Methods of Experimental Physics

VOLUME 3

MOLECULAR PHYSICS

SECOND EDITION

PART A

METHODS OF EXPERIMENTAL PHYSICS:

L. Marton, *Editor-in-Chief*

Claire Marton, *Assistant Editor*

Volume 3

Molecular Physics

Second Edition

PART A

Edited by

DUDLEY WILLIAMS

Physics Department
Kansas State University
Manhattan, Kansas

1974

ACADEMIC PRESS · New York and London
A Subsidiary of Harcourt Brace Jovanovich, Publishers

ACADEMIC PRESS, INC.
111 Fifth Avenue, New York, New York 10003

United Kingdom Edition published by
ACADEMIC PRESS, INC. (LONDON) LTD.
24/28 Oval Road, London NW1

Library of Congress Cataloging in Publication Data

Williams, Dudley, DATE   ed.
    Molecular physics.

    (Methods of experimental physics, V. 3)
    Includes bibliographical references.
    1.   Molecules,   2.   Molecular theory.   I.   Title.
II.   Series.
QC175.16.M6W553   539'.12   73-8905
ISBN 0-12-476003-1

CONTENTS OF VOLUME 3, PART A

CONTRIBUTORS TO VOLUME 3, PART A

Numbers in parentheses indicate the pages on which the authors' contributions begin.

W. E. BLASS, *Department of Physics and Astronomy, The University of Tennessee, Knoxville, Tennessee* (126)

GEORGE W. CHANTRY, *Department of Trade and Industry, National Physical Laboratory, Teddington, Middlesex, England* (302)

GEOFFREY DUXBURY, *School of Chemistry, University of Bristol, Bristol, England* (302)

DAVID R. LIDE, Jr., *National Bureau of Standards, Washington, D. C.* (11)

C. WELDON MATHEWS, *Department of Chemistry, Ohio State University, Columbus, Ohio* (203)

A. H. NIELSEN, *Department of Physics and Astronomy, The University of Tennessee, Knoxville, Tennessee* (216)

D. H. RANK, *Physics Department, The Pennsylvania State University, University Park, Pennsylvania* (395)

T. A. WIGGINS, *Physics Department, The Pennsylvania State University, University Park, Pennsylvania* (395)

DUDLEY WILLIAMS, *Department of Physics, Kansas State University, Manhattan, Kansas* (1)

FOREWORD

Close to 12 years have elapsed since I wrote the foreword to the first edition of the volume on Molecular Physics in our series. At that time I estimated to be about halfway in our task to present a concise survey of the methods used by experimental physicists. The original concept of six volumes has since grown to ten published volumes, with several of them split into double volumes. At this time I can report on advanced plans for further additions: a volume on polymer physics, another on fluid dynamics, and a third on environmental studies.

In publishing this revised edition of the volume on Molecular Physics I would like to trace briefly the reason for issuing it. My fellow editors and I were pleased with the reception of our series by our readers, as manifested by book reviews and by the distribution of the books. Several volumes were reprinted, but when it came to reprinting the molecular physics (as well as the electronics) volumes, we concluded that the time had come for major revisions. The results achieved by Professor Williams are presented herewith and both he and I hope physicists will find this edition as valuable as the first version.

In this new edition you will find the subjects rearranged, with a considerable amount of new material added and some of the old omitted. You will recognize also some of the authors from the first edition, with a number of new authors added. Professor Williams' introduction gives an excellent survey of the organization of the new "Molecular Physics."

It remains a pleasant duty to thank Professor Dudley Williams and all the authors for their untiring labors. The cooperation of the publishers is gratefully acknowledged, as well as all the contributions to the editorial work by Mrs. Claire Marton.

L. Marton

CONTENTS OF VOLUME 3, PART B

CONTRIBUTORS TO VOLUME 3, PART B

THOMAS C. ENGLISH, *Department of Physics and Astrophysics, Duane Physical Laboratories, University of Colorado, Boulder, Colorado*

EDWIN N. LASSETTRE, *Center for Special Studies and the Department of Chemistry, Mellon Institute of Science, Carnegie-Mellon University, Pittsburgh, Pennsylvania*

P. S. LEUNG, *Union Carbide Corporation, Sterling Forest Research Center, Tuxedo, New York*

C. A. McDOWELL, *Department of Chemistry, University of British Columbia, Vancouver, British Columbia, Canada*

J. D. MEMORY, *Department of Physics, North Carolina State University, Raleigh, North Carolina*

G. W. PARKER, *Department of Physics, North Carolina State University, Raleigh, North Carolina*

AUSMA SKERBELE, *Center for Special Studies and the Department of Chemistry, Mellon Institute of Science, Carnegie-Mellon University, Pittsburgh, Pennsylvania*

G. J. SAFFORD, *Union Carbide Corporation, Sterling Forest Research Center, Tuxedo, New York*

JENS C. ZORN, *Department of Physics, University of Michigan, Ann Arbor, Michigan*

1. INTRODUCTION*

1.1. Introduction

Although there are certain very general methods and general principles that constitute the science known as "physics," the methods become so specialized in modern research that detailed treatment of those involved in the various fields of current research seems highly desirable. The present volume, devoted to the methods of *molecular physics*, has been prepared with the purpose of providing an introduction to this subject for young scientists such as graduate students in physics and chemistry and for more mature scientists who are already specialists in other research fields.

Recognizing that it is very difficult to arrive at formal definitions of specific areas of research, we merely point out that research in molecular physics is largely conducted by physicists and physical chemists. In fact, the list of contributors to the present volume is nearly equally divided between professional physicists and professional chemists; the references include articles in the official journals of physical societies and chemical societies as well as in more specialized journals devoted to such subjects as chemical physics, molecular physics, and molecular spectroscopy.

In this volume we attempt to present a survey of the experimental methods employed in certain selected, currently active, fields of molecular physics. We place emphasis on general *methods* of investigation as opposed to detailed description of currently popular *techniques* and *gadgetry*, which are subject to rapid change. In most parts of the book we supply examples of the ways in which each experimental method is applied to a specific molecular problem, and we summarize the results obtained for related problems; however, we do not attempt to be encyclopedic.

Recognizing that theory forms the *rationale* of experiments and is necessary to the interpretation of the results of experimental measurements in terms of meaningful molecular parameters, we present basic

---

* Part 1 is by Dudley Williams.

1

theoretical relationships at appropriate points in the text and give references to original papers, review articles, and definitive theoretical treatises; in some cases, for which suitable references are not readily available, it has been necessary to insert somewhat lengthy chapters on theory. In general, however, we have concentrated on *general experimental methods of molecular physics*.

Before beginning our treatment of this subject, we present first a brief review of the origins of the molecular theory of matter, and then a brief survey of the subjects we shall regard as constituting molecular physics.

1.2. Origins of the Molecular Theory

The molecular theory of matter is a relatively modern elaboration of the atomic theory. Without this elaboration, the atomic theory led to such serious inconsistencies and contradictions that it never achieved universal acceptance.

The "atomic philosophy" can be traced from the time of Democritus in the fifth century B.C. Although accepted by the Epicureans, it was discredited by Aristotle and was in abeyance during the Middle Ages. During the Renaissance there was a revival of interest in atomism. Galileo regarded it with favor, and P. Gassendi presented a restatement of the views of Epicurus and Lucretius.

During the latter half of the seventeenth century, atomism gradually changed from a philosophical speculation to a fruitful concept employed widely, but by no means universally, by the scientists of the time. The atomic nature of matter was treated by Robert Boyle at various intervals between 1661 and his death in 1691. In the "Principia," 1687, Isaac Newton pictured matter as consisting of atoms surrounded by vacuum. Boyle and Newton both freely employed atomism in their chemical and physical speculations; the atomic view was also supported by Christian Huyghens. Despite the opposition of Gottfried von Leibniz, the constitution of matter was generally considered as atomic by many physical investigators. Atomic ideas were employed in a highly sophisticated manner by Daniel Bernoulli in 1738 in his development of the kinetic theory of gases; although his theory of gases was quantitatively correct in certain respects, Bernoulli's work appears to have been largely unappreciated for more than a century after its publication.

Although the assumption of the existence of atoms was attractive, there was an understandable lack of clarity in the atomic concept until the beginning of the nineteenth century; prior to that time the terms *atom, corpuscle, particle,* and the like were used more or less interchangeably to refer to invisibly small and indivisible parts of objects. However, no precise and unambiguous conceptual scheme of the structure of matter had been developed. Many choices of detailed models were possible; the choice of fruitful models was hampered by a lack of understanding of the nature of heat. It was possible to adopt a *kinetic theory* of gases of the kind proposed by Bernoulli; on the basis of the kinetic model, the particles of the gas interact only during collisions and the speed of the particles increases with temperature. According to the opposing *static theory*, supported initially by Robert Hooke and later by John Dalton himself, the addition of heat or caloric fluid produces repulsion between the particles of a gas. At the end of the eighteenth century, the static model was generally accepted. According to both theories, mutual attraction between particles in solids and in liquids was assumed.

The modern atomic theory was first set forth in John Dalton's treatise, "A New System of Chemical Philosophy," which appeared in 1808. Many of the ideas presented can be traced to Robert Boyle, who gave a modern operational definition of chemical element in "The Sceptical Chymist" in 1661 and to Antoine Lavoisier's quantitative treatment of chemical reactions in his "Traité Elémentaire de Chimie," published in 1789. According to Dalton's theory, matter is composed of indivisible atoms; there are as many varieties of atoms as there are chemical elements. In chemical reactions, atoms are neither created nor destroyed but only rearranged; the smallest portion of a compound was assumed to consist of a definite number of elementary atoms and was called a *compound atom.* In the elaboration of his theory, Dalton employed what has been called the *rule of greatest simplicity.* According to this rule, elements entered reactions atom by atom and, when a compound atom was formed, it was assumed to be *binary,* consisting of two elementary atoms, "unless some other cause appears to the contrary"; when "other causes" appeared, *ternary* or higher combinations of elementary atoms were assumed. Dalton's theory gave a satisfactory account of the gravimetric law of multiple proportions but ultimately led to ambiguities in the determinations of relative atomic weights.

In the very year in which Dalton's book was published (1808), J. L. Gay-Lussac announced the discovery of certain important *volumetric* relations applying to gaseous reactions: reactive gases, under conditions of equal pressure and temperature, combine in simple volumetric proportions. Gay-Lussac's experimental results could not be explained in terms of Dalton's theory. Some additional amendment of Dalton's theory was needed.

Such an amendment was supplied in 1811 by the Italian physicist Amedeo Avogadro. With Avogadro's amendment, the atomic theory ultimately emerged in its present-day form. Avogadro pointed out that if there is a simple numerical relation between combining volumes of gases and if they combine into uniform "atomic groups," then there must be some simple connection between the actual numbers of these "atomic groups" in equal volumes of combining gases. These "atomic groups" behave like discrete particles prior to reaction but are divisible during reaction. These discrete groups of atoms Avogadro named *molecules* or "little masses." Avogadro's hypothesis—now called *Avogadro's law*—asserted that equal volumes of all gases under identical conditions of temperature and pressure contain equal numbers of molecules; the recognition that the molecules of elements as well as those of compounds may be diatomic or polyatomic proved to be the salvation of the atomic theory.

Avogadro's ideas were not immediately acceptable to his colleagues and his papers lay neglected for many years. However, the vigorous development of chemistry during succeeding decades disclosed so many shortcomings in Dalton's simple theory that following 1840 there was a general loss of faith in the atomic theory of chemistry. The solution to the confusion of the chemists came in 1858 when the Italian chemist Stanislao Cannizzaro called for a revival of Avogadro's theory and showed that it provided a reasonable basis for chemistry.

By the time of Cannizzaro's suggestion in 1858, the work of Benjamin Thompson (Count Rumford) and J. P. Joule had led to an understanding of the nature of heat. The abandonment of the caloric theory of heat and the growing acceptance of the kinetic theory of gases made the existence of molecules acceptable to the physicists of the time. The rapid progress of chemistry involving the development of the concept of valence, D. I. Mendeleev's recognition of "periodic" variations of chemical properties of the elements, and a clearer understanding of organic compounds was firmly based on a molecular–atomic theory of chemistry by the end of the nineteenth century. Similarly, the elaboration of the kinetic–molecular theory of gases by such men as H. von Helmholtz, J. C. Maxwell, L. Boltzmann, and J. W. Gibbs eventually became one of the major triumphs of nineteenth century physics.

There was one final attack on the kinetic–molecular theory of matter between 1890 and 1908 during the period when the extensive uses of thermodynamics were being realized. Wilhelm Ostwald and his "Energetics School" attacked the molecular theory on the grounds that it was based on unjustified hypotheses involving particles whose existence had never been demonstrated. From the point of view of the Energetics School, the mere statement of the thermodynamic relations describing physical processes or chemical reactions should be the goal of science and any discussion of detailed mechanisms is not only needless but also undesirable. The attacks of Ostwald were suddenly brought to a halt in 1908 as a result of the brilliant experiments of Jean Perrin on the Brownian motion of particles large enough to be seen by means of a microscope. The observed behavior of the visible particles agreed quantitatively with Einstein's predictions based on kinetic–molecular theory.

Since the time of Perrin's definitive experiments, there has been no serious reason to doubt the existence of atoms and molecules. The physicists have proceeded with their studies of the properties and structure of atoms. Chemists have successfully used the resulting quantum-mechanical models in understanding chemical reactions.

1.3. Molecular Physics

In our treatment of molecular physics in this book, we devote our primary attention to stable molecules, i.e., to atomic groups that maintain their identity under ambient laboratory conditions. The internuclear separation of atoms in stable molecules is typically of the order of one angstrom (1 Å = 0.1 nm = 10^{-10} m); dissociation energies normally fall within the range 1–5 eV. This restriction of the subject eliminates from consideration such entities as *collision pairs*, for which there is no energetically stable configuration, and *van der Waals molecules*, in which internuclear separations are typically from 3 to 5 Å and for which dissociation energies are small as compared with those of stable molecules.

Most of the molecular physics to be presented deals with relatively small molecules. We devote little attention to *macromolecules*, although these entities are of major importance to the polymer chemist and to the molecular biologist, and are being investigated intensively and extensively by physical methods. Similarly, our treatment of large arrays of molecules will be rather limited, since the detailed studies of such arrays more properly falls within the realm of solid-state physics.

The general approach to molecular physics in the present volume consists of a presentation of the methods currently employed to determine:

(a) the sizes and shapes of molecules; (b) the electric and magnetic properties of molecules; (c) the internal energy levels of molecules; and (d) molecular ionization and dissociation energies.

Some of the earliest information regarding the sizes of molecules was obtained in studies of the thermodynamic properties of various substances. From the equations of state established in such studies, information regarding intermolecular forces could also be obtained. Early measurements of dielectric constants provided valuable information regarding the electric polarizabilities and dipole moments of molecules. Acoustical studies provided early information regarding the equipartition of energy between translational, rotational, and vibrational degrees of freedom; extension of these studies to the ultrasonic frequency range also provided information regarding molecular relaxation phenomena. Early studies of dielectric breakdown and thermal decomposition provided useful information regarding molecular ionization and dissociation phenomena. Although some of these "classical methods" of investigation have been modernized and are still being used in various laboratories, the basic methods involved are familiar and have been described in text books. Thus we omit them from detailed consideration in the present book.

Similarly, in the interests of brevity, it has been necessary for us to omit any detailed treatment of the important diffraction methods of studying molecular structure. X-ray diffraction studies have provided valuable information on internuclear distances and are particularly effective for structures containing atoms of elements of high atomic number. In contrast to X-ray diffraction, neutron diffraction studies are particularly effective in the analysis of structures containing hydrogen and other elements of low mass number. X-ray and neutron diffraction studies are thus complementary. Early electron diffraction studies gave valuable information regarding the internuclear separation of atoms in the molecules of gases; more recently electron diffraction techniques have been used very effectively in studies of surface effects in solids. Although the experimental methods employed in diffraction studies continue to be improved, the greatest and most significant advance in recent years has been the application of high-speed digital computers to the formidable task of obtaining structural parameters from the initial diffraction data obtained experimentally. The modern computer has, in a sense, completely revolutionized the whole field of diffraction studies.

Part 2 of the book, immediately following this brief introduction, is devoted to *molecular spectroscopy* and covers the requisite theory together with a treatment of recent methods. Microwave spectroscopy provides a knowledge of molecular rotational energy levels and therefore moments of inertia of molecules; in cases where isotopic substitutions are possible, highly precise values of internuclear separations can be obtained. From measurements of the Stark effect, molecular electric dipole moments can be obtained. Although much of the early work in the microwave region was done with components designed primarily for communication systems, excellent microwave spectrographs are becoming available from commercial sources and will doubtless be widely used. Infrared spectroscopy, which provides us with detailed knowledge of vibrational and rotational energy levels, has been improved markedly in the near and intermediate infrared ranges by the development of greatly improved gratings, filters, and solid-state detectors. An important advance of primary importance in the far infrared has been the development of interferometers; from the resulting interferograms, absorption and emission spectra can be obtained as Fourier transforms. The development of *Fourier transform spectroscopy* has become of great practical importance as a result of the availability of modern high-speed computers. Computers are also now almost indispensable in the field of electronic spectroscopy, where improvements in resolution have made it possible to map literally thousands of lines in the spectra of certain molecules. The molecular parameters obtained from electronic spectroscopy include excited electronic levels together with the vibrational levels in the excited electronic states. In some cases, moments of inertia of molecules in excited states can be obtained; energies of dissociation can also be determined.

An important development of the past decade has been the development of lasers, which provide nearly monochromatic radiation in various parts of the electromagnetic spectrum. We include a chapter on molecular lasing systems, which shows how existing knowledge of molecular energy levels can be used in this new field of optical technology. Although the CO_2 laser with wavelengths in the 10-μm region is perhaps the best known of the molecular lasers, other molecular lasers provide sources of coherent radiation at wavelengths in many other spectral regions out to wavelengths as long as 300 μm. Of course, it will be recalled that Townes' original NH_3 laser was a molecular device operating in the microwave region!

Part 3 of the book is devoted to *light-scattering* methods of studying phenomena. The development of strong sources of coherent radiation has revolutionized techniques of studying the Raman effect. With an argon-ion laser as a source, it is actually possible under favorable conditions to make *visual* observations of Raman scattering! Because of the negligible spectral spread of the source radiation, it is possible to observe Raman lines within 1 cm^{-1} of the exciting line. New photon-counting techniques employing circuits previously developed for use in nuclear physics can be used with newly developed photodetectors to observe weak Raman lines that would have hitherto been unobservable. With pulsed lasers having high energy output, many new previously unobserved optical phenomena have been discovered. The molecular information derivable from studies of Rayleigh, Raman, and Brillouin scattering are discussed.

The fourth part of the volume (which begins Part B of this volume) deals chiefly with *magnetic-resonance* phenomena. The frequency shifts of the nuclear resonance signals from the positions they would have for a free nucleus in the same magnetic field provide a sensitive means of probing the local electronic environments of each type of nucleus with nonzero spin in a given molecule. These "chemical shifts" along with certain small spin–spin interactions provide the organic chemist with a sensitive method of "finger-printing" various compounds and provide new insights into molecular structure. Part 4 of this volume gives the experimental methods of determining the pertinent molecular parameters from nuclear magnetic resonance, electron-spin resonance, and nuclear-quadrupole resonance phenomena.

Part 5 of the volume is devoted to the *mass spectrometry* of molecules. This method provides valuable insights into such molecular properties as ionization and dissociation energies. Within the last decade, there has been much progress in this field. Mass-spectrographic analysis is proving to be a useful technique in most well-equipped chemical research laboratories. After the general methods have been introduced, the recent developments are treated in considerable detail.

Part 6 of this volume is devoted to *molecular beams*. Nearly three decades ago one of the pioneers in this field remarked, "It is extremely important that our experiments be done properly, since the techniques involved are so difficult that in all probability the experiments will never be repeated." The present résumé describes the general methods and shows how the initially difficult techniques have been refined and extended. Some of the resulting molecular parameters cannot be obtained by other methods.

The present volume closes with Part 7, dealing with "special methods," which includes three brief chapters on three different recently developed methods. The first of these deals with *photoelectric studies*; analyses of the energies of electrons ejected from a compound by ultraviolet photons shows that a portion of the initial photon energy goes into the excitation of the molecules of the compound; this provides a method of determining molecular energy levels. The second and third chapters in Part 7 deal with the *inelastic scattering of electrons and neutrons*, respectively; energy losses by these particles during collision is transferred selectively to internal energy of the target molecules. The molecular energy changes involved are not subject to selection rules like those involved in photon absorption or scattering. The special methods described in the final chapter thus supplement those described in Parts 2 and 3 of the book.

2. MOLECULAR SPECTROSCOPY

2.1. Microwave Spectroscopy*†‡

2.1.1. Introduction

The microwave region of the electromagnetic spectrum extends from about 1000 MHz (or 1 GHz) to perhaps 300,000 MHz (300 GHz) or, in wavelength terms, from about 30 cm to 1 mm. The upper limit is somewhat arbitrary, but conventional microwave techniques become increasingly difficult at wavelengths below 1 mm. This submillimeter region merges into the far infrared region of the spectrum, where optical techniques are more efficient.

Microwave measurements are characterized by the extremely high frequency resolution that can be achieved. Most microwave oscillators contain a resonant element that ensures a monochromatic output to a high approximation. Even greater frequency stability can be obtained by locking the oscillator to a suitable frequency standard. Thus a degree of resolution and accuracy of measurement (in absolute terms) is possible in the microwave region that was completely out of reach of optical spectroscopy until the advent of lasers.

Most materials absorb microwave radiation to some measurable degree. In condensed phases and in gases at moderate pressure, the absorption tends to vary rather slowly with frequency. In gases at very low pressures, however, sharply resonant features begin to appear in the microwave absorption spectrum. As used in this chapter, the term "microwave spectroscopy" means the study of the sharp-line spectra that are characteristic of low-pressure gases.

The spectra observed in gases at microwave frequencies are primarily the result of transitions between molecular rotational energy levels. Thus

---

† See also Volume 2, Part 10.

‡ The opinions expressed in this article are those of the author and do not necessarily reflect the official position of the National Bureau of Standards.

* Chapter 2.1 is by David R. Lide, Jr.

the geometry and mass distribution of a molecule usually determine the gross features of its microwave spectrum. However, many other structural features make a secondary (and sometimes a major) contribution. Detailed studies of microwave spectra have thus provided a wealth of information on many features of molecular structure.

This chapter is concerned to only a minor extent with the experimental details of measuring microwave spectra. The measurement techniques have become relatively standardized by this time (see Section 2.1.14), and suitable apparatus for most conventional purposes is available from commercial sources. Instead, the principal emphasis is on the interpretation of the spectra and the deduction of molecular information therefrom. After a discussion of the spectrum of the rigid rotor, the various effects present in real molecules that cause deviations from this ideal case are considered. An effort is made to present these in a systematic manner that will facilitate their recognition in observed spectra.

It is difficult to cover all facets of the subject in a chapter of this length. More space has been devoted to certain topics, such as vibration–rotation interactions and the classification of asymmetric-rotor patterns, which have not been treated in great detail in other microwave texts. Subjects such as nuclear-quadrupole hyperfine structure and centrifugal distortion, which have been thoroughly discussed elsewhere, are outlined only briefly here. Some areas where research has so far been fairly limited have been omitted entirely; these include Zeeman effects, magnetic hyperfine structure, and molecules with electronic angular momentum. There are several comprehensive texts on microwave spectroscopy that include discussions of these subjects.[1-4]

2.1.2. The Rigid Rotor

The starting point in the analysis of most microwave spectra is the approximation of a molecule as a rigid body rotating freely in space. It was originally shown by Born and Oppenheimer that the Hamiltonian

[1] C. H. Townes and A. L. Schawlow, "Microwave Spectroscopy." McGraw Hill, New York, 1955.

[2] J. E. Wollrab, "Rotational Spectra and Molecular Structure." Academic Press, New York, 1967.

[3] T. M. Sugden and C. N. Kenney, "Microwave Spectroscopy of Gases." Van Nostrand-Reinhold, Princeton, New Jersey, 1965.

[4] W. Gordy and R. L. Cook, "Microwave Molecular Spectra." Wiley (Interscience), New York, 1970.

of an isolated molecule can be separated, to a high approximation, into contributions from the electronic motion, the vibration of the nuclei, and the overall rotation of the molecule. The rotational contribution to the energy is generally much less than the vibrational and electronic contributions, and the time scale associated with the rotation is correspondingly slower than that of the vibrational and electronic motion. Thus the rotation may be described by approximating the molecule as a rigid body in which each atomic mass is placed at a position determined by averaging over the vibrational and electronic motion. While the exact nature of this average is important in the interpretation of the finer details found in microwave spectra, and situations do arise in which the separation of rotation from vibration is not such a good approximation, the gross features of the majority of the microwave spectra so far observed can be explained on the basis of the rigid-rotating molecule. We shall therefore develop first the theory of the rigid rotor, and subsequently consider the effects that must be included in higher orders of approximation.

2.1.2.1. Classical Kinetic Energy of a Rigid Rotor. Any rigid body may be represented by a collection of point masses m_i ($i = 1, 2, \ldots$) whose positions relative to each other are fixed. We wish to describe the motion of this group of masses relative to some inertial frame of reference XYZ, which will be referred to as the *space-fixed frame*. It is convenient to define another Cartesian coordinate system xyz which is fixed in the rigid body; this *body-fixed frame* will be chosen so that its origin is at the center of mass of the body. If ω is the angular velocity of the xyz frame relative to the XYZ frame, the kinetic energy is given by

$$T = \tfrac{1}{2} \sum_{\alpha,\beta} I_{\alpha\beta}\omega_\alpha\omega_\beta, \qquad \alpha, \beta = x, y, z. \qquad (2.1.1)$$

The $I_{\alpha\beta}$ are components of the inertia tensor defined by

$$I_{xx} = \sum_i m_i(y_i^2 + z_i^2), \qquad I_{xy} = -\sum_i m_i x_i y_i, \qquad \text{etc.}$$

The expression for the kinetic energy may be simplified by choosing the xyz frame as the principal axes of inertia, which diagonalize the inertia tensor. This is done by solving the secular equation

$$\begin{vmatrix} I_{xx} - I & I_{xy} & I_{xz} \\ I_{xy} & I_{yy} - I & I_{yz} \\ I_{xz} & I_{yz} & I_{zz} - I \end{vmatrix} = 0. \qquad (2.1.2)$$

The introduction of angular momenta, defined by

$$P_\alpha = \partial T/\partial \omega_\alpha, \qquad \alpha = x, y, z,$$

then gives the kinetic energy as

$$T = (P_x^2/2I_x) + (P_y^2/2I_y) + (P_z^2/2I_z). \qquad (2.1.3)$$

The rotation of the rigid body is thus completely described by the three principal moments of inertia, I_x, I_y, I_z.

Several special cases are apparent. In the *linear rotor* one of the principal moments (usually designated as I_z) vanishes, and the other two are equal. The *symmetric rotor* has two of the principal moments equal, but the third is different from zero. The *spherical rotor*, in which all three moments are equal, is of little interest in microwave spectroscopy. The most general case, with all three moments different, is called the *asymmetric rotor*.

A well-known theorem for a planar rigid body states that

$$I_z = I_x + I_y,$$

where xy is the plane of the body. This may be generalized to a body in which xy is a plane of symmetry. We then have

$$I_x + I_y - I_z = \sum_i m_i z_i^2.$$

2.1.2.2. Quantum-Mechanical Hamiltonian of the Rigid Rotor. The classical kinetic energy of the general rigid rotor, expressed in terms of the components of angular momentum in the body-fixed frame, is given by Eq. (2.1.3). In the absence of external fields the potential energy is zero, so that the Hamiltonian is given by the same equation. Since the coefficients in Eq. (2.1.3) are constant, the transformation from classical to quantum-mechanical Hamiltonian is straightforward. It is convenient to express the angular momentum components in units of $\hbar$ by writing the quantum-mechanical Hamiltonian as

$$H = (\hbar^2/2I_x)P_x^2 + (\hbar^2/2I_y)P_y^2 + (\hbar^2/2I_z)P_z^2, \qquad (2.1.4)$$

where the angular momentum operators P_x, P_y, and P_z are now dimensionless.

2.1.2.3. Units. We shall later introduce rotational constants, or reciprocal moments of inertia, for the coefficients that appear in the Hamiltonian of Eq. (2.1.4). It is customary in microwave spectroscopy to specify the rotational constants in frequency rather than energy units, because the basic spectroscopic measurement is one of frequency. This is equivalent to replacing Eq. (2.1.4) with a reduced Hamiltonian given by

$$H/h = (h/8\pi^2 I_x)P_x{}^2 + (h/8\pi^2 I_y)P_y{}^2 + (h/8\pi^2 I_z)P_z{}^2. \quad (2.1.5)$$

A typical rotational constant, for example, the constant B for a linear molecule, is then defined in frequency units by

$$B = h/8\pi^2 I_x.$$

The eigenvalues are therefore designated by E/h and have units of frequency.

If the moments of inertia are measured in atomic mass units times angstroms squared (amu Å^2), the numerical relation becomes

$$B \quad (\text{MHz}) = 505391/I_x \quad (\text{amu Å}^2).$$

This factor is based on the currently accepted atomic mass scale in which $m(C^{12}) = 12$ and on the fundamental constants of Taylor et al.[5]

2.1.2.4. Angular Momentum Operators. While the solution to Eq. (2.1.4) may be found by wave mechanics, it is more customary to use matrix methods. The matrix representation of angular momentum operators is discussed in all standard texts on quantum mechanics. The specific application to the rigid-rotor problem is treated in some detail by Allen and Cross.[6] We shall not repeat the discussion here, but will only outline the calculation and quote the results.

It is well known that the operators representing the angular momentum components in the body-fixed frame follow the commutation rules

$$[P_x, P_y] = P_x P_y - P_y P_x = -iP_z$$
$$[P_y, P_z] = -iP_x, \quad (2.1.6)$$
$$[P_z, P_x] = -iP_y.$$

A similar set of equations applies to the components P_X, P_Y, P_Z in the space-fixed frame, except that the signs on the right-hand side are positive.[6] It may also be shown that each component, in either body or

[5] B. N. Taylor, W. H. Parker, and D. N. Langenberg, *Rev. Mod. Phys.* **41**, 375 (1969).
[6] H. C. Allen and P. C. Cross, "Molecular Vib-Rotors." Wiley, New York, 1963.

space-fixed frame, commutes with P^2. We may thus add to Eq. (2.1.6) two further relations

$$[P^2, P_Z] = 0, \qquad [P^2, P_z] = 0.$$

Therefore, we can choose a matrix representation in which P^2, P_Z, and P_z are diagonal matrices (although the Hamiltonian will, in general, not be diagonal). This representation is labeled by quantum numbers J, K, M; the well-known result is

$$(JKM \mid P^2 \mid JKM) = J(J + 1),$$
$$(JKM \mid P_Z \mid JKM) = M, \qquad (2.1.7)$$
$$(JKM \mid P_z \mid JKM) = K.$$

The quantum numbers M and K are each restricted to values

$$J, J - 1, J - 2, \ldots, -J,$$

while J must be a positive integer or zero. Thus we can interpret M as the projection of the total angular momentum on the space-fixed Z axis, and K as the projection on the body-fixed z axis.

With the representation chosen in this way, Eq. (2.1.6) may be used to derive the other matrix elements that are required. The result is

$$(J, K, M \mid P_x^2 \mid J, K, M)$$
$$= (J, K, M \mid P_y^2 \mid J, K, M)$$
$$= \tfrac{1}{2}[J(J + 1) - K^2],$$
$$(J, K, M \mid P_x^2 \mid J, K \pm 2, M) \qquad (2.1.8)$$
$$= -(J, K, M \mid P_y^2 \mid J, K \pm 2, M)$$
$$= -\tfrac{1}{4}[(J \mp K)(J \pm K + 1)(J \mp K - 1)(J \pm K + 2)]^{1/2}.$$

All other matrix elements vanish.

We now have all the elements necessary to set up the matrix for the Hamiltonian of Eq. (2.1.4). Before considering the most general case, which has been designated as the asymmetric rotor, we shall discuss two special situations, the linear rotor and the symmetric top, where the Hamiltonian reduces to simpler form.

2.1.2.5. The Linear Rotor.

The linear rotor has already been defined as a rigid body in which all of the mass points lie on a straight line. If this line is designated as the z axis, then

$$I_x = I_y, \qquad I_z = 0.$$

The fact that $I_z = 0$ in a linear rotor implies also that $P_z = 0$. The Hamiltonian for a linear rotor becomes

$$H/h = B(P_x^2 + P_y^2) = BP^2, \qquad (2.1.9)$$

where

$$B = h/8\pi^2 I_x.$$

We see immediately that H is diagonal in the representation defined by Eq. (2.1.7); the energy eigenvalues are

$$E/h = BJ(J + 1). \qquad (2.1.10)$$

Thus the energy levels of a rigid linear rotor have a very simple form, varying quadratically with the quantum number J. The levels have the usual spacial degeneracy of $2J + 1$, corresponding to the allowed values of M.

The eigenfunctions of the rigid linear rotor are also easily derived (although in practice the explicit expression for the eigenfunctions is not often required). These eigenfunctions are most conveniently expressed as functions of the Eulerian angles θ, ϕ, χ, which define the relative orientation of the body-fixed and space-fixed frames. In the linear rotor the x and y axes may be fixed arbitrarily, so that we can specify $\chi = 0$. Furthermore, the angles θ and ϕ are identical with the angular coordinates of the usual spherical polar coordinate system. Since the Hamiltonian is proportional to the square of the total angular momentum, the energy eigenfunctions are simply angular momentum eigenfunctions, which can be written as

$$\Psi(\theta, \phi) = \Theta(\theta)\Phi(\phi),$$

where, aside from constant factors,

$$\Theta(\theta) = P_J^{|M|}(\cos\theta), \qquad \Phi(\phi) = \exp(iM\phi).$$

Here $P_J^{|M|}(\cos\theta)$ is an associated Legendre function. When the normalization factor is properly chosen, we finally have

$$\Psi_{JM} = \left[\frac{(2J+1)(J-|M|)!}{4\pi(J+|M|)!}\right]^{1/2} e^{iM\phi} P_J^{|M|}(\cos\theta). \qquad (2.1.11)$$

The only significant symmetry property of the linear rotor is the parity of the levels; i.e., the behavior of the eigenfunctions upon inversion of all particles through the origin of the xyz frame. This inversion operation is described by

$$\theta \to \pi - \theta, \qquad \phi \to \phi.$$

An examination of the properties of the associated Legendre functions shows that the eigenfunctions Ψ_{JM} have even parity $(+)$ when J is even and odd parity $(-)$ when J is odd. We shall use this result later for deriving selection rules.

2.1.2.6. The Symmetric Rotor. In a symmetric rotor two of the principal moments of inertia are equal, while the third is different. We shall designate the axis corresponding to the unique moment as z (frequently referred to as the "figure" axis). Let us define rotational constants A and B by

$$B = h/8\pi^2 I_x = h/8\pi^2 I_y,$$
$$A = h/8\pi^2 I_z.$$

To maintain consistency with the notation for asymmetric rotors the unique rotational constant is sometimes designated by A when $I_z < I_x$ but by C when $I_z > I_x$. However, we shall use the symbol A for both situations. Our general Hamiltonian in Eq. (2.1.5) now becomes

$$H/h = B(P_x^2 + P_y^2) + AP_z^2 = BP^2 + (A - B)P_z^2. \qquad (2.1.12)$$

As was true for a linear rotor, the representation in which P^2 and P_z are diagonal also diagonalizes H, so that the energy eigenvalues may be obtained directly from Eq. (2.1.7). The result is

$$E/h = BJ(J + 1) + (A - B)K^2. \qquad (2.1.13)$$

We may note that in a *prolate symmetric rotor*, where $A > B$, the energy increases with increasing K (for fixed J), while in an *oblate rotor* the reverse is true. The quantum number K describes the projection of the total angular momentum on the figure axis, which is a constant of motion for a symmetric rotor.

The derivation of the eigenfunctions for a symmetric rotor is somewhat more complicated than was true for a linear rotor, since now the Eulerian angle χ, which describes the rotation of the body about its figure axis, must be taken into account. The calculation is outlined by

Townes and Schawlow.[1] The resulting eigenfunction can be written as

$$\Psi_{JKM} = N_{JKM}e^{iM\phi}e^{iK\chi}x^{|K-M|/2}(1-x)^{|K+M|/2}F(x), \qquad (2.1.14)$$

where

$$x = \tfrac{1}{2}(1 - \cos\theta).$$

$F(x)$ is the hypergeometric function and N_{JKM} is a normalization factor. It may be shown that this function reduces to the linear-rotor eigenfunction when $K = 0$.

In addition to the spacial degeneracy of $2J + 1$, each level of the rigid symmetric rotor, except those with $K = 0$, has a twofold degeneracy because of the fact that the energy does not depend on the sign of K. The net degeneracy is $2(2J + 1)$ when $K \neq 0$ and $2J + 1$ when $K = 0$. One further type of degeneracy should be mentioned here, although it will not be considered in detail until a later chapter. This is the so-called "inversion degeneracy." Unless all of the mass of a body lines in the same plane, an inversion of the internal coordinate system (i.e., the operation $x \to -x$, $y \to -y$, $z \to -z$) carries the body into a new form that cannot be derived from the original by a rotation of the xyz frame. The moments and products of inertia, however, are clearly unchanged by this operation. Thus we must recognize the existence of two forms of the rigid body, one right-handed and one left-handed, and these two forms will have identical energy. The resulting twofold inversion degeneracy applies to all nonplanar rigid bodies, whether symmetric rotors or not. However, as long as the body is completely rigid, the inversion degeneracy is not experimentally detectable and can be ignored. It becomes important only when nonrigidity effects are taken into account and removal of the degeneracy becomes possible. This situation will be discussed in Section 2.1.7.

We must now discuss the symmetry properties of the symmetric-rotor levels. If the rotor is symmetric because two of the principal moments happen accidentally to be equal, the only pertinent symmetry operation is inversion. However, it may be shown[7] that one of the two levels resulting from the inversion degeneracy is positive with respect to inversion while the other is negative. Therefore the $+$ or $-$ character of the levels of a rigid symmetric rotor is of no practical importance, since the levels occur in $+$ and $-$ pairs that cannot be separated. On the other hand, if

[7] G. Herzberg, "Infrared and Raman Spectra of Polyatomic Molecules." Van Nostrand-Reinhold, Princeton, New Jersey, 1945.

the body is a symmetric rotor because of its inherent symmetry properties, then additional symmetry considerations must be taken into account. The symmetry properties of a molecule are conventionally discussed in terms of the *point group* to which the rigid body representing the molecule belongs. Each point group consists of all the symmetry elements possessed by the body. The important point groups are discussed in many places[7,8] and will not be tabulated here. Several of these point groups include elements (in particular, a threefold or higher axis of symmetry) which necessarily imply that the body is a symmetric rotor. However, if we are concerned solely with microwave spectra, the only point groups of importance in discussing symmetric rotors are C_{nv}, with $n \geq 3$. The reason is the requirement of a permanent dipole moment in order for a molecule to exhibit a microwave spectrum; other point groups to which symmetric rotors can belong (D_{3d}, D_{3h}, D_{2d}, etc.) have such high symmetry that no pure rotational spectrum is allowed. Furthermore, the point group C_{3v} is by far the most important in practice, and we shall restrict our attention to it.

In describing the symmetry properties of rotational energy levels we need only concern ourselves with the rotations that are included in the point group in question. In the case of C_{3v}, the rotational subgroup is C_3, which includes rotation by $2\pi/3$ about the figure (z) axis plus the identity operation. The symmetry species in C_3 are the totally symmetric species A and the degenerate species E. Now a rotation about the z axis affects only the Eulerian angle χ. The symmetric-rotor eigenfunction given in Eq. (2.1.14) involves χ only in the factor $\exp(iK\chi)$, so that the effect of a rotation by $2\pi/3$ is

$$\Psi_{JKM} \rightarrow e^{i2\pi K/3}\Psi_{JKM}.$$

It is seen immediately that when $K = 0$ the eigenfunction is unchanged by the rotation. The levels with $K = 0$ therefore belong to the totally symmetric species A. The same is true when K is an integral multiple of 3. The two levels which exist for $K = \pm 3p$, where p is a positive integer, are each of species A, although they are degenerate in the rigid-rotor limit that we are now discussing. However, when K is neither zero nor an integral multiple of 3, the above transformation carries the eigenfunction into a linear combination of the degenerate eigenfunctions Ψ_{JKM} and $\Psi_{J,K,-M}$. These levels therefore belong to the degenerate species E.

[8] E. B. Wilson, J. C. Decius, and P. C. Cross, "Molecular Vibrations." McGraw Hill, New York, 1955.

2.1.2.7. The Asymmetric Rotor. We now consider the most general type of rigid rotor, that in which all three principal moments of inertia are different. The energy eigenvalues of the asymmetric rotor cannot be expressed in closed form, and the pattern of levels is much more complex than that of the linear and symmetric rotors. A secular equation must be solved to obtain the eigenvalues and eigenfunctions. We shall discuss several formulations that are commonly used for expressing the Hamiltonian of the asymmetric rotor. The choice for a particular problem is to some extent a matter of taste, but is also influenced by the computational methods that are to be used.

Let us first rewrite the Hamiltonian of Eq. (2.1.4) is an equivalent form

$$H = \tfrac{1}{2}[(\hbar^2/2I_x) + (\hbar^2/2I_y)](P_x{}^2 + P_y{}^2) + (\hbar^2/2I_z)P_z{}^2$$
$$+ \tfrac{1}{2}[(\hbar^2/2I_x) - (\hbar^2/2I_y)](P_x{}^2 - P_y{}^2). \tag{2.1.15}$$

The first two terms describe a rigid symmetric rotor with an effective nonunique moment $I_x I_y/(I_x + I_y)$. This part of the Hamiltonian is therefore diagonal in the JKM representation that we have used previously. The last term is diagonal in J and M, but not in K, as may be seen from the matrix elements of $P_x{}^2$ and $P_y{}^2$ in Eq. (2.1.8). The total angular momentum J is thus a constant of the motion, as is its projection M on the space-fixed axis, but there are no other good quantum numbers. However, the form of Eq. (2.1.15), in which the nondiagonal part of the Hamiltonian has been separated, suggests the obvious use of a symmetric rotor (J, K, M) basis for constructing the energy matrix of the asymmetric rotor.

Before proceeding further we must consider the labeling of the axes. There is no unique axis in the asymmetric rotor that can be chosen as z; in fact, the only way of distinguishing the axes is by the relative magnitudes of the principal moments. The conventional notation is to label the principal axes as a, b, and c, with the understanding that

$$I_a < I_b < I_c.$$

We may also define rotational constants A, B, C by

$$A = h/8\pi^2 I_a, \qquad B = h/8\pi^2 I_b, \qquad C = h/8\pi^2 I_c.$$

There are then six ways of identifying a, b, c with the axes x, y, z. These six choices have been designated by King, Hainer, and Cross[9] with the representation symbols given in Table I. The representations provide six distinct (but equivalent) ways of setting up the asymmetric rotor Hamiltonian. The type I and type III representations turn out to be more convenient than type II in most cases. We shall make use here of the I^r and III^l representations.

TABLE I. Identification of Axes in the Asymmetric-Rotor Problem

Representation	x	y	z
I^r	b	c	a
I^l	c	b	a
II^r	c	a	b
II^l	a	c	b
III^r	a	b	c
III^l	b	a	c

A very useful parameter that describes the degree of asymmetry of the rotor may be defined by

$$\varkappa = (2B - A - C)/(A - C).$$

The range of $\varkappa$ is -1 to $+1$. It is seen that $\varkappa = -1$ when $B = C$, which is the condition for a prolate symmetric rotor; likewise $\varkappa = +1$ when $A = B$, corresponding to the limiting oblate symmetric rotor.

In terms of the rotational constants which we have defined, Eq. (2.1.15) becomes

$$(H/h)[I^r] = \tfrac{1}{2}(B+C)(P_b{}^2+P_c{}^2)+AP_a{}^2+\tfrac{1}{2}(B-C)(P_b{}^2-P_c{}^2), \quad (2.1.16)$$

$$(H/h)[III^l] = \tfrac{1}{2}(A+B)(P_a{}^2+P_b{}^2)+CP_c{}^2+\tfrac{1}{2}(A-B)(P_a{}^2-P_b{}^2). \quad (2.1.17)$$

If the last term of Eq. (2.1.16) could be ignored, the Hamiltonian would describe a prolate symmetric rotor whose energy levels are determined by quantum numbers J and K. We shall label this K as K_{-1}, where the subscript indicates the value of the asymmetry parameter $\varkappa$ in the prolate symmetric limit; K_{-1} is to be taken as an unsigned quantum number. Likewise, the neglect of the last term of Eq. (2.1.17) leaves the Hamil-

[9] G. W. King, R. M. Hainer, and P. C. Cross, *J. Chem. Phys.* **11**, 27 (1943).

tonian of an oblate symmetric rotor ($\varkappa = 1$) whose levels are described by quantum numbers J and K_1.

In both limiting cases the levels are doubly degenerate, except for those where K_{-1} or K_1 are zero. However, the presence of the last term in Eqs. (2.1.16) and (2.1.17) removes the symmetric rotor degeneracy, giving $2J + 1$ distinct levels for each value of J. From symmetry considerations, levels of different K but the same J cannot cross. Therefore a simple correlation can be made between the levels of the limiting oblate and prolate symmetric rotors. This correlation diagram is shown in Fig. 1 for the case $J = 3$. A similar diagram may be drawn for each value of J.

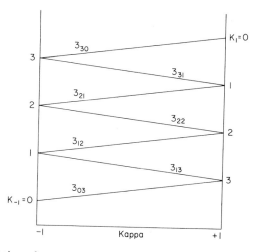

FIG. 1. Correlation of asymmetric-rotor levels between limiting prolate and oblate rotors ($J = 3$).

The correlation between limiting prolate and oblate symmetric rotors also provides a useful way of labeling the levels of the asymmetric rotor. A level is uniquely specified if the K values of the limiting symmetric rotors, K_{-1} and K_1, are given, together with the value of J. The conventional way of writing this (due to King et al.[9]) is

$$J_{K_{-1}K_1}.$$

The levels in Fig. 1 are labeled in this manner. It should be emphasized that K_{-1} and K_1 are simply indices used to label the asymmetric-rotor levels and are not good quantum numbers except in the symmetric rotor limits.

Another method of labeling the asymmetric-rotor levels is to number them with an index τ. For a given value of J, the level of lowest energy is designated by $\tau = -J$, the next by $\tau = -J + 1$, and so on, to $\tau = J$ for the highest level. The relation between the two schemes is given by

$$\tau = K_{-1} - K_1.$$

We can now investigate the form of the Hamiltonian matrix. It is convenient to rewrite Eqs. (2.1.16) and (2.1.17) as

$$(H/h)[\mathrm{I^r}] = \tfrac{1}{2}(B + C)P^2 + [A - \tfrac{1}{2}(B + C)]h(b_\mathrm{p}), \quad (2.1.18)$$

$$(H/h)[\mathrm{III^l}] = \tfrac{1}{2}(A + B)P^2 + [C - \tfrac{1}{2}(A + B)]h(b_\mathrm{o}), \quad (2.1.19)$$

where $h(b)$ is a dimensionless operator defined by

$$h(b) = P_z{}^2 - b(P_x{}^2 - P_y{}^2). \quad (2.1.20)$$

It may be noted that we have reverted to xyz labels for the axes in order to be able to handle both the $\mathrm{I^r}$ and $\mathrm{III^l}$ cases with a single expression. The quantity b is an asymmetry parameter originally defined by Wang. In the $\mathrm{I^r}$ case it is given by

$$b = b_\mathrm{p} = (C - B)/(2A - B - C),$$

while for the $\mathrm{III^l}$ representation

$$b = b_\mathrm{o} = (A - B)/(2C - A - B).$$

We may note the relations between b and $\varkappa$

$$b_\mathrm{p} = (\varkappa + 1)/(\varkappa - 3), \qquad b_\mathrm{o} = (\varkappa - 1)/(\varkappa + 3).$$

It is seen that $b_\mathrm{p} = 0$ in the prolate limit, while $b_\mathrm{o} = 0$ for the limiting oblate symmetric rotor.

Since the first terms in Eqs. (2.1.18) and (2.1.19) are diagonal, our problem reduces to finding the eigenvalues of the dimensionless matrix $h(b)$. If these eigenvalues are designated by $w(b)$, the asymmetric-rotor levels are given by

$$(E/h)[\mathrm{I^r}] = \tfrac{1}{2}(B + C)J(J + 1) + [A - \tfrac{1}{2}(B + C)]w(b_\mathrm{p}), \quad (2.1.21)$$

$$(E/h)[\mathrm{III^l}] = \tfrac{1}{2}(A + B)J(J + 1) + [C - \tfrac{1}{2}(A + B)]w(b_\mathrm{o}). \quad (2.1.22)$$

It should be emphasized at this point that Eqs. (2.1.21) and (2.1.22) are completely equivalent and will yield identical energy levels for a given set of rotational constants. However, from the standpoint of practical computations, the determination of $w(b)$ is easier when b is small. Therefore, one normally chooses the I^r expression when the rotor is nearer the prolate limit and the III^l expression near the oblate limit.

The matrix elements of $h(b)$ in the representation with P^2 and P_z diagonal may be written with the aid of Eqs. (2.1.7) and (2.1.8). The result is

$$(J, K \mid h \mid J, K) = K^2$$

$$(J, K \mid h \mid J, K \pm 2) = bf^{1/2}(J, K \pm 1)$$

where

$$f(J, n) = \tfrac{1}{4}[J(J + 1) - n(n + 1)][J(J + 1) - n(n - 1)]. \quad (2.1.23)$$

We have suppressed the quantum number M, since it does not affect the energy. There is one such matrix, of rank $2J + 1$, for each value of J; the rows and columns are labeled by K.

The matrix of $h(b)$ can be factored by applying the Wang transformation[9]

$$X = X^\dagger = 2^{-1/2} \begin{vmatrix} \cdots & & & & & \cdots \\ & -1 & & & 1 & \\ & & -1 & & 1 & \\ & & & 2^{1/2} & & \\ & & 1 & & 1 & \\ & 1 & & & & 1 \\ \cdots & & & & & \cdots \end{vmatrix}. \quad (2.1.24)$$

For a given J value the transformation X is of rank $2J + 1$, and the labeling is such that the first row and column correspond to $K = -J$, the second to $K = -J + 1$, and so on. It will be recognized that the application of this transformation is equivalent to the use of linear combinations of symmetric rotor functions

$$2^{-1/2}(\Psi_{JKM} \pm \Psi_{J,-K,M})$$

as basis functions. In addition to the twofold factoring produced by the Wang transformation, a further factoring into even and odd K is possible because the level K is connected only with the levels $K \pm 2$. Application of the Wang transformation thus gives

$$X^\dagger h X = h(E^+) + h(E^-) + h(O^+) + h(O^-)$$

where

$$h(E^\pm) = \begin{vmatrix} 0 & 2^{1/2}bf^{1/2}(J,1) & 0 & 0 & \cdots \\ 2^{1/2}bf^{1/2}(J,1) & 4 & bf^{1/2}(J,3) & 0 & \cdots \\ 0 & bf^{1/2}(J,3) & 16 & bf^{1/2}(J,5) & \cdots \\ 0 & 0 & bf^{1/2}(J,5) & 36 & \cdots \\ \cdots & \cdots & \cdots & \cdots & \cdots \end{vmatrix}$$

$$(2.1.25)$$

$$h(O^\pm) = \begin{vmatrix} 1 \pm bf^{1/2}(J,0) & bf^{1/2}(J,2) & 0 & 0 & \cdots \\ bf^{1/2}(J,2) & 9 & bf^{1/2}(J,4) & 0 & \cdots \\ 0 & bf^{1/2}(J,4) & 25 & bf^{1/2}(J,6) & \cdots \\ 0 & 0 & bf^{1/2}(J,6) & 49 & \cdots \\ \cdots & \cdots & \cdots & \cdots & \cdots \end{vmatrix} .$$

The matrix for E^- is identical to that for E^+ except that the first row and first column are omitted, as indicated by the dotted line. The rows and columns are labeled as

$$E^+: \quad |K| = 0, 2, 4, 6, \ldots ,$$
$$E^-: \quad |K| = 2, 4, 6, \ldots ,$$
$$O^+: \quad |K| = 1, 3, 5, \ldots ,$$
$$O^-: \quad |K| = 1, 3, 5, \ldots .$$

The maximum value of K is $2J + 1$. This is as far as symmetry considerations will take us, and the matrices in Eq. (2.1.25) must be diagonalized in order to obtain the eigenvalues $w(b)$.

Let us note some qualitative conclusions that can be drawn from Eq. (2.1.25). When b is very small, the diagonal elements, which are equal to K^2, yield a zeroth approximation to the energy. This gives the energy of the limiting symmetric rotor, as expected. From the O^+ and O^- matrices we see that the degeneracy of the $K = 1$ level is removed in first order; the splitting is symmetric in this order, and the separation of the split levels is just

$$2bf^{1/2}(J,0) = bJ(J+1).$$

Likewise, we see from the $E^\pm$ matrices that the $K = 2$ level will split in second order, but in an unsymmetric manner. It may be shown in general that the Kth level will split in the Kth order of perturbation, although all levels are shifted even in second order.

The other commonly used formulation of the asymmetric-rotor Hamiltonian makes use of the asymmetry parameter $\varkappa$. With some algebraic manipulation we may write the Hamiltonian as

$$H/h = AP_a^2 + BP_b^2 + CP_c^2$$
$$= \tfrac{1}{2}(A + C)P^2 + \tfrac{1}{2}(A - C)(P_a^2 + \varkappa P_b^2 - P_c^2).$$

If the eigenvalues of the dimensionless operator $P_a^2 + \varkappa P_b^2 - P_c^2$ are designed by $E(\varkappa)$, the eigenvalues of H become

$$E/h = \tfrac{1}{2}(A + C)J(J + 1) + \tfrac{1}{2}(A - C)E(\varkappa). \qquad (2.1.26)$$

This is a particularly convenient expression because it may be shown that

$$E_\tau^J(\varkappa) = -E_{-\tau}^J(-\varkappa), \qquad (2.1.27)$$

where we have identified the levels by the J_τ notation described previously. Thus in the tabulation of the reduced energy $E(\varkappa)$ it is only necessary to include positive values of $\varkappa$.

The matrix elements of the operator $P_a^2 + \varkappa P_b^2 - P_c^2$ are given for the various representations by King et al.[9] Application of the Wang transformation leads again to a factoring into four submatrices E^+, E^-, O^+, and O^-. The form of these matrices is discussed by King et al.[9]

To obtain the energy levels of the asymmetric rotor we must calculate the eigenvalues of the reduced Hamiltonian matrix. Explicit algebraic expressions for the eigenvalues can be obtained only for low values of J, where the matrices are small. The expressions for $w(b)$ and $E(\varkappa)$ have been tabulated by Allen and Cross (Table 2.1).[6] It is sometimes convenient to have the eigenvalues expressed directly in terms of the rotational constants: these are given in Table II for levels of $J \leq 3$.

When J is larger, a direct algebraic solution for the eigenvalues is not practical, and some suitable numerical method must be used. The most successful method for hand calculation is the continued fraction technique, which has been fully described.[9] However, the routine availability of high-speed digital computers has made such hand calculations unnecessary. There are now many programs for calculating asymmetric-rotor eigenvalues to any desired accuracy, even for high values of J.

TABLE II. Asymmetric-Rotor Energy Levels for $J \leq 3$ Expressed in Terms of the Rotational Constants

Level	Energy
0_{00}	0
1_{10}	$A + B$
1_{11}	$A + C$
1_{01}	$B + C$
2_{20}	$2A + 2B + 2C + 2[(B - C)^2 + (A - C)(A - B)]^{1/2}$
2_{21}	$4A + B + C$
2_{11}	$A + 4B + C$
2_{12}	$A + B + 4C$
2_{02}	$2A + 2B + 2C - 2[(B - C)^2 + (A - C)(A - B)]^{1/2}$
3_{30}	$5A + 5B + 2C + 2[4(A - B)^2 + (A - C)(B - C)]^{1/2}$
3_{31}	$5A + 2B + 5C + 2[4(A - C)^2 - (A - B)(B - C)]^{1/2}$
3_{21}	$2A + 5B + 5C + 2[4(B - C)^2 + (A - B)(A - C)]^{1/2}$
3_{22}	$4A + 4B + 4C$
3_{12}	$5A + 5B + 2C - 2[4(A - B)^2 + (A - C)(B - C)]^{1/2}$
3_{13}	$5A + 2B + 5C - 2[4(A - C)^2 - (A - B)(B - C)]^{1/2}$
3_{03}	$2A + 5B + 5C - 2[4(B - C)^2 + (A - B)(A - C)]^{1/2}$

Tables of the reduced eigenvalues $E(\varkappa)$ are often useful for preliminary calculations and predictions of spectra. A set of tables covering the range $J = 0$–12 in intervals of 0.01 in $\varkappa$ has been given by Townes and Schawlow.[1] A table for $J = 0$–40, $\varkappa(0.1)$ may be found in Allen and Cross.[6]

A number of approximate methods for calculating asymmetric-rotor eigenvalues were developed before large digital computers were available, but most of these now have only historical interest. An exception is the perturbation expansion of the eigenvalues about the symmetric-rotor limit, which still finds use. The expansion is best done in terms of the asymmetry parameter b. The function $w(b)$ which appears in Eqs. (2.1.21) and (2.1.22) can be expressed as a power series in b, which is rapidly convergent when b is small. Thus we can write

$$w(b) = K^2 + C_1 b + C_2 b^2 + \cdots, \qquad (2.1.28)$$

where K stands for K_{-1} if a I^r representation is used ($b = b_p$), and $K = K_1$ for a III^l representation ($b = b_0$). It may be shown that only even powers of b appear in the series when K is even. The formulas for

C_1 through C_5 are given by Allen and Cross (Appendix VI).[6] A tabulation of these coefficients for $J = 1$–12 may be found in Townes and Schaw-low.[1] The power series approximation is clearly most useful near the symmetric rotor limits, where b is small. The convergence is slow for levels of high J and low K.

The eigenfunctions of the rigid asymmetric rotor cannot be written explicitly. The most convenient way of expressing them is as linear combinations of symmetric-rotor eigenfunctions. In setting up the asymmetric-rotor Hamiltonian we applied the Wang transformation, which means that the appropriate basis functions are symmetric and antisymmetric linear combinations of the symmetric rotor functions for $+K$ and $-K$. We can write these as

$$S_{JKM\gamma} = 2^{-1/2}[\Psi_{JKM} + (-1)^\gamma \Psi_{J,-K,M}],$$

where Ψ_{JKM} is given by Eq. (2.1.14) and γ takes values 0 and 1. The value $\gamma = 0$ applies to the E^+ and O^+ submatrices, while $\gamma = 1$ corresponds to the E^- and O^- submatrices. If we designate the asymmetric-rotor eigenfunctions as $A_{J\tau M}$, they can be expressed as

$$A_{J\tau M} = \sum_K t_{K\tau} S_{JKM\gamma}. \qquad (2.1.29)$$

The summation extends only over those values of K contained in the submatrix in question. The coefficients $t_{K\tau}$ are the elements of the trans-formation matrix that diagonalizes the asymmetric-rotor Hamiltonian; they are functions of J and γ (but not of M).

The direct calculation of the $t_{K\tau}$ is discussed by King, Hainer, and Cross.[9] However, since most calculations of asymmetric-rotor energy levels are now done by computer, the transformation coefficients are easily obtained as a byproduct of the energy calculation. A table of $t_{K\tau}$ for $J = 1$–15 and at intervals of 0.1 in $\varkappa$ has been presented by Schwen-deman.[10] Perturbation expressions for the $t_{K\tau}$ are sometimes useful in situations near the symmetric rotor limits.[11]

We have seen that the asymmetric-rotor Hamiltonian undergoes a fourfold factoring, which implies that intrinsic symmetry properties are present. The Hamiltonian (2.1.15) is easily seen to be invariant under the operations of the so-called **V**-group, which consists of rotations of

[10] R. H. Schwendeman, Tables for the Rigid Asymmetric Rotor. *Nat. Std. Ref. Data Ser., Nat. Bur. Std. (U.S.)* **12** (1968).

[11] D. R. Lide, *J. Chem. Phys.* **20**, 1761 (1952).

$180°$ (C_2) about each of the principal axes. The four irreducible representations of the **V**-group are labeled A, B_a, B_b, B_c; the subscript on B indicates the axis for which the character is $+1$. The submatrices E^+, E^-, O^+, O^- thus belong to the species A, B_a, B_b, B_c, but the specific identification depends upon the parity of J and γ and upon the representation used. The full details of this identification are given by Allen and Cross.[6]

The symmetry properties of the asymmetric-rotor levels may be expressed in a simple form by making use of the parity of the indices K_{-1} and K_1 which appear in the notation $J_{K_{-1}K_1}$. Using the symbol eo to indicate a level in which K_{-1} is even and K_1 is odd, and similarly for the other combinations, we find the following unique correspondence with the species of the **V**-group

$$ee \leftrightarrow A, \qquad oo \leftrightarrow B_b,$$
$$eo \leftrightarrow B_a, \qquad oe \leftrightarrow B_c. \qquad (2.1.30)$$

This result is seen immediately from the form of the symmetric-rotor eigenfunctions, Eq. (2.1.14), which shows that the behavior with respect to a twofold rotation $(\chi \rightarrow \chi + \pi)$ is determined at the prolate and oblate limits by the parity of K. Thus the effect of a twofold rotation about the a axis is determined by the parity of K_{-1}, that of a rotation about the c axis by K_1, and that of a rotation about the b axis by the parity of the product $K_{-1}K_1$.

2.1.3. Selection Rules and Relative Intensities

The spectra that are observed in the microwave region are, with few exceptions, of the electric dipole type.* A necessary requirement for observing pure rotational transitions is the existence of a permanent electric dipole moment. The characteristics of the rotational eigenfunctions determine what transitions are allowed and the relative probability of each. In this section we shall first discuss the selection rules that can be derived from symmetry considerations and then give the procedure for quantitative calculation of relative intensities of rotational transitions.

2.1.3.1. Selection Rules for Linear, Symmetric, and Asymmetric Rotors.
Since we are concerned with electric dipole transitions, only those transi-

---

* An important exception is the magnetic dipole spectrum of O_2.

tions that follow the general selection rule

$$\varDelta J = 0, \pm 1$$

are allowed. In a rigid linear rotor, $\varDelta J = 0$ implies no change of state, and the only allowed transitions in the absorption spectrum are those in which J increases by one unit.

In the symmetric rotor an additional selection rule on the quantum number K must be established. In the most common case where the rotor is symmetric because of its threefold (or higher) symmetry, the dipole moment must necessarily coincide with the symmetry (figure) axis. It is thus independent of the Eulerian angle χ. However, the symmetric-rotor eigenfunction of Eq. (2.1.14) is seen to involve this angle through the factor $\exp(iK\chi)$. Therefore, all matrix elements of the dipole moment between states of different K must vanish, and we have the selection rules

$$\varDelta J = \pm 1, \qquad \varDelta K = 0$$

for the rigid symmetric rotor.

The situation is more complicated in the asymmetric rotor, since the dipole moment $\boldsymbol{\mu}$ may have components along each of the principal axes. From the discussion of symmetry properties in Section 2.1.2.7 it is seen that the dipole components μ_a, μ_b, μ_c belong to symmetry species B_a, B_b, B_c, respectively, of the V-group. Thus the dipole component μ_a gives rise to transitions between levels of species A and B_a and between levels of species B_b and B_c, and similarly for the other components. Making use of the correspondences of Eq. (2.1.30), we thus find the selection rules

$$\mu_a: \quad \text{ee} \leftrightarrow \text{eo} \quad \text{and} \quad \text{oo} \leftrightarrow \text{oe},$$

$$\mu_b: \quad \text{ee} \leftrightarrow \text{oo} \quad \text{and} \quad \text{eo} \leftrightarrow \text{oe},$$

$$\mu_c: \quad \text{ee} \leftrightarrow \text{oe} \quad \text{and} \quad \text{oo} \leftrightarrow \text{eo}.$$

Another way of stating these selection rules is to note that K_{-1} does not change parity for a μ_a-type transition, K_1 does not change parity for a μ_c-type transition, while both indices change parity for a μ_b-type transition. In addition, the selection rule $\varDelta J = 0, \pm 1$ still applies, and $\varDelta J = 0$ now represents a nontrivial type of transition.

2.1.3.2. Rotational Line Strengths. It will be shown later (Section 2.1.12) that the absolute intensity of a rotational absorption line is proportional to the square of a dipole matrix element computed in a space-fixed axis frame. The instantaneous value of the dipole component along the space-fixed Z axis may be expressed as

$$\mu_Z = \mu_x \Phi_{Zx} + \mu_y \Phi_{Zy} + \mu_z \Phi_{Zz}, \qquad (2.1.31)$$

where Φ_{Zx} is the cosine of the angle between the space-fixed Z axis and the molecule-fixed x axis. Similar expressions hold for μ_X and μ_Y. Thus the set of direction cosines Φ_{Fg}, with $F = X, Y, Z$ and $g = x, y, z$, are required for the quantitative calculation of intensities. The matrix elements of these direction cosines give an immediate indication of the relative intensities of the various allowed transitions.

The direction cosine matrix elements in a symmetric rotor representation may be expressed in closed form. The results are shown in Table III. Details of the calculation may be found in Allen and Cross.[6] The direction cosine elements of the asymmetric rotor cannot in general be calculated explicitly, but are conveniently described as linear combinations of symmetric-rotor matrix elements by using Eq. (2.1.29).

In order to calculate the intensity of a rotational transition between the levels R and R' in the absence of external fields, we require the squares of the direction cosine matrix elements summed over the three spacial directions and over all of the spacially degenerate states. This quantity is conventionally defined as the *line strength* S of the transition. It is given explicitly by

$$S = \sum_{F,M,M'} |(R | \Phi_{Fg} | R')|^2, \qquad (2.1.32)$$

where $F = X, Y, Z$. It should be noted that this line strength refers to the particular dipole component μ_g that is responsible for the transition $R \leftrightarrow R'$, i.e., the intensity of the transition is proportional to $S\mu_g^2$.

The line strengths of the allowed transitions in linear and symmetric rotors are easily calculated from the matrix elements of Table III. For the $J \rightarrow J + 1$ transition of a linear rotor the result is

$$S = J + 1 \qquad (2.1.33a)$$

while for the transition $J \rightarrow J + 1$, $K \rightarrow K$ of a symmetric rotor we obtain

$$S = (J + 1)[1 - K^2/(J + 1)^2]. \qquad (2.1.33b)$$

The line strengths for asymmetric-rotor transition are most readily discussed by reference to the limiting symmetric-rotor cases. Explicit expressions can be derived for asymmetric-rotor transitions in the sym-

TABLE III. Direction Cosine Matrix Elements in a Symmetric-Rotor Representation[a]

	$J' = J + 1$	$J' = J$	$J' = J - 1$
$(\Phi^0_{Fy})_{J,K;J',K\pm1} = \mp i(\Phi^0_{Fx})_{J,K;J',K\pm1}$	$\mp[(J \pm K + 1)(J \pm K + 2)]^{1/2}$	$[(J \mp K)(J \pm K + 1)]^{1/2}$	$\mp[(J \mp K)(J \mp K - 1)]^{1/2}$
$(\Phi^0_{Fz})_{J,K;J',K}$	$2[(J + K + 1)(J - K + 1)]^{1/2}$	$2K$	$-2[J^2 - K^2]^{1/2}$
$(\Phi^0_{Fy})_{J,0;J',1} = -i(\Phi^0_{Fx})_{J,0;J',1}$	$-[2(J + 1)(J + 2)]^{1/2}$	$[2J(J + 1)]^{1/2}$	$-[2J(J - 1)]^{1/2}$

[a] Line 1 applies to $K > 0$; line 2 to all K.

metric-rotor limits by the use of Eq. (2.1.32) and Table III.[12] These results are summarized in Table IV. The line strength may be calculated for arbitrary asymmetry by using the limiting direction cosine elements of Table III with the transformation of Eq. (2.1.29), followed by summation according to Eq. (2.1.32). Most computer programs for asymmetric-rotor energy eigenvalues may be extended to calculate the line strengths as well. Useful tabulations of asymmetric rotor line strengths may be found in Townes and Schawlow[1] [$J \leq 12; \varkappa = -1$ (0.5) 1], Laurie and Schwendeman[13] [$J \leq 12; \varkappa = -1$ (0.1) 1], and Wacker and Pratto[14] [$J \leq = 35, \varkappa = \pm(0, 0.3, 0.6, 0.8, 0.95, 1.0)$]. Perturbation formulas for the near-symmetric limit have been given by Lide.[11]

Some general comments on asymmetric-rotor line strengths can be made, although specific values are often sensitive to the degree of asymmetry and to the J values involved in the transition. Transitions that are allowed in both the prolate ($\varkappa = -1$) and oblate ($\varkappa = +1$) symmetric-rotor limits have reasonably large line strengths for all values of the asymmetry parameter. Transition of the μ_b-type in which both K_{-1} and K_1 change by one unit fall into this category, as do μ_a-type transitions with $\Delta K_{-1} = 0$, $\Delta K_1 = \pm 1$ and μ_c-type transitions with $\Delta K_{-1} = \pm 1$, $\Delta K_1 = 0$. The following classes of transitions are strong in a near-prolate rotor but vanish in strength as the oblate limit is approached:

$$\mu_b: \quad \Delta K_{-1} = \pm 1, \quad \Delta K_1 = \pm 3,$$
$$\mu_c: \quad \Delta K_{-1} = \pm 1, \quad \Delta K_1 = \pm 2,$$

while the converse is true for the transitions

$$\mu_b: \quad \Delta K_{-1} = \pm 3, \quad \Delta K_1 = \pm 1,$$
$$\mu_a: \quad \Delta K_{-1} = \pm 2, \quad \Delta K_1 = \pm 1.$$

Other transitions are forbidden in both prolate and oblate limits but may have significant strength at intermediate values of asymmetry. It should be emphasized, however, that the degree to which the prolate or oblate limit is approached depends not only on the value of $\varkappa$ but also on the J values involved.

[12] P. C. Cross, R. M. Hainer, and G. W. King, *J. Chem. Phys.* **12**, 210 (1944).

[13] R. H. Schwendeman and V. W. Laurie, "Tables of Line Strengths." Pergamon, Oxford, 1958.

[14] P. F. Wacker and M. R. Pratto, Microwave Spectral Tables: Line Strengths of Asymmetric Rotors. *Nat. Bur. Std. (U.S.)* Monograph **70**, Vol. II (1964).

TABLE IV. Rotational Line Strengths in the Symmetric-Rotor Limit[a]

	$J' = J+1$	$J' = J$	$J' = J-1$
$S^0_{J,K;J',K+1}$	$(J+K+2)(J+K+1)/4(J+1)$	$(2J+1)(J+K+1)(J-K)/4J(J+1)$	$(J-K-1)(J-K)/4J$
$S^0_{J,K;J',K}$	$(J+K+1)(J-K+1)/(J+1)$	$K^2(2J+1)/J(J+1)$	$(J^2-K^2)/J$
$S^0_{J,0;J',1}$	$(J+2)/2$	$(2J+1)/2$	$(J-1)/2$

[a] Line 1 applies to $K > 0$; line 2 to all K.

2.1.3.3. Nuclear Spin Statistical Weights. A molecule that belongs to a point group containing a finite rotation as one of its symmetry elements must necessarily have two or more equivalent nuclei. The total wave function, including nuclear spin, will either remain the same or change sign, depending on the statistics followed by the nuclei, under a rotation which interchanges equivalent nuclei. This leads to statistical weight factors for the rotational levels which affect the relative intensities of transitions.

The symmetry properties of a symmetric rotor with a threefold symmetry axis were discussed in Section 2.1.2.6. A consideration of the symmetry properties of the spin functions for a nucleus of $I = \frac{1}{2}$ shows that levels of $K = 0, 1, 2, 3, 4, 5, 6, \ldots$ have relative statistical weights 2, 1, 1, 2, 1, 1, 2, $\ldots$, respectively. When the nuclear spin is zero, levels in which K is not a multiple of three are missing entirely, while the remainder (including $K = 0$) have equal weights. These considerations assume that no inversion splitting is resolved.

Nuclear spin weights are also important in asymmetric rotors belonging to point group C_2 or C_{2v}. The rotational wave functions must either remain the same or change sign for a rotation of 180° about the symmetry axis. The same is true for the nuclear spin functions. The rotational and spin functions must be combined in such a way that the behavior of the overall wave function conforms to the net statistics of the indistinguishable nuclei.

In the simple case where there are only two indistinguishable nuclei of spin $\frac{1}{2}$ (as in H_2O), there are four spin functions, three of which are unchanged (symmetric) while the fourth is antisymmetric to an interchange of nuclei. Since the total wave function must be antisymmetric for spin $\frac{1}{2}$ nuclei, we see that the three symmetric-spin functions must be

TABLE V. Statistical Weights for a C_{2v} or C_2 Molecule Containing One Pair of Indistinguishable Nuclei of Spin $\frac{1}{2}$

C_2 symmetry axis	Statistical weights			
	ee	eo	oe	oo
a	1	1	3	3
b	1	3	3	1
c	1	3	1	3

combined with an antisymmetric-rotational function and vice versa. The behavior of the rotational functions is easily determined from the $K_{-1}K$ indices with which the levels are labeled [see Eq. (2.1.30)]. Thus we obtain the results in Table V for the weights applicable to the three possible identifications of the symmetry axis with a principal axis.

2.1.4. Rigid-Rotor Spectral Patterns

The rotational fine structure of infrared vibrational bands can be classified in a logical manner which is very useful in spectral analysis. Thus one can draw certain conclusions from the overall shape of the band and then refine the interpretation step by step until all details are explained. In principle, the same classification scheme can be applied to the pure-rotational spectra observed in the microwave region. However, a practical problem arises. A typical vibration–rotation band in the infrared might extend over a region of 50 to 100 cm^{-1}. The pure rotational spectrum covers a similar range, but this is much larger than the entire microwave region. In fact, experimental limitations often restrict the available microwave data to a range of 1 to 2 cm^{-1}. Even when microwave measurements are available for a broader range, differences in sensitivity and detector response make it difficult to compare intensities in different parts of the spectrum. Thus the analysis of a microwave spectrum must often be based on data obtained for a small "window" in the total spectrum, a window whose location is dictated by experimental considerations.

For this reason it is not easy to devise a comprehensive classification scheme that will be useful for all microwave spectra. Nevertheless, certain patterns do tend to recur whose recognition greatly simplifies the interpretation of the spectrum.

2.1.4.1. Linear Molecules and Symmetric Rotors.
The rotational spectra of rigid linear molecules and symmetric rotors require little discussion. We have seen that the pertinent selection rule for pure rotational transitions is $\Delta J = +1$, with the additional restriction that $\Delta K = 0$ for symmetric rotors. In both cases the frequency is given by

$$\nu(J \to J + 1) = 2B(J + 1). \tag{2.1.34}$$

The spectrum thus consists of lines separated by an interval $2B$, as shown in Fig. 2. The line strength increases with increasing J, according to Eq. (2.1.33a), and a further increase results from the ν^2–dependence of the absorption coefficient (Section 2.1.12). However, the rotational

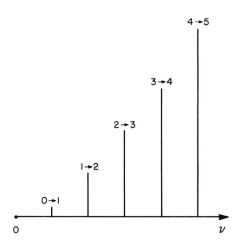

FIG. 2. Spectral pattern for rigid linear or symmetric rotor. The transitions are labeled $J \to J + 1$.

Boltzmann factor eventually becomes more significant and leads to decreasing intensities. The peak intensity comes roughly at $J \approx (3kT/2B)^{1/2}$. Measurement of two or three lines is generally sufficient to establish the J assignments with certainty. The rotational constant B is then calculated immediately from Eq. (2.1.34).

Deviations from this simple rigid-rotor pattern are caused by centrifugal distortion and by vibration–rotation interactions. As is shown in Section 2.1.6, centrifugal distortion changes slightly the intervals between successive lines. Also, in symmetric rotors the lines of different K become slightly separated. When centrifugal distortion is included, Eq. (2.1.34) must be replaced by

$$\nu(J \to J + 1) = 2B(J + 1) - 4D_J(J + 1)^3 - 2D_{JK}K^2, \quad (2.1.35)$$

where D_J and D_{JK} are centrifugal distortion constants. The D_{JK} term should be ignored for linear molecules. In symmetric rotors D_{JK} is readily determined if the K-structure is resolved. Accurate measurement of two transitions, preferably with rather different J values, then permits calculation of B and D_J. If only one transition can be measured, it is possible (Section 2.1.6) to estimate D_J from vibrational data and thereby to obtain a somewhat more accurate value of B.

The other complication found in the spectra of real molecules is the presence of satellite lines from excited vibrational states. These generally fall very close to the ground-state transition, so that a large region between two successive rotational transitions is still free of lines. Vibrational satellite patterns are discussed in Section 2.1.6.

2.1.4.2. Asymmetric-Rotor Patterns. The first point to distinguish in the classification of asymmetric rotor spectra is the dipole component that is responsible for a particular set of transitions. Thus we can refer to a-type, b-type, and c-type transitions resulting from the dipole components μ_a, μ_b, and μ_c, respectively. If the orientation of the dipole moment is fixed by molecular symmetry, only one or two of these types will be allowed. In particular, if the molecule contains an axis of symmetry, there will be only one type of transition represented in the microwave spectrum. In such cases a rough knowledge of the molecular structure usually (though not always) permits the type of spectrum to be predicted. In other cases one type of transition may dominate the spectrum even though the others are allowed. Since the absorption intensity depends upon the square of the dipole component, one type of transition will be far stronger than the others if the dipole moment vector is even roughly parallel to a principal axis. If, for example, in a molecule in which the a and b axes lie in a plane of symmetry, the dipole moment falls $20°$ from the a axis, the μ_a transitions will be seven times as strong, on the average, as the μ_b transitions. Therefore, even very crude estimates of the orientation of the dipole moment are useful in predicting the type of transitions to be expected.

In the vibration–rotation spectra of symmetric rotors a distinction is made between *parallel* and *perpendicular* bands, i.e., between bands in which the vibrational transition moment is parallel or perpendicular to the symmetry axis of the rotor. The rotational selection rule for parallel bands is $\Delta K = 0$, while for perpendicular bands $\Delta K = \pm 1$. Though not rigorous for asymmetric rotors, this classification is still convenient if the degree of asymmetry is small. The same classification is very useful in the pure-rotational spectra of slightly asymmetric rotors. That is, if any meaning can be attached to the quantum number K in an asymmetric rotor, we can sort the principal transitions according to whether $\Delta K = 0$ or $\Delta K = \pm 1$. The $\Delta K = 0$ transitions are produced by the component of the permanent dipole moment that is parallel to the pseudo-symmetry axis, while the $\Delta K = \pm 1$ transitions result from the perpendicular dipole component. One might think that such a classification would be meaningful only for cases very close to the symmetric-rotor limit; however, it often turns out to be a useful concept even for quite asymmetric molecules.

2.1.4.3. Parallel Transitions. It is clear that parallel transitions can be associated with either the μ_a dipole component in a near-prolate rotor or the μ_c component in a near-oblate rotor. In the former case we identify $K \equiv K_{-1}$, and the rigorous selection rules of Section 2.1.3.1 state that the parity of K must not change, i.e., that $\Delta K = 0, \pm 2, \pm 4, \ldots$. If the rotor is only slightly asymmetric, the line strengths of the $\Delta K = 0$ transitions will be much greater than those for the transitions in which $\Delta K \neq 0$. Therefore, what we have called the "parallel transitions" comprise the strongest group of transitions produced by the μ_a component. Exactly the same considerations apply to a near-oblate rotor, where we identify K with K_1.

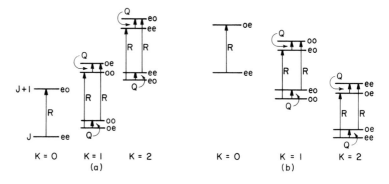

FIG. 3. Energy-level patterns for slightly asymmetric rotors, with parallel transitions indicated; μ_a transitions for the (a) near-prolate case and μ_c for the (b) near-oblate case. The labels are appropriate for an even value of J.

The pattern of levels for near-prolate and near-oblate rotors is illustrated in Fig. 3, with the parallel transitions indicated. The general selection rule $\Delta J = 0, \pm 1$ implies two types of transitions which are conventionally designated as Q-branch ($\Delta J = 0$) and R-branch ($\Delta J = 1$). The nature of the spectral patterns may be deduced from the power series expression for the asymmetric rotor energy, Eq. (2.1.28). Considering first the R-branch transitions, we see that a pair of lines appears for each value of K (except $K = 0$). However, if the asymmetry is small, all of the $2J + 1$ lines for a given $J \rightarrow J + 1$ transition fall close to a mean value given by $(B + C)(J + 1)$ in the near-prolate case and $(A + B)(J + 1)$ in the oblate case. The resulting spectral patterns are shown in Fig. 4. Here we have labeled the transitions ν_K^{h} and ν_K^{l}, with the superscript indicating whether a line is the higher or lower frequency member of a K-doublet. The $K = 1$ doublet undergoes the largest

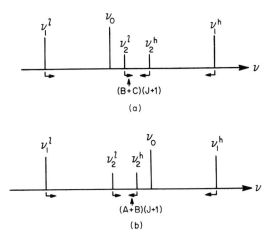

FIG. 4. Spectral patterns for parallel transitions in a slightly asymmetric rotor: (a) near-prolate, (b) near-oblate. The transition is taken as $J \rightarrow J + 1$, but lines with $K > 2$ are not shown. The arrows indicate the predominant direction of the Stark effect.

splitting, and the remaining lines fall between this pair (at least for small asymmetry). It will be noted from Fig. 4 that the patterns for near-prolate and near-oblate rotors have a mirror-image relation; the ν_0 line, for example, falls at a lower frequency than the mean of the ν_1^h and ν_1^l lines in a near-prolate rotor, but at a higher frequency in the near-oblate case. Thus it is easy to distinguish the two cases from the qualitative appearance of the pattern.

The parallel R-branch spectrum of a slightly asymmetric rotor can therefore be visualized as a symmetric-rotor pattern (Fig. 2) in which each line has split into a group of the type shown in Fig. 4. The centers of gravity of successive groups are separated by an amount $B + C$ in the prolate case and $A + B$ in the oblate case. Both the number of lines in each group $(2J + 1)$ and the frequency spread of the group increases with increasing J. At a sufficiently large value of J the adjacent groups will begin to overlap, and the characteristic appearance of a parallel-type spectrum will then be lost. This point will be reached at a lower value of J as the asymmetry becomes greater. Furthermore, other transitions appear with detectable intensity as the near-symmetric approximation breaks down. Nevertheless, it is often possible to recognize groups of the type in Fig. 4 even when there is considerable overlapping.

The characteristic Stark effects in a parallel-type spectrum are helpful in making assignments. The Stark interaction through the μ_a component in a near-prolate rotor (or μ_c in a near-oblate rotor) leads to a strong repulsion between the two levels that have been split by asymmetry (Section 2.1.9.3). Reference to Fig. 3 shows that the two transitions of given K value will therefore tend to be brought together by the Stark effect. Thus the line $\nu_K{}^h$ should show a Stark shift predominately to low frequency, while $\nu_K{}^l$ should be shifted to high frequency by a comparable amount. This behavior is indicated by the arrows in Fig. 4. The magnitude of the Stark shifts becomes larger with increasing K, since the degeneracy is then more nearly exact. The $K = 0$ line is easily distinguished from the rest of the pattern because of its much weaker Stark effect. It should be remarked, however, that these observations are valid only if the dipole component that produces the transition also makes the dominant contribution to the Stark effect. Rather different Stark effects may appear if another dipole component is significant.

When the lines in a parallel-type pattern have been identified, approximate values of the rotational constants are easily derived. We can show that $B + C$ (or $A + B$ in the oblate case) is given, to first order, by

$$B + C = (\nu_1{}^h + \nu_1{}^l)/2(J + 1). \qquad (2.1.36)$$

Similarly, we obtain for $B - C$ (or $A - B$)

$$B - C = (\nu_1{}^h - \nu_1{}^l)/(J + 1). \qquad (2.1.37)$$

The A constant can be estimated from the splitting $\nu_2{}^h - \nu_0$ (or $\nu_0 - \nu_2{}^l$ in the oblate case), or from $\nu_0 - \frac{1}{2}(\nu_1{}^h + \nu_1{}^l)$. These splittings are given to a first approximation by the quadratic terms in the power series expansion of Eq. (2.1.28).

It is also possible to have Q-branch ($\Delta J = 0$) transitions between the two components of a K-doublet, as shown in Fig. 3. These transitions form a series for each value of K. The first series, for $K = 1$, follows the approximate formula

$$\nu(J) = \frac{1}{2}(B - C)J(J + 1),$$

and the expressions for other values of K are easily obtained from the power series coefficients. Since the asymmetry splitting decreases rapidly with increasing K [see Eq. (2.1.28)], the Q-branch transitions of higher K will fall at much lower frequencies (for comparable values of J). The line strengths for the $\Delta J = 0$ transitions tend to be small and to decrease with increasing J. Therefore, these transitions are usually not as prominent as those with $\Delta J = 1$.

2.1.4.4. Perpendicular Transitions. The types of perpendicular transitions to be considered are

near-prolate rotor: μ_b or μ_c transitions,

near-oblate rotor: μ_a or μ_b transitions.

In each case the rigorous selection rules require that the parity of K change during a transition. As before, we consider only the transitions that predominate near the symmetric limits, namely those with $\Delta K = \pm 1$. We then recognize the following possible changes in the quantum numbers J and K

$$\Delta J = 0, \qquad \Delta K = \pm 1 \qquad (Q\text{-branch}),$$
$$\Delta J = \pm 1, \qquad \Delta K = \pm 1 \qquad (R_+\text{-branch}),$$
$$\Delta J = \pm 1, \qquad \Delta K = \mp 1 \qquad (R_-\text{-branch}).$$

Here we use a rather condensed notation in which the symbol R indicates that J changes in the transition (by either $+1$ or -1). The subscript $+$ or $-$ on R indicates whether the change in K has the same or opposite sign as the change in J. This is more convenient for microwave spectroscopy than the more elaborate notation of Cross et $al.$[12]

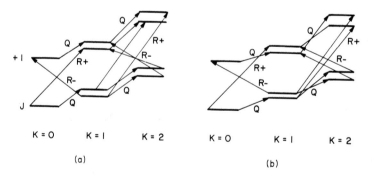

Fig. 5. Allowed perpendicular transitions in a near-prolate rotor: (a) μ_b-type, (b) μ_c-type.

The level pattern and the allowed transitions are illustrated in Fig. 5 for the near-prolate case. A similar diagram is easily derived for the near-oblate limit. The Q-branch transitions will be discussed first, because they are most likely to lead to recognizable spectral patterns. In Fig. 6 a plot of Q-branch transitions is given for a moderately asymmetric

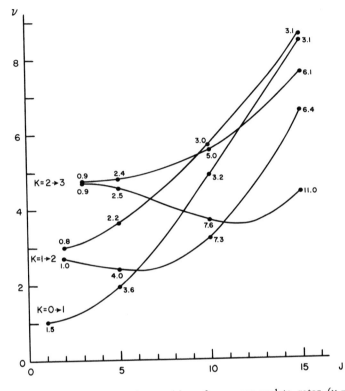

FIG. 6. Perpendicular Q-branch transitions for a near-prolate rotor ($\varkappa = -0.80$) with μ_b selection rules. The ordinate is in units of $A - C$. Line strengths are given at selected values of J.

near-prolate rotor ($\varkappa = -0.80$) with μ_b selection rules. For each value of K (except $K = 0$) the transitions $K \rightarrow K + 1$ form two series that diverge from each other as J increases. If the curves are extrapolated to $J = 0$, the origin of the series $K \rightarrow K + 1$ is given to a first approximation by

$$\nu_0(K \rightarrow K + 1) \approx [A - \tfrac{1}{2}(B + C)](2K + 1).$$

The two series for a given K have comparable intensities, although the line strengths are somewhat greater for the lower-frequency branch. This branch initially moves to low frequency with increasing J, but eventually passes through a minimum, and the frequencies begin to increase. As J increases the appearance of well-defined series tends to be lost. At very high J values, however, a new series of doublets may appear as the level pattern approaches the oblate limit. Thus it may be seen from Fig. 6 that the $K = 0 \to 1$ series merges at high J with the upper branch of the $K = 1 \to 2$ series (but here our $K = K_{-1}$ is no longer a meaningful quantum number).

The corresponding patterns for μ_c-type transitions are rather different (see Fig. 7). Here most of the intensity goes into the low-frequency branch, and at high values of J the line strength in the high-frequency branch become very small. This is basicly because the transitions of the lower branch (involving $\Delta K_{-1} = 1$, $\Delta K_1 = 0$) are permitted at both the prolate and oblate limits, while the upper branch, with $\Delta K_{-1} = 1$, $\Delta K_1 = 2$, is forbidden at the oblate limit. The μ_b transitions, on the contrary, are allowed at both limits for both branches. Another important difference is that the lower branch in the μ_c case approaches zero frequency as J increases. Furthermore, the $K = 0 \to 1$ series, which is often the easiest to recognize, moves to lower frequency in the μ_c case, in contrast to its behavior with μ_b selection rules.

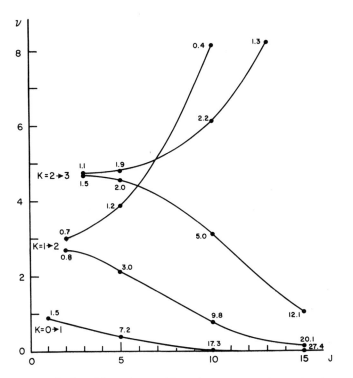

FIG. 7. Perpendicular Q-branch transitions for a near-prolate rotor ($\varkappa = -0.80$) with μ_c selection rules. The ordinate is in units of $A - C$. Line strengths are given for selected values of J.

The pattern for other values of $\varkappa$ will be qualitatively similar to Fig. 6 and 7. If the asymmetry of the rotor is smaller, the branches will not begin to diverge or the series overlap until higher values of J are reached. In case of greater asymmetry there may be no easily recognizable series because of the strong overlapping of the branches even at low J values.

Of course, it must be recalled that other transitions are possible besides the $\varDelta K = 1$ transitions considered here, and these will tend to become more important as the rotor deviates from the symmetric limit. However, the transitions of $\varDelta K = 3, 5$, etc., will almost always be much weaker than the transitions plotted in Figs. 6 and 7. At $\varkappa = 0.80$ and $J = 10$, for example, the primary transitions are seen to have line strengths in the range 3–17, while the strongest of the secondary transitions has a line strength of only 0.16. Generally similar intensity relations are found even in more asymmetric cases.

Although the labels on Figs. 6 and 7 refer to a near-prolate rotor, the same curves can be applied to the near-oblate case. This can be seen from the relation

$$E_\tau{}^J(\varkappa) = -E_{-\tau}^J(-\varkappa), \qquad (2.1.27)$$

which implies that the transition $J_{mn} \rightarrow J_{st}$ at a given value of $\varkappa$ has the same frequency (and, as may be shown, the same line strength) as the transition $J_{ts} \rightarrow J_{nm}$ at an asymmetry value of $-\varkappa$. Therefore, series of the type in Figs. 6 and 7 will be found for a near-oblate rotor, but the appropriate quantum-number change is $\varDelta K = -1$ rather than $\varDelta K = +1$.

The Q-branch series of this type frequently provide the most useful clues for analyzing a perpendicular-type spectrum. A rather convenient technique for checking the assignment of such series is the so-called "Q-branch plot." The $\varDelta J = 0$ transitions in a rigid rotor are functions of only two parameters, which may be taken, for example, as $A - C$ and $\varkappa$ [see Eq. (2.1.26)]. Each assigned transition then provides a relationship between $A - C$ and $\varkappa$. Over a small range of $\varkappa$ these curves can be approximated as straight lines which are readily calculated from a table of $E(\varkappa)$. If all assignments are correct, the curves should have a mutual point of intersection. Thus it is easy to test any tentative assignment with this graphical procedure, and in the same process to obtain approximate values of $A - C$ and $\varkappa$. It must be remembered, of course, that in a real molecule centrifugal distortion will spoil the mutual intersection of the curves to a slight extent.

The R-branch transitions generally produce less distinctive spectral features than the Q-branch patterns discussed above. In the zeroth approximation of the limiting symmetric rotor, the R-branch transitions $J, K \rightarrow J + 1, K \pm 1$, which we designate as $R_\pm$, have frequencies given by

$$\nu(R_\pm) \approx (B + C)(J + 1)$$
$$+ [A - \tfrac{1}{2}(B + C)](\pm 2K + 1) \quad \text{(near-prolate rotor)},$$
$$\nu(R_\pm) \approx (A + B)(J + 1)$$
$$+ [C - \tfrac{1}{2}(A + B)](\pm 2K + 1) \quad \text{(near-oblate rotor)}.$$

$$(2.1.38)$$

We note first that $A - \tfrac{1}{2}(B + C)$ is necessarily positive, while $C - \tfrac{1}{2}(A + B)$ is negative. Thus in the near-prolate case the two terms in Eq. (2.1.38) have the same sign in $\nu(R_+)$ but opposite signs in $\nu(R_-)$. The reverse is true in the near-oblate case. We may therefore recognize one class of perpendicular transitions, R_+ for the near-prolate rotor and R_- for the near-oblate, in which the frequency tends to increase systematically with increasing J and K (although there may be regions where the J-dependence is not monotonic). These transitions form series which we may regard as being derived from the Q-branch series in Figs. 6 and 7. That is, to a first approximation the R_+ $(K \rightarrow K + 1)$ series in the near-prolate case and the R_- $(K \rightarrow K - 1)$ series in the near-oblate case can be obtained by adding a linear function of $J + 1$ to the curves in Figs. 6 and 7. The resulting curves have much steeper slopes than the Q-branch curves and are thus not so likely to produce recognizable series. In light molecules most transitions of this type will fall at rather high frequencies.

The other class of R-branch perpendicular transitions is that in which the two terms in Eq. (2.1.38) have opposite signs, i.e., R_- for a near-prolate rotor and R_+ for the near-oblate case.* Here it is possible for the two terms to cancel approximately, so that transitions with quite high values of J and K may appear at low frequencies. Transitions of this type are perhaps the most difficult of all to identify. Since the frequency is given by a small difference between two large numbers, predictions

---

* It is obvious that Eq. (2.1.38) can lead to negative as well as positive frequencies. To follow conventional notation, we might reverse the signs in Eq. (2.1.38) and designate these as $P_\pm$ transitions ($\Delta J = -1$). However, such a notation tends to be cumbersome in microwave spectroscopy, so that we shall use the symbol R for any $|\Delta J| = 1$ transition.

based upon estimated rotational constants tend to be poor. Also, such transitions are subject to a large centrifugal distortion correction (expressed as a percentage of the frequency).

Transitions of this type tend to be distributed rather randomly throughout the spectrum, and recognizable patterns occur only from accidental factors such as special relationships among the rotational constants. For example, in a near-prolate rotor if $A - \frac{1}{2}(B + C)$ happens to be almost equal to an integral multiple, say N, of $\frac{1}{2}(B + C)$, it is easily shown from Eq. (2.1.38) that the transition $J, K \rightarrow J + 1, K - 1$ will have approximately the same frequency as $J + N, K + 1 \rightarrow J + N + 1 \rightarrow K$. Of course, this approximation is very crude, and the statement means only that the two transitions will fall in roughly the same region of the spectrum. However, it is possible in this way to get a series of lines related by the fact that successive members involve an increase of 1 in K and of N in J. Series of this type have been found in CH_2F_2 [15] and propylene oxide.[16] There will be two such series because of the asymmetry splitting, which may or may not fall close together, depending on the precise relationship among the rotational constants. If it happens that the asymmetry splitting is small, a recognizable pattern of doublets may result. This was found to be the case in CH_2F_2.

The R_- transitions are on the average somewhat weaker than the R_+ transitions. This may be seen by noting that in the symmetric-rotor limit the line strengths of the transitions $J + 1, K \rightarrow J, K + 1$ and $J, K \rightarrow J + 1, K + 1$ are in the ratio

$$(J - K - 2)(J - K - 1)/(J + K + 2)(J + K + 1).$$

Thus in the near-prolate rotor the rather irregular R_- transitions discussed in the last paragraph tend to be weaker than the R_+ and Q transitions. In the near-oblate case, on the other hand, the irregular transitions are R_+ and thus are comparable in intensity to the Q-branch spectrum.

A summary of the possible R-branch transitions for near-prolate and near-oblate rotors is given in Table VI. This table includes only the primary transitions, i.e., those in which $\Delta K = \pm 1$. While other transitions may certainly appear with detectable intensity, they generally do not contribute systematic patterns to the spectrum.

[15] D. R. Lide, *J. Amer. Chem. Soc.* **74**, 3548 (1952).
[16] D. R. Herschbach and J. D. Swalen, *J. Chem. Phys.* **29**, 761 (1958).

TABLE VI. Classification of Principal R-Branch Perpendicular Transitions

	ΔJ	ΔK_{-1}	ΔK_1	$\Delta \tau$	Type
Near-Prolate Rotor:					
	$+1$	$+1$	$+1$	0	$R_+(\mu_b)$
	$+1$	$+1$	-1	$+2$	$R_+(\mu_b)$
	$+1$	-1	$+1$	-2	$R_-(\mu_b)$
	$+1$	-1	$+3$	-4	$R_-(\mu_b)$
	$+1$	$+1$	0	$+1$	$R_+(\mu_c)$
	$+1$	-1	$+2$	-3	$R_-(\mu_c)$
Near-Oblate Rotor:					
	$+1$	$+1$	$+1$	0	$R_+(\mu_b)$
	$+1$	-1	$+1$	-2	$R_+(\mu_b)$
	$+1$	$+1$	-1	$+2$	$R_-(\mu_b)$
	$+1$	$+3$	-1	$+4$	$R_-(\mu_b)$
	$+1$	0	$+1$	-1	$R_+(\mu_a)$
	$+1$	2	-1	$+3$	$R_-(\mu_a)$

The correlation between the various types of transitions is summarized in Table VII. It is seen, for example, that a parallel spectrum of a near-oblate rotor becomes a perpendicular spectrum as the asymmetry is varied toward the prolate limit, and vice versa. There is obviously an intermediate region in which our terminology loses meaning. This is

TABLE VII. Correlation between Transitions of Limiting Prolate and Oblate Rotors

Type	$\lvert \Delta K_{-1} \rvert$	$\lvert \Delta K_1 \rvert$	Near prolate	Near oblate
μ_a	0	1	$\parallel (Q, R)$	$\perp (Q, R_+)$
	2	1	—	$\perp (Q, R_-)$
μ_b	1	1	$\perp (Q, R_+, R_-)$	$\perp (Q, R_+, R_-)$
	1	3	$\perp (R_-)$	—
	3	1	—	$\perp (R_-)$
μ_c	1	0	$\perp (Q, R_+)$	$\parallel (Q, R)$
	1	2	$\perp (Q, R_-)$	—

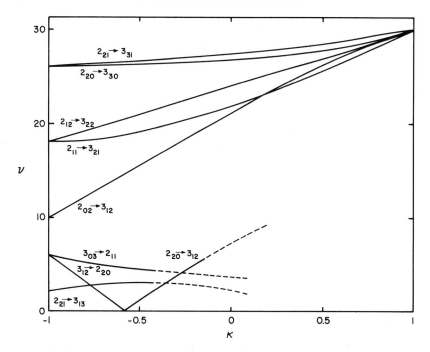

FIG. 8. Correlation of μ_c-type R-branch transitions between prolate and oblate limits. The ordinate is in units of $A - C$. The curves are calculated with $(A+C)/(A-C)$ = 1.5.

illustrated in Fig. 8 for a specific set of transitions involving $J = 2 \rightarrow 3$ with a μ_c selection rule. At the left-hand side of the diagram the characteristic perpendicular pattern of a near-prolate rotor is apparent, i.e., the R_+ transitions at higher frequency and the weaker, less regular, R_- transitions which, for this particular choice of constants, fall at lower frequency. As one moves toward the oblate limit, the R_- transitions lose intensity and eventually become negligible, while the R_+ transitions merge to form the characteristic pattern of a parallel spectrum of a near-oblate rotor (compare Fig. 4). For this particular example one can see that the parallel pattern is well defined in the region $1 > \varkappa > 0.5$, while the perpendicular pattern remains recognizable through a roughly equivalent region near the prolate limit. In the intermediate region, say $0.5 > \varkappa > -0.5$, neither the perpendicular nor the parallel classification is very appropriate, although it may still be useful to visualize the pattern as being derived from one of the limiting cases.

In attempting to generalize on the conditions under which a near-prolate or near-oblate pattern is to be expected, one must remember that not only the asymmetry but also the values of J and K are important. This point has been discussed in a quantitative way by Hainer et al.[17] in connection with the calculation of asymmetric rotor energy eigenvalues. It is convenient to plot a quantity $\lambda = K_{-1}/[J(J+1)]^{1/2}$ against $\varkappa$, as shown in Fig. 9. Then the upper-left half of the diagram represents a rotor which is more prolate in character, while the lower-right half represents a more oblate rotor. Near the dividing line, which is defined by $\varkappa = E_{\tau}^{J}(\varkappa)/J(J+1)$, any generalization based upon the limiting symmetric cases is practically useless. As one moves further away from this line the behavior becomes more symmetric in character. Thus the higher K_{-1} transitions (i.e., $\lambda \to 1$) may form recognizable near-prolate patterns, while at the same value of $\varkappa$ the transitions of low K_{-1} form patterns which are more oblate in character.

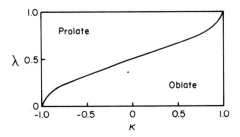

FIG. 9. Prolate–oblate character of asymmetric-rotor levels. (See text for details.)

2.1.5. Vibration–Rotation Interactions

The general features of the microwave spectra of most molecules can be explained fairly well on the basis of the rigid-rotor model. However, a quantitative fit of the spectrum to the accuracy attainable from modern microwave measurements usually requires the refinement of this model to include the influence of internal vibrations on the rotation of the molecule. In most simple molecules the vibration–rotation interactions have a fairly mild effect, but in some cases the spectrum is modified in a major way.

In this section we shall discuss three of these effects: (a) the dependence of effective rotational constants on vibrational state; (b) Coriolis resonance; and (c) Fermi resonance. The first of these is responsible for the observation of satellite lines corresponding to rotational transitions in

[17] R. M. Hainer, P. C. Cross, and G. W. King, *J. Chem. Phys.* **17**, 826 (1949).

excited vibrational states. The second and third lead to the frequently observed deviation of this satellite spectrum from a simple, regular pattern. Three other effects, centrifugal distortion, internal rotation splitting, and inversion doubling, also are due to the interaction of internal and overall motions. However, these will be treated in separate sections.

2.1.5.1. Hamiltonian for the Vibrating Rotor. The internal motions in "normal" molecules can be treated by the theory of small vibrations. In this formulation one introduces normal coordinates Q_s that are linear combinations of the Cartesian displacements of the atoms from their equilibrium positions. These normal coordinates are chosen such that the kinetic energy, as well as the potential energy when expressed through terms quadratic in the coordinates, are diagonal quadratic forms (see Wilson et al.[8] or Herzberg[7] for details). The purely vibrational part of the Hamilton can thus be written

$$H_v = \tfrac{1}{2} \sum_s P_s + \lambda_s Q_s,$$ (2.1.39)

where P_s is the momentum conjugate to Q_s and

$$\lambda_s = (2\pi c \omega_s)^2.$$

Here ω_s is the characteristic vibrational fundamental, expressed in units of reciprocal centimeters, and the summation extends from $s = 1$ to $3N - 6$ ($3N - 5$ in the case of a linear molecule), where N is the number of atoms. The solution of Eq. (2.1.39) gives the vibrational levels (in reciprocal centimeters) as

$$(E_v/hc) = \sum_s (v_s + \tfrac{1}{2})\omega_s.$$ (2.1.40)

The general Hamiltonian for a molecule that is both vibrating and rotating was first derived by Wilson and Howard.[18] The vibrations affect the rotational part of the kinetic energy in two principal ways. First, the moments and products of inertia are no longer constants but become functions of the internal coordinates; and second, the vibrational motion generates an internal angular momentum which is coupled to the angular momentum associated with overall rotation. In order to describe these effects, a suitable molecule-fixed coordinate system must be introduced. The most convenient choice is that defined by the Eckart conditions,

[18] E. B. Wilson and J. B. Howard, *J. Chem. Phys.* **4**, 260 (1936).

which specify that (a) the origin of the coordinate system lies at the instantaneous center of mass and (b) the vibrational angular momentum vanishes as all atoms pass through their equilibrium positions. With this choice the classical kinetic energy, which was given by Eq. (2.1.1) for the rigid rotor, now becomes

$$2T = I_{xx}\omega_x{}^2 + I_{yy}\omega_y{}^2 + I_{zz}\omega_z{}^2 + 2I_{xy}\omega_x\omega_y$$

$$+ 2I_{xz}\omega_x\omega_z + 2I_{yz}\omega_y\omega_z + 2\omega_x \sum_{st} \zeta_{st}^x Q_s \dot{Q}_t$$

$$+ 2\omega_y \sum_{st} \zeta_{st}^y Q_s \dot{Q}_t + 2\omega_z \sum_{st} \zeta_{st}^z Q_s \dot{Q}_t + \sum_t \dot{Q}_t{}^2. \qquad (2.1.41)$$

Here ζ_{st}^x, etc., are Coriolis coupling constants which depend upon the transformation from Cartesian to normal coordinates.

The potential energy can be expressed as a power series in the normal coordinates, as long as the displacements from equilibrium are not too large. It is convenient at this point to introduce dimensionless normal coordinates q_s defined by

$$q_s = (\hbar^2/\lambda_s)^{-1/4} Q_s = (\hbar/2\pi c\omega_s)^{1/2} Q_s.$$

The potential energy then becomes

$$V = \tfrac{1}{2}hc \sum_s \omega_s q_s{}^2 + hc \sum_{stu} k_{stu} q_s q_t q_u$$

$$+ hc \sum_{stuv} k_{stuv} q_s q_t q_u q_v + \cdots. \qquad (2.1.42)$$

We shall not carry this expansion beyond the fourth-order term.

The derivation of the quantum-mechanical Hamiltonian from Eqs. (2.1.41) and (2.1.42) involves a rather tedious calculation (see Wilson et al.,[8] pp. 279–284). Furthermore, the resulting Hamiltonian is so general that an exact calculation of its eigenvalues is not feasible. Therefore, simplifying approximations, which are suitable for the particular problem under consideration, must be made. When the vibrational amplitudes are small, several terms in the general Hamiltonian can be neglected in the first approximation. In addition, we can expand the moments and products of inertia as power series in the normal coordinates

$$I_{\alpha\alpha} = I_{\alpha\alpha}^e + \sum_s a_s^{\alpha\alpha} Q_s + \sum_{st} A_{st}^{\alpha\alpha} Q_s Q_t + \cdots,$$

$$I_{\alpha\beta} = -\sum_s a_s^{\alpha\beta} Q_s - \sum_{st} A_{st}^{\alpha\beta} Q_s Q_t + \cdots, \qquad (2.1.43)$$

where $I_{\alpha\alpha}^{\rm e}$ is the equilibrium moment and

$$a_s^{\alpha\alpha} = (\partial I_{\alpha\alpha}/\partial Q_s)_{\rm e},$$

$$A_{st} = (\partial^2 I_{\alpha\alpha}/\partial Q_s\,\partial Q_t)_{\rm e}.$$

With these approximations we are led to a working Hamiltonian of the form

$$H = H_0 + H_1 + H_2 + H_3 + H_4 + H_5 + H_6,$$

where

$$H_0 = \tfrac{1}{2} \sum_s (p_s^2 + \lambda_s Q_s^2) + (P_x^2/2I_{xx}^{\rm e}) + (P_y^2/2I_{yy}^{\rm e}) + (P_z^2/2I_{zz}^{\rm e}),$$

$$H_1 = -[(p_x P_x/I_{xx}^{\rm e}) + (p_y P_y/I_{yy}^{\rm e}) + (p_z P_z/I_{zz}^{\rm e})],$$

$$H_2 = -\tfrac{1}{2} \sum_s Q_s \sum_{\alpha,\beta} (a_s^{\alpha\beta}/I_{\alpha\alpha}^{\rm e} I_{\beta\beta}^{\rm e}) P_\alpha P_\beta,$$

$$H_3 = hc \sum_{stu} k_{stu} q_s q_t q_u, \tag{2.1.44}$$

$$H_4 = hc \sum_{stuv} k_{stuv} q_s q_t q_u q_v,$$

$$H_5 = -\tfrac{1}{2} \sum_s \sum_t Q_s Q_t \sum_{\alpha\beta} C_{st}^{\alpha\beta} P_\alpha P_\beta (I_{\alpha\alpha}^{\rm e} I_{\beta\beta}^{\rm e})^{-1},$$

$$H_6 = \sum_\alpha p_\alpha^2/I_{\alpha\alpha}^{\rm e}.$$

The p_x, p_y, and p_z which appear in H are components of the net vibrational angular momentum; e.g.,

$$p_x = \sum_{st} \zeta_{st}^x Q_s p_t, \tag{2.1.45}$$

where p_t is the momentum conjugate to the normal coordinate Q_t.
In H_5 the new symbol $C_{st}^{\alpha\beta}$ is an abbreviation for

$$C_{st}^{\alpha\beta} = A_{st}^{\alpha\beta} - \sum_u \zeta_{su}^\alpha \zeta_{tu}^\beta - \sum_\gamma a_s^{\alpha\gamma} a_t^{\gamma\beta}/I_{\gamma\gamma}^{\rm e}.$$

The numbering of terms in Eqs. (2.1.44) is somewhat different from the conventions introduced by Nielsen,[19] but is more convenient for microwave spectroscopy. Since no degeneracy index has been indicated,

[19] H. H. Nielsen, *Handbuch der Phys.* **37/1**, 173 (1959).

it is understood that the summations run over all components of a degenerate vibrational mode.

The leading term of Eqs. (2.1.44) describes a rigid rotor, harmonic oscillator model whose solution has already been discussed. The complete Hamiltonian may be set up in that basis and perturbation techniques applied. However, it is convenient to retain the angular momentum components P_x, P_y, P_z in operator form while carrying out a perturbation treatment on the vibrational part of the problem. In this way we are left with an effective rotational Hamiltonian for each vibrational state (or, in case of degeneracy, for each pair of states). A general procedure for doing this, through the use of a contact transformation, has been developed by Nielsen.[19] However, in this section we shall not give the general solution but shall simply apply second-order perturbation theory to calculate those terms which have a particularly important influence on the microwave spectrum.

Before considering the Hamiltonian of Eqs. (2.1.44) in detail, it is helpful to summarize the effects of the various terms when second-order perturbation theory is applied (we defer consideration of essential degeneracies). H_1 has matrix elements which are off-diagonal in the vibrational quantum numbers. In second-order these elements will contribute terms in $P_x{}^2$, $P_y{}^2$, etc., whose coefficients involve the vibrational parameters. These terms may be absorbed into the rigid-rotor part of the Hamiltonian, yielding effective rotational constants A_v, B_v, C_v which depend upon the vibrational state. Also H_2 has off-diagonal elements, but the square of H_2 gives a term that is quartic in P_x, P_y, and P_z. This is the origin of centrifugal distortion, which is discussed in Section 2.1.6. Some of the off-diagonal elements of H_3 have similar form to H_2. The cross term in the square of $H_2 + H_3$ is quadratic in the overall angular momentum and will thus contribute to the effective rotational constants (note, however, that this contribution involves the anharmonic potential constants, while H_1 is a strictly harmonic term). H_5 has diagonal elements that are quadratic in the angular momenta, and so makes a further contribution to the effective rotational constants. In cases of accidental near-degeneracies of excited vibrational levels, the elements from H_1 may lead to rotational (Coriolis) resonances. Similarly, H_3 and H_4 are responsible for Fermi resonance between nearly degenerate levels.

2.1.5.2. Dependence of Rotational Constants on Vibration State. In the absence of vibrational degeneracy or near-degeneracy it is found experimentally that the rotational spectrum in an excited vibrational state usually follows a rigid-rotor pattern, but with rotational constants that differ slightly from those of the ground state. In most cases the dependence of the rotational constants on vibrational state may be expressed as a rapidly convergent power series in the vibrational quantum numbers

$$B_v = B_e - \sum_s \alpha_s^{(B)}(v_s + \tfrac{1}{2}) + \sum_{st} \gamma_{st}^{(B)}(v_s + \tfrac{1}{2})(v_t + \tfrac{1}{2}) + \cdots . \quad (2.1.46)$$

Similar expressions apply to A_v and C_v. B_e is the hypothetical rotational constant corresponding to all atoms being frozen at their equilibrium positions. The coefficients α_s, γ_s, etc., may in principle be calculated by a perturbation treatment of the Hamiltonian of Eqs. (2.1.44). Since this calculation becomes very complicated when one goes beyond the first term, we shall limit our attention to the α_s coefficients.

The problem is to find those perturbation terms that are quadratic in P_x, P_y, P_z and linear in the quantity $(v + \tfrac{1}{2})$. As discussed in the last section, terms of this type arise from $H_1, H_2 + H_3$, and H_5. A rather tedious calculation yields the following expression for α_s.

$$\alpha_s^{(B)} = -\frac{2B_e^2}{c\omega_s} \left[3A_{ss}^{xx} + 4\sum_t (\zeta_{st}^x)^2 \frac{\omega_t^2}{\omega_s^2 - \omega_t^2} \right]$$
$$- 2B_e^2 \left[3ck_{sss}(c\omega_s)^{-3/2} \frac{2\pi a_s^{xx}}{h^{1/2}} + \sum_{t \neq s} ck_{sst}(c\omega_t)^{-3/2} \frac{2\pi a_t^{xx}}{h^{1/2}} \right].$$
$$(2.1.47)$$

Here we have arbitrarily identified the rotational constant B with the x axis. It should be noted that ω_s and k_{sss} are measured in units of reciprocal centimeters. Thus $c\omega_s$ and ck_{sss}, as well as B_e and $\alpha_s^{(B)}$, have units of frequency.

The term in the first square brackets of Eq. (2.1.47) depends only on the harmonic part of the vibrational force field. If this is known with sufficient accuracy from other measurements, the harmonic contribution to $\alpha_s^{(B)}$ can be calculated. Explicit formulas for several types of molecules have been given by Laurie and Herschbach.[20] The second term involves the cubic anharmonic constants as well, which are known for only a limited number of molecules. Therefore, while it is difficult to predict the anharmonic contribution, the procedure can be reversed and the cubic potential constants calculated from the observed α_s. This has been

[20] V. W. Laurie and D. R. Herschbach, *J. Chem. Phys.* **37**, 1687 (1962).

done, for example, by Morino *et al.* for SO_2.[21] In some cases an estimate of the anharmonic part of α_s can be made by transferring potential constants from related molecules.

2.1.5.3. Degenerate States in Linear and Symmetric Rotors. The expressions for the vibration–rotation interaction constants which were derived in the last section did not take into account possible degeneracies among the vibrational states. We first consider essential degeneracies arising from the high symmetry of a molecule. From the viewpoint of interpreting microwave spectra, we need only treat the doubly degenerate vibrations that occur in linear molecules and symmetric rotors. In such a molecule the zeroth-order Hamiltonian H_0 contains one or more terms representing a two-dimensional harmonic oscillator. Each of these terms will contribute to the vibrational energy an amount

$$hc\omega_r(v_r + 1),$$

where the subscript r is used to label the quantities associated with the degenerate level. The solution of the two-dimensional harmonic oscillator (see Townes and Schawlow,[1] pp. 31–32) shows that the degeneracy of the level v_r is equal to $v_r + 1$. The individual states are conventionally labeled by a quantum number l_r, which takes values

$$l_r = v_r, v_r - 2, v_r - 4, \ldots, -v_r.$$

Furthermore, it can be shown that the vibrational angular momentum has a nonvanishing component on the figure axis of the rotor, given by

$$\langle p_z \rangle = \sum_r \zeta_r l_r,$$

where ζ_r is the Coriolis coupling constant associated with the two components of the degenerate vibration.

The presence of a nonvanishing component of vibrational angular momentum in a symmetric rotor leads to a first-order vibration–rotation interaction. The component of total angular momentum along the figure axis is still described by a quantum number K, but this is composed of two parts, a vibrational contribution $\sum \zeta_r l_r$ and a pure rotational contribution $K - \sum \zeta_r l_r$. The rotational energy is given in first order by

$$E/h = B_e J(J + 1) + (A_e - B_e)K^2 - 2A_e K \sum \zeta_r l_r, \qquad (2.1.48)$$

[21] Y. Morino, Y. Kikuchi, S. Saito, and E. Hirota, *J. Mol. Spectrosc.* **13**, 95 (1964).

where the last term comes from the first-order perturbation contribution of the term H_1 in Eqs. (2.1.44).

When a degenerate mode of a symmetric rotor is singly excited ($v_r = 1$, $l_r = \pm 1$), Eq. (2.1.48) shows that the initial fourfold degeneracy (twofold vibrational plus twofold K-degeneracy of the rigid rotor) is split; the result is a pair of doubly-degenerate levels displaced by $\pm 2A_e\zeta_r \mid K \mid$ from the unperturbed position. When $K = 0$ there is one doubly-degenerate level corresponding to $l_r = \pm 1$. The symmetry properties of these levels may be established from the rules of Section 2.1.2.6 by replacing K with $K - l_r$.

The first-order splitting in degenerate vibrational states does not influence the microwave spectrum directly because of the selection rule $\Delta K = 0$. However, in higher orders the rotational energy is affected by vibrational degeneracy in an observable way. We consider first the case of a linear molecule. The perturbation term H_1 in Eq. (2.1.44) has matrix elements connecting the state $(v_s v_r l_r K)$ with states $(v_s + 1, v_r \pm 1, l_r \pm 1, K \pm 1)$. A perturbation calculation shows that the degeneracy in l_r is split in second order when $l = \pm 1$, yielding a rotational energy

$$E/h = B_v J(J + 1) \pm (q_l/4)(v_r + 1)J(J + 1), \qquad (2.1.49)$$

where

$$q_l = \frac{2B_e^2}{c\omega_r}\left[1 + 4\sum_s \frac{\zeta_{sr}^2 \omega_r^2}{\omega_s^2 - \omega_r^2}\right]. \qquad (2.1.50)$$

This effect is commonly referred to as "l-type doubling." It should be noted that q_l depends only on the harmonic part of the force field. The levels with $\mid l \mid > 1$ remain degenerate in this order, but they in turn will split in higher orders.[19,22]

The degeneracy in a linear rotor also has an effect on the expression for the α constants. To obtain $\alpha_r^{(B)}$ for a degenerate mode, Eq. (2.1.47) must be modified by insertion of a factor $\frac{1}{2}$ before the first bracket.

In the most familiar case of a linear triatomic molecule it is customary to designate the degenerate mode, which describes a bending motion, with the quantum number v_2. The rotational constant is then expressed by

$$B_v = B_e - \alpha_1(v_1 + \tfrac{1}{2}) - \alpha_2(v_2 + 1) - \alpha_3(v_3 + \tfrac{1}{2}). \qquad (2.1.51)$$

[22] A. G. Maki and D. R. Lide, *J. Chem. Phys.* **47**, 3206 (1967).

The complete expressions for the α's are

$$\alpha_1 = -(2B_e^2/c\omega_1)[3 + 4\zeta_{21}^2\omega_2^2(\omega_1^2 - \omega_2^2)^{-1}]$$
$$-(2B_e)^{3/2}[(3ck_{111}\zeta_{23})(c\omega_1)^{-3/2} + (ck_{113}\zeta_{21})(c\omega_3)^{-3/2}],$$

$$\alpha_2 = (B_e^2/c\omega_2)[1 + 4\zeta_{21}^2\omega_2^2(\omega_1^2 - \omega_2^2)^{-1} + 4\zeta_{23}^2\omega_2^2(\omega_3^2 - \omega_2^2)]$$
$$-(2B_e)^{3/2}[(ck_{221}\zeta_{23})(c\omega_1)^{-3/2} + (ck_{223}\zeta_{21})(c\omega_3)^{-3/2}], \qquad (2.1.52)$$

$$\alpha_3 = -(2B_e^2/c\omega_3)[3 + 4\zeta_{23}^2\omega_2^2(\omega_3^2 - \omega_2^2)^{-1}]$$
$$-(2B_e)^{3/2}[(ck_{331}\zeta_{23})(c\omega_1)^{-3/2} + (3ck_{333}\zeta_{21})(c\omega_3)^{-3/2}].$$

In the above expressions for the α's the first bracket involves only harmonic terms, while the second includes cubic anharmonic constants. It should be pointed out that this division is somewhat arbitrary. A different formulation, which takes into account the curvilinear nature of the bending motion, leads to a rather different partitioning of harmonic and anharmonic contributions in the bending mode.[23]

The symmetry species of an excited vibrational state of a linear molecule is determined by the value of l. States with $l = 0, \pm 1, \pm 2, \pm 3$, etc., are classified as $\Sigma, \Pi, \Delta, \Phi$, etc., respectively. An important symmetry property of the levels is the behavior of the eigenfunctions upon inversion. It may be shown that in Σ states (including the ground state) the rotational levels of even J are symmetric $(+)$ with respect to inversion, while those of odd J are antisymmetric $(-)$. In Π states, which are split by l-type doubling, the level of lower energy is $(+)$, while the upper level is $(-)$, when J is even; the opposite is true when J is odd. It should be noted that $J \geq |l|$, so that no $J = 0$ level exists in a Π state.

The selection rules for electric dipole transitions in linear molecules are

$$\Delta J = 0, \pm 1, \qquad (+) \leftrightarrow (-).$$

Therefore, in a $J \rightarrow J + 1$ transition in a Π state the lower component of level J combines with the lower component of level $J + 1$, and similarly for the upper components. The transition thus appears as a doublet with a frequency separation of

$$q_l(v + 1)(J + 1).$$

When $|l| \neq 1$ the transitions for states of the same v but different l

[23] D. R. Lide and C. Matsumura, *J. Chem. Phys.* **50**, 3080 (1969).

all have the same frequency in the present approximation. In practice, however, perturbations or higher terms are usually present which lead to a resolvable separation of components with different $|\,l\,|$.

A typical pattern of vibrational satellite lines from excited states of a degenerate mode of a linear molecule is indicated in Fig. 10. The pattern has been drawn with a negative α_2, which is the most common case. It should be remembered that the lines from Δ states $(l = \pm 2)$ will first appear in the $J = 2 \rightarrow 3$ transition, Φ states will appear first in $J = 3 \rightarrow 4$, and so on.

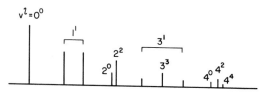

FIG. 10. Vibrational satellite pattern for a linear triatomic molecule in which the bending mode ν_2 is excited.

Another type of transition is possible in a Π state, namely $\Delta J = 0$, $(+) \rightarrow (-)$. From Eq. (2.1.49) the frequency of such a transition is seen to be

$$\nu = (q_l/2)(v_r + 1)J(J + 1).$$

These direct transitions between the two components of an l-doublet have been observed in several molecules.[22,24]

The treatment of degenerate vibrational levels in symmetric rotors is more complicated. The phenomenon of l-type doubling occurs here also, but the quantum number K must be taken into account. We recall that l describes the internal angular momentum while K gives the total angular momentum along the figure axis. The rotational angular momentum of the top is thus $K - l$. When $K - l = 0$, an l-type doubling exactly analogous to that in linear molecules occurs. In the case where one degenerate mode is singly excited, so that $l = \pm 1$, the states $K = l = 1$ and $K = l = -1$ (i.e., those with $Kl = +1$) are split in first order. The rotational energy for these states is given by

$$E/h = B_v J(J+1) + (A_v - B_v) - 2A_v \zeta_r \pm (q_l/4)(v_r+1)J(J+1). \quad (2.1.53)$$

When K takes other values (including the states $Kl = -1$) a second-order

[24] A. G. Maki, J. Mol. Spectrosc. **23**, 110 (1967).

contribution proportional to

$$q_l^2/(1 - \zeta_r)(A_v - B_v)$$

occurs. The net result on the energy levels is indicated schematically in Fig. 11. However, it has been found that centrifugal distortion and other higher-order terms usually make contributions of comparable magnitude to the second-order terms involving q_l. Detailed discussions of the level patterns in excited degenerate states of symmetric rotors have been given by Weber[25] and by Grenier-Besson and Amat.[26]

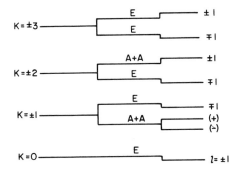

FIG. 11. Rotational energy-level structure in a singly excited degenerate vibrational state of a symmetric rotor (schematic). The left-hand side represents the level structure for a rigid rotor. In the center the first-order Coriolis splitting is taken into account; this is independent of J. The higher-order, J-dependent interactions are included on the right. The symbols $(+)$ and $(-)$ refer to symmetric and antisymmetric linear combinations of states with $K = l = 1$ and $K = l = -1$.

Although the quantitative interpretation may be difficult, it is usually possible to assign the spectra in singly excited degenerate states on the basis of Stark effects and relative intensities. The common pattern consists of two widely-split lines (with $Kl = +1$) with characteristic Stark effects of the type discussed in Section 2.1.9.5 (i.e., the Stark components of the higher-frequency transition move to lower frequency, and viceversa, as the field increases). The remaining lines are clustered near the mean of the l-doublets. These have symmetric first-order Stark effects, except for the $K = 0, l = \pm 1$ transition, which is second order. Typical patterns have been illustrated by Lide and Mann.[27]

[25] G. G. Weber, *J. Mol. Spectrosc.* **10**, 321 (1963).
[26] M. L. Grenier Besson and G. Amat, *J. Mol. Spectrosc.* **8**, 22 (1962).
[27] D. R. Lide and D. E. Mann, *J. Chem. Phys.* **29**, 914 (1958).

2.1.5.4. Diatomic Molecules. Although the general formulas derived above include the diatomic rotor, it is worthwhile to give the explicit results for this case. The vibration–rotation interaction constant α is given by

$$\alpha = -6(B_e{}^3/c\omega)^{1/2}[(2^{1/2}k_3/\omega) + (B_e/c\omega)^{1/2}], \qquad (2.1.54)$$

where k_3 is the cubic anharmonic constant. It has been found that k_3 is almost always negative and that the anharmonic term in Eq. (2.1.54) is generally from two to four times as large as the harmonic term.[28] Thus the value of α in diatomic molecules is usually positive, so that the effective rotational constant B_v is less than B_e.

Vibration–rotation interaction constants in diatomic molecules have been calculated to a much higher approximation by Dunham. The Dunham treatment is discussed by in considerable detail by Townes and Schawlow[1] (pp. 9–11). The vibrational potential is expressed in the form

$$V = a_0\xi^2(1 + a_1\xi + a_2\xi^2 + \cdots),$$

where $\xi = (r - r_e)/r_e$. The energy levels are written

$$E_{vJ} = \sum_{lj} Y_{lj}(v + \tfrac{1}{2})^l[J(J + 1)]^j. \qquad (2.1.55)$$

The constant Y_{01} is identical, to a high approximation, with B_e, and Y_{11} may be identified with $-\alpha$. The full expressions for the Y_{lj} in terms of the a_i coefficients are given by Townes and Schawlow.[1]

2.1.5.5. Coriolis Resonance. In the theory which has been developed so far the assumption has been made that all vibrational states, except those which are degenerate because of the molecular symmetry, are sufficiently well separated in energy that second-order perturbation theory can be applied. In practice this assumption is often found to be invalid. When two vibrational levels happen accidentally to have almost equal energies, significant deviations from the predictions of the theory may be found. We refer to such situations as examples of resonance or accidental vibrational degeneracy. An observable resonance requires not only a near-coincidence of energy but also a nonvanishing matrix element connecting the states. Since the vibration–rotation Hamiltonian is totally symmetric, the first requirement is that the overall symmetry species of the two vibration–rotation states be the same. Furthermore, resonance

---

[28] D. R. Herschbach and V. W. Laurie, *J. Chem. Phys.* **35**, 458 (1961).

can occur only between states of the same J value, since the Hamiltonian is diagonal in J.

We may recognize two general types of accidental degeneracy, those that produce perturbations of specific rotational levels and those that only result in a mixing of vibrational states. The former type is referred to as *Coriolis resonance* and is associated with the term H_1 of the general Hamiltonian. The second type is discussed in the next section.

We shall treat only the case of Coriolis resonance between vibrational fundamentals (i.e., $v_s = 0$, $v_t = 1$ nearly coincident with $v_s = 1$, $v_t = 0$), since this is the most common situation encountered in microwave spectra. The internal angular momentum operators p_α (with $\alpha = x, y, z$) that appear in the term H_1 of the general Hamiltonian are proportional to the Coriolis coupling constants ζ_{st}^α [see Eq. (2.1.45)]. Therefore, a Coriolis interaction between two fundamentals requires that one of the ζ_{st}^α be different from zero. It was shown by Jahn (see Herzberg,[7] p. 376) that an interaction is permitted only if the product of the symmetry species of the two vibrational states contains the species of a rotation. The allowed interactions for those point groups that are most important in microwave spectroscopy are summarized in Table VIII. An entry in the table, say x, indicates that $\zeta_{st}^x \neq 0$ for two modes ω_s and ω_t belonging to the species indicated in the appropriate row and column.

The matrix element of H_1 connecting two nondegenerate fundamentals may be shown to be

$$(v_s = 0, \ v_t = 1 \mid H_1 \mid v_s = 1, \ v_t = 0) = \beta_{st}^x P_x + \beta_{st}^y P_y + \beta_{st}^z P_z, \quad (2.1.56)$$

where

$$\beta_{st}^\alpha = -iB_e^{\alpha\alpha}\zeta_{st}^\alpha(\omega_s + \omega_t)(\omega_s\omega_t)^{-1/2}.$$

In the simplest case there is only one nonvanishing term in Eq. (2.1.56), although reference to Table VIII shows that for C_s molecules, for example, both an x and y term may be present. The factor $(\omega_s + \omega_t) \times (\omega_s\omega_t)^{-1/2}$ has been retained in the definition of β_{st}^α although this will be very nearly 2 if the coincidence of levels is close. When the magnitude of $\beta_{st}^z p_z$ is significant in comparison with the vibrational energy difference $\omega_s - \omega_t$, the effective rotational Hamiltonians for the two states will be coupled, and both states will show deviations from the normal rigid-rotor pattern. The coupling is absent when $J = 0$ and so does not affect the vibrational energy. The deviations generally become larger as J increases, since the coupling term is proportional to P_α.

TABLE VIII. Allowed Coriolis Interactions in Some Important Point Groups

C_s (C_2)	A′	A″		
A′	z			
A″	x, y	z		
C_{2v}	A_1	A_2	B_1	B_2
A_1	—			
A_2	z	—		
B_1	y	x	—	
B_2	x	y	z	—
C_{3v}	A_1	A_2	E	
A_1	—			
A_2	z	—		
E	x, y	x, y	x, y, z	
$C_{\infty v}$	Σ	Π		
Σ	—			
Π	x, y	z		

The specific form of the coupling depends on the type of molecule. We shall consider a few simple examples. In a C_{2v} molecule two given levels can be coupled through only one of the terms β_{st}^{α} (see Table VIII). Let us consider a resonance between two fundamentals of symmetry species A_1 and A_2 (or of species B_1 and B_2), which are coupled by ζ_{st}^{z}. The z axis is defined as the twofold symmetry axis of the molecule, but this can be identified with either the a, b, or c axis of the rigid asymmetric rotor which we use as the first approximation for the rotational problem. To be definite, we shall take a molecule where $z \leftrightarrow a$ (formaldehyde is an example). Then it is natural to choose a I^r representation ($x, y, z \leftrightarrow b, c, a$), and the matrix elements of $P_z \equiv P_a$ in a symmetric-rotor basis are just (in units of $\hbar$)

$$(K \mid P_z \mid K) = K.$$

The interaction matrix element is

$$(v_s = 0, \quad v_t = 1, J, K \mid H_1 \mid v_s = 1, \quad v_t = 0, J, K) = \beta_{st}^{a} K. \quad (2.1.57)$$

The asymmetric-rotor matrix elements in the representation are, from Section 2.1.2.7,*

$$(J, K \mid H_R \mid J, K) = \tfrac{1}{2}(B_v + C_v)J(J + 1) + [A_v - \tfrac{1}{2}(B_v + C_v)]K^2,$$
$$(J, K \mid H_R \mid J, K \pm 2) = \tfrac{1}{4}(C_v - B_v)[(J - K)(J - K - 1)$$
$$\times (J + K + 1)(J + K + 2)]^{1/2}$$

where the subscript v stands for s or t. Since the interaction element, Eq. (2.1.57), is diagonal in K, the Hamiltonian still factors into sub-matrices for even and odd K. Furthermore, if we apply the Wang transformation, Eq. (2.1.24), as is done for the rigid asymmetric rotor, it is easy to show that the matrix elements of Eq. (2.1.57) connect the states

$$E_s{}^+(J, K) \leftrightarrow E_t{}^-(J, K), \qquad O_s{}^+(J, K) \leftrightarrow O_t{}^-(J, K),$$
$$E_s{}^-(J, K) \leftrightarrow E_t{}^+(J, K), \qquad O_s{}^-(J, K) \leftrightarrow O_t{}^+(J, K).$$

Here $E^\pm$ and $O^\pm$ are the symbols for the factored asymmetric rotor submatrices [see Eq. (2.1.25)], and the subscript s or t indicates the vibrational state.

The exact calculation of the Coriolis perturbation therefore requires the diagonalization of a set of matrices, each of which is twice as large as the corresponding rigid-rotor matrix. The results cannot be expressed in simple form. We shall look only at the solution for $J = 1$. Here we have two 2×2 matrices whose eigenvalues correlate with the asymmetric-rotor levels 1_{10} and 1_{11},

$$
\begin{matrix}
v_s = 0, & v_t = 1 \\
v_s = 1, & v_t = 0
\end{matrix}
\begin{pmatrix}
\omega_t + A_t + B_t & \beta_{st}^a \\
(\beta_{st}^a)^* & \omega_s + A_s + C_s
\end{pmatrix},
$$
$$
\begin{matrix}
v_s = 0, & v_t = 1 \\
v_s = 1, & v_t = 0
\end{matrix}
\begin{pmatrix}
\omega_t + A_t + C_t & \beta_{st}^a \\
(\beta_{st}^a)^* & \omega_s + A_s + B_s
\end{pmatrix}.
$$

$$(2.1.58)$$

The 1_{01} levels are not affected by the resonance, since the interaction matrix element of Eq. (2.1.57) vanishes for $K = 0$. A solution in closed form is therefore possible for the levels with $J = 1$.

---

* The use of effective rotational constants A_v, B_v, C_v in the diagonal elements implies that we have carried out the usual second-order treatment for all the other (nonresonant) interactions. Strictly speaking, the interaction element should be modified accordingly. In most cases this correction is trivial, and we can simply use the ground-state constant (A_0 in this case) in the definition of β_{st}^a.

The exact calculation of eigenvalues becomes tedious when J is large. In making preliminary assignments, however, the Coriolis matrix elements may sometimes be treated in an approximate fashion by absorbing their effects into a set of modified rotational constants.* Examples of this procedure have been given for SO_2F_2 [29] and F_2CO.[30] A somewhat more complicated situation is found in CH_3CH_2CN,[31] where it is possible to fit the spectrum by introducing empirical terms of a centrifugal distortion-like character.

Coriolis resonance can also occur in symmetric rotors. In particular, Table VIII shows that a species E fundamental of a C_{3v} molecule can interact with an A_1 or A_2 mode. In this case one obtains a 3×3 matrix for each value of J and K, but this matrix factors further for the special cases $K = 0$ and $K = \pm J$. The complete formulation is given by Kuczkowski and Lide,[32] who analyzed a resonance of this type in PF_3BH_3. Such resonances can be very strong, shifting lines far from the region of the ground-state transitions, and in some cases causing a breakdown of the normal symmetric-rotor selection rules.

Some general comments can be made on the analysis of a spectrum in which Coriolis interactions are present. Fitting the spectrum to the appropriate energy matrix is complicated by the fact that so many parameters are involved. In addition to the effective rotational constants in the two vibrational states, one must introduce the unknown energy difference $\omega_s - \omega_t$ plus one or more Coriolis constants ζ_{st}^a. Furthermore, in symmetric rotors the resonance is sensitive to ζ_r, the Coriolis constant in the degenerate state. It is helpful whenever possible to identify low J lines or certain lines not affected by the perturbation, since these may provide values for some of the effective rotational constants. It is also possible, in a preliminary analysis, to ignore the difference between effective rotational constants in the two states. Finally, simple perturbation formulas that can be derived from the matrices given above are useful in predicting which transitions will be most affected by the resonance.

2.1.5.6. Fermi Resonance.

The other important type of resonance is the anharmonic or Fermi resonance. This type of interaction can occur

[29] D. R. Lide, D. E. Mann, and R. M. Fristrom, *J. Chem. Phys.* **26**, 734 (1957).
[30] V. W. Laurie and D. T. Pence, *J. Mol. Spectrosc.* **10**, 155 (1963).
[31] V. W. Laurie, *J. Chem. Phys.* **31**, 1500 (1959).
[32] R. L. Kuczkowski and D. R. Lide, *J. Chem. Phys.* **46**, 357 (1967).

---

* These same matrix elements contribute to the vibration–rotation interaction constant α_s in situations where there is no accidental degeneracy.

only between vibrational states of the same symmetry species. It may be significant whenever a state $v_s = n$, $v_t = 0$ is accidentally degenerate with the state $v_s = 0$, $v_t = m$, i.e. when $n\omega_s \approx m\omega_t$ (and at least one of the integers n and m is different from unity). If the vibrational species of the two nearly-degenerate levels is the same, the anharmonic terms H_3 and H_4 in the general Hamiltonian (and higher terms which have been neglected) may connect the levels. These are purely vibrational terms, so that the interaction occurs only between states with the same rotational quantum numbers. In general, the interaction becomes weaker the larger the difference in vibrational quantum numbers between the states, because higher-order anharmonic terms in the potential function must be invoked.

The most common type of Fermi resonance involves H_3, which contains the cubic terms in the potential. If a term $k_{stt}q_sq_t^2$ is present in H_3, it will have nonvanishing matrix elements connecting the state v_s, v_t with the states $v_s \pm 1$, v_t and $v_s \pm 1$, $v_t \pm 2$. The first type of matrix element cannot produce resonant interactions, since it connects two states which differ in energy by ω_s. However, a matrix element connecting v_s, v_t with $v_s - 1$, $v_t + 2$ (or with $v_s + 1$, $v_t - 2$) can have a large effect if the energy difference $\omega_s - 2\omega_t$ happens to be close to zero. This is the classical example of Fermi resonance, first observed in CO_2, in which the first overtone of one mode coincides accidentally with another fundamental. The cubic potential term $k_{stt}q_sq_t^2$ causes a mixing of the wave functions of these two states and a modification of their energies. This type of interaction has not been taken into account so far in our calculation, since it ordinarily does not affect the rotational constants in the second order of perturbation. In case of accidental degeneracy, however, it must be included.

The higher anharmonic terms can also produce resonant interactions. If the symmetry requirements are satisfied, the term $k_{sstt}q_s^2q_t^2$ (which occurs in the term H_4 in our general Hamiltonian) has nonvanishing matrix elements between the states v_s, v_t and $v_s \pm 2$, $v_t \mp 2$. Thus if we have two nearly coincident fundamentals of different symmetry, whose first overtones have the same symmetry, these overtones (v_s, v_t = 0, 2 and 2, 0) can interact through the quartic term. A well-known example is the resonance between the overtones of the two stretching fundamentals of H_2O, which was first analyzed by Darling and Dennison.[7] Similar interactions may be produced by higher-order anharmonic terms. If it happens that one fundamental is accidentally an integral multiple of another (say, $\omega_s = n\omega_t$), an interaction between the states

v_s, $v_t = 0$, n and $1, 0$ may be observed, even though n is quite large (assuming, of course, that the symmetry requirements are satisfied). It is sometimes observed in microwave spectra that the rotational constants of successive excited states of a single vibrational mode ($v_s = 0, 1, 2, 3, \ldots$) vary smoothly with v_s except for one state which is out of line. This is an indication of the probable existence of a higher-order resonance of the above type.

We shall derive a general expression for the effect of Fermi resonance involving two accidentally coincident levels, without specifying which anharmonic term is responsible. We label the levels 1 and 2 and designate their unperturbed vibrational energies as W_1^0 and W_2^0 and unperturbed wave functions as Ψ_1^0 and Ψ_2^0. These unperturbed values may be taken as harmonic oscillator eigenvalues and eigenfunctions; or, for a somewhat better approximation, they may be corrected for the second-order contributions of all the other (nonresonant) interactions. We assume that the levels 1 and 2 are connected by a matrix element W_{12}. Since this matrix element connects states with the same rotational quantum numbers, we shall ignore the small difference in rotational energy in comparison to $W_1^0 - W_2^0$. This is almost always a safe approximation, since the differences in rotational constants, $A_1 - A_2$, etc., are so small.

The portion of the Hamiltonian involving these two levels is just

$$\begin{pmatrix} W_1^0 & W_{12} \\ W_{12} & W_2^0 \end{pmatrix}.$$

To be definite, we take $W_1^0 \geq W_2^0$. Since W_{12} involves matrix elements of powers of the normal coordinates, we can take it to be real, so that $W_{21} = W_{12}$. The eigenvalues of this matrix are

$$W_1 = \tfrac{1}{2}(W_1^0 + W_2^0) + [\tfrac{1}{4}(W_1^0 - W_2^0)^2 + W_{12}^2]^{1/2},$$
$$W_2 = \tfrac{1}{2}(W_1^0 + W_2^0) - [\tfrac{1}{4}(W_1^0 - W_2^0)^2 + W_{12}^2]^{1/2}. \tag{2.1.59}$$

The eigenfunctions may be written

$$\Psi_1 = a\Psi_1^0 - b\Psi_2^0, \qquad \Psi_2 = b\Psi_1^0 + a\Psi_2^0, \tag{2.1.60}$$

where $b = +(1 - a^2)^{1/2}$ and

$$\begin{aligned} a &= 2^{-1/2}\{1 + [1 + 4W_{12}^2/(W_1^0 - W_2^0)^2]^{-1/2}\}^{1/2} \\ &= 2^{-1/2}\{1 + [1 - 4W_{12}^2/(W_1 - W_2)^2]^{1/2}\}^{1/2} \\ &= 2^{-1/2}\{1 + (W_1^0 - W_2^0)/(W_1 - W_2)\}^{1/2}. \end{aligned} \tag{2.1.61}$$

We have listed several equivalent formulas for the transformation coefficient a which may be useful under different circumstances. The positive square root is always to be taken in these formulas.

One effect of the Fermi resonance is to change the vibrational energies. It is seen from Eqs. (2.1.59) that the level of higher initial energy moves up while the lower moves down in energy. The mean energy is unchanged, since

$$W_1 + W_2 = W_1^0 + W_2^0.$$

The eigenfunction of each resulting state is a mixture of the eigenfunctions of the two initial states, as given by Eqs. (2.1.60). Therefore the effective rotational Hamiltonian for each state (which consists of the rigid-rotor Hamiltonian plus the vibration–rotation interaction terms resulting from the usual second-order perturbation calculation) becomes a weighted average of the expressions for the unperturbed states. If no other resonances affect these states, we therefore expect a rigid-rotor pattern of levels with effective rotational constants A_v, B_v, C_v given by

$$A_1 = a^2 A_1^0 + b^2 A_2^0, \qquad A_2 = b^2 A_1^0 + a^2 A_2^0, \qquad (2.1.62)$$

where A_1^0 and A_2^0 are the values that would apply in the absence of the resonance. Similar expressions hold for B_v and C_v. Since $a^2 + b^2 = 1$ we have the relations

$$A_1 + A_2 = A_1^0 + A_2^0.$$

If the Fermi resonance has been analyzed from its effect on the vibrational spectrum, so that W_{12} and $W_1 - W_2$ are known, the calculation of the perturbed rotational constants is straightforward. Alternatively, the interaction constant W_{12} may be determined from the observed rotational constants if the unperturbed constants can be estimated and if $W_1 - W_2$ or $W_1^0 - W_2^0$ is known.

The specific form of W_{12} depends upon the type of Fermi resonance. In the most important case, where the cubic potential terms are responsible, the levels 1 and 2 can be identified with states v_s, v_t and $v_s - 1$, $v_t + 2$. Then if both s and t are nondegenerate modes, W_{12} is just

$$\begin{aligned}
W_{12} &= (v_s, v_t \mid H_3 \mid v_s - 1, v_t + 2) \\
&= k_{stt}(v_s \mid q_s \mid v_s - 1)(v_t \mid q_t^2 \mid v_t + 2) \\
&= 2^{-1/2} k_{stt}[v_s(v_t + 1)(v_t + 2)]^{1/2}. \qquad (2.1.63)
\end{aligned}$$

We have used harmonic oscillator matrix elements (see, e.g., Wilson *et al.*[8]) and are again expressing all constants in reciprocal centimeters. When one of the modes (say, ω_t) is doubly degenerate, we obtain

$$W_{12} = (v_s, v_t, l_t \mid H_3 \mid v_s - 1, v_t + 2, l_t)$$
$$= 2^{-1/2}k_{stt}\{v_s[(v_t + 2)^2 - l_t^2]\}^{1/2}. \tag{2.1.64}$$

It is important to note that W_{12} is nonvanishing only if the two states have the same l_t. For this reason Fermi resonance is only possible between two states in which the quantum number of the degenerate mode differs by 2 (or by a larger even number if higher-order potential terms are involved).

The expressions for W_{12} can be calculated in the same manner for other cases. For example, in a resonance between states $v_s v_t$ and $v_s - 2$, $v_t + 2$ (both modes nondegenerate), we have

$$W_{12} = (w_s v_t \mid H_4 \mid v_s - 2, v_t + 2)$$
$$= \tfrac{1}{4}k_{sstt}[v_s(v_s - 1)(v_t + 1)(v_t + 2)]^{1/2}.$$

In linear triatomic molecules Fermi resonance between the lower-frequency stretching fundamental and the overtone of the bending mode is very common; in fact it occurs to some extent in almost all known molecules of this type. Using the common notation for such molecules, in which the vibrational state is written $v_1 v_2 v_3$, this is a resonance between the 10^00 and 02^00 states. The value of W_{12} for this interaction is found from Eq. (2.1.64), since ω_2 is a degenerate mode:

$$W_{12} = 2^{1/2}k_{122}.$$

The resonance will change the rotational constants of the two states, giving

$$B_{10^00} = a^2 B^0_{10^00} + b^2 B^0_{02^00}, \qquad B_{02^00} = b^2 B^0_{10^00} + a^2 B^0_{02^00}. \tag{2.1.65}$$

In analyzing such a resonance we usually do not have available the values of W_{12} and the unperturbed constant $B^0_{10^00}$. However, we can obtain a good estimate of $B^0_{02^00}$ (i.e., of the unperturbed value of α_2) from the observed rotational constant in either the 01^10 or 02^20 level, since these states are not influenced by the resonance. Thus if only terms in $(v_2 + 1)$ are retained in the expression for B_v, we may calculate α_2 from either of the relations

$$B_{02^20} = B_0 - 2\alpha_2, \qquad B_{01^10} = B_0 - \alpha_2.$$

Therefore, with the observed values of B_{10^00} and B_{02^00}, we can solve the pair of equations (2.1.65) simultaneously to yield a^2 and $B^0_{10^00}$. The explicit result is

$$a^2 = (B_{10^00} - B^0_{02^00})/(B_{10^00} + B_{20^00} - 2B^0_{02^00}),$$

$$B^0_{10^00} = B_{10^00} + B_{02^00} - B^0_{02^0}. \qquad (2.1.66)$$

The mixing coefficient a^2 involves both k_{122} and the vibrational energy difference. If the latter is known from the infrared spectrum, the cubic constant k_{122} may be determined.

One of the strongest Fermi resonances of this type that has been studied is in ClCN.[33] Here the unperturbed vibrational energy difference is only 9 cm^{-1}, while the observed levels are separated by about 70 cm^{-1}. The coefficients determined from the analysis are $a^2 = 0.59$, $b^2 = 0.41$, which approaches the limit of $a^2 = b^2 = 0.50$ corresponding to complete mixing of the states. The spectral pattern for a typical transition is shown in Fig. 12. This figure illustrates the quite general result that the effective rotational constants of the resonating states become more nearly equal as a result of the interaction.

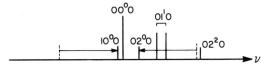

FIG. 12. Effect of Fermi resonance on the vibrational satellite pattern of ClCN.

The treatment of Fermi resonance in nonlinear molecules is not essentially different from the above procedure. In a resonance between v_s, $v_t = 1, 0$ and $0, 2$ states, measurements on the v_s, $v_t = 0, 1$ state can provide unperturbed values of the vibration–rotation interaction constants $\alpha_t^{(A)}$, $\alpha_t^{(B)}$, $\alpha_t^{(C)}$. With these constants we can calculate $A_v{}^0$, $B_v{}^0$, $C_v{}^0$ for the 0, 2 state. Then the observed A_v, B_v, C_v in the two resonating states permit us to calculate the mixing coefficient a^2 and the unperturbed rotational constants for the 1, 0 state. For the general asymmetric rotor there is some redundancy in this calculation, since we have six observed rotational constants available to calculate four unknowns. The extent to which the data can be fitted with a single value of a^2 provides a check on the adequacy of the approximations. An example of such a treatment

[33] W. J. Lafferty, D. R. Lide, and R. A. Toth, *J. Chem. Phys.* **43**, 2063 (1965).

is the rather strong Fermi resonance between the 100 and 020 states in F_2O, which has been satisfactorily analyzed by Morino and Saito.[34]

Our treatment of Fermi resonance has been based on the assumption that the two resonating levels are so close in energy, compared to their separation from all other levels, that their interactions with other levels can be handled by the standard second-order perturbation formulas. In a very strong resonance, such as occurs in the example of ClCN mentioned above, this is likely to be a good assumption. In case of a weaker resonance, however, this assumption may not be valid. In the OCS molecule, for example, the 02^00 level is separated from the 10^00 level by about 170 cm^{-1}. This leads to a relatively mild, though easily detectable Fermi resonance. The next-nearest level that can interact with 02^00 is the 00^01 level, which lies about 1050 cm^{-1} away; this interaction occurs through the potential constant k_{322}. We can see that the latter interaction cannot necessarily be neglected in comparison with the former, especially if we allow for the possibility that k_{322} may be larger than k_{122}. A more correct treatment of the Fermi resonance should include the three levels 10^00, 02^00, and 00^01, which will involve two potential constants, k_{122} and k_{322}. Morino and Matsumura[35] have treated the data on OCS in this way. An analysis of Fermi resonance in other linear triatomic molecules has been given by Lide.[36]

2.1.6. Centrifugal Distortion

The rotational spectrum of a real molecule, even in its ground vibrational state, always deviates slightly from the ideal rigid rotor pattern. This effect is known as *centrifugal distortion*. In a classical picture one can think of the stretching of a real, nonrigid molecule as the rate of rotation increases. This stretching tends to increase the moment of inertia of the molecule, and the interval between successive energy levels is thus reduced from the rigid-rotor value.

The proper quantum-mechanical treatment starts with the general Hamiltonian of Eq. (2.1.44). The term that is responsible for centrifugal distortion is H_2

$$H_2 = -\tfrac{1}{2} \sum_s Q_s \sum_{\alpha,\beta} a_s^{\alpha\beta} (I_{\alpha\alpha}^e I_{\beta\beta}^e)^{-1} P_\alpha P_\beta.$$

The normal coordinate Q_s has a matrix element connecting the states

[34] Y. Morino and S. Saito, *J. Mol. Spectrosc.* **19**, 435 (1966).

[35] Y. Morino and C. Matsumura, *Bull. Chem. Soc. Japan* **40**, 1095 (1967).

[36] D. R. Lide, *J. Mol. Spectrosc.* **33**, 448 (1970).

v_s and $v_s \pm 1$

$$(v_s \mid Q_s \mid v_s \pm 1) = (h/8\pi c\omega_s)^{1/2}(v_s + 1)^{1/2}.$$

Thus the ground-vibrational state is connected to the first excited state of each normal mode by an element of H_2, and this element is quadratic in the overall angular momentum. A second-order perturbation calculation yields terms that are quartic in the angular momentum components. It is easily seen that these terms will be inversely proportional to ω_s^2 and will involve inverse fourth powers of the moments of inertia; furthermore, they will depend quadratically on $a_s^{\alpha\beta}$, the first derivatives of the moments of inertia [Eqs. (2.1.43)].

The sum of all these terms constitutes the centrifugal distortion operator, which may be considered as a perturbation to the rigid-rotor Hamiltonian. In order to obtain the first-order solution, the expectation values $\langle P_x^4 \rangle$, $\langle P_x^2 P_y^2 \rangle$, etc., must be evaluated in a rigid-rotor basis. This approximation is usually adequate for moderate J values, although a more accurate calculation may be required for very high rotational levels.

2.1.6.1. Diatomic and Linear Rotors. The centrifugal distortion term for a diatomic molecule is easily derived. Noting that the moment of inertia derivative is given by

$$a_1^{xx} = a_1^{yy} = 2I_e^{1/2},$$

one finds that the contribution to the rotational energy is

$$-(4B_e^3/c^2\omega^2)\langle P^4 \rangle = -(4B_e^3/c^2\omega^2)J^2(J + 1)^2.$$

It is conventional to write the energy as

$$E/h = B_0 J(J + 1) - D_J J^2(J + 1)^2, \qquad (2.1.67)$$

where

$$D_J = 4B_e^3/(c\omega)^2. \qquad (2.1.68)$$

The inclusion of the centrifugal distortion correction leads to a formula for the frequency of allowed transition

$$\nu(J \rightarrow J + 1) = 2B_0(J + 1) - 4D_J(J + 1)^3. \qquad (2.1.69)$$

Measurement of two transitions therefore permits the calculation of B_0 and D_J.

The calculation follows a similar pattern for linear rotors. Both Eqs. (2.1.67) and (2.1.69) are valid for the linear rotor. In the case of a linear triatomic molecule, D_J is given by

$$D_J = 4B_e{}^3[\zeta_{23}^2/(c\omega_1)^2 + \zeta_{12}^2/(c\omega_3)^2], \tag{2.1.70}$$

where ζ_{23} and ζ_{12} are the Coriolis coupling constants introduced in Eq. (2.1.41).

The effect of centrifugal distortion in diatomic and linear molecules is to decrease the intervals between successive rotation transitions. Since D_J/B_e is typically of the order of 10^{-6}, this is a small effect, but it is easily observable, even at low J values. If the harmonic force field is known, D_J may be calculated from the equations given above. Alternatively, the measured D_J gives information that can be used in the determination of the force field.[37]

2.1.6.2. Symmetric Rotors. When centrifugal distortion is taken into account, the expression for the energy levels of a symmetric rotor becomes

$$E/h = B_0 J(J+1) - D_J J^2(J+1)^2 - D_{JK}K^2 J(J+1) - D_K K^4. \tag{2.1.71}$$

The frequency of a rotational transition is then given by

$$\nu = 2(B_0 - D_{JK}K^2)(J+1) - 4D_J(J+1)^3. \tag{2.1.72}$$

The D_{JK} distortion term thus tends to separate transitions of different K values, which would be coincident in the rigid-rotor approximation. The D_J term has the same effect as in linear rotors, while D_K does not produce observable effects on the microwave spectrum.

Expressions for D_J, D_{JK}, and D_K for molecules of point group $\mathbf{C}_{3v}$ have been given by Wilson[38] and by Dowling et $al.$[39] These are much more complicated than the examples discussed previously. Using the notation which will be introduced in the next section for asymmetric rotors, we may write the distortion constants as

$$\begin{aligned}
D_J &= -\tfrac{1}{4}\tau_{xxxx}, \\
D_{JK} &= -2D_J - \tfrac{1}{2}(\tau_{xxzz} + 2\tau_{xzxz}), \\
D_K &= -D_J - D_{JK} - \tfrac{1}{4}\tau_{zzzz}.
\end{aligned} \tag{2.1.73}$$

The quantities τ_{xxxx}, etc., are defined in Eq. (2.1.75).

[37] Y. Morino and E. Hirota, $Bull.$ $Chem.$ $Soc.$ $Japan$ **31**, 423 (1958).
[38] E. B. Wilson, $J.$ $Chem.$ $Phys.$ **27**, 986 (1957).
[39] J. M. Dowling, R. Gold, and A. G. Meister, $J.$ $Mol.$ $Spectrosc.$ **1**, 265 (1957); **2**, 411 (1958).

Explicit expressions for the distortion constants in X_3YZ molecules are given by Dowling *et al.*[39] Examples of the application of distortion constants to determine the force field in X_3Y molecules have been given by Mirri.[40]

2.1.6.3. Asymmetric Rotors.

The treatment of centrifugal distortion in asymmetric rotors was first developed by Kivelson and Wilson.[41] The effective Hamiltonian including centrifugal distortion may be written in the form

$$H = AP_a^2 + BP_b^2 + CP_c^2 + \tfrac{1}{4} \sum_{\alpha\beta\gamma\delta} \tau_{\alpha\beta\gamma\delta} P_\alpha P_\beta P_\gamma P_\delta. \qquad (2.1.74)$$

Here $A = h/8\pi^2 I_a$, etc., where I_a is the effective moment of inertia in the particular vibrational state; i.e., I_a includes the vibration–rotation interaction corrections discussed in Section 2.1.5.2. The quantities $\tau_{\alpha\beta\gamma\delta}$ (which, like A, B, C, are expressed in frequency units) are defined by

$$\tau_{\alpha\beta\gamma\delta} = \frac{(h^3/8\pi^3)}{I_\alpha I_\beta I_\gamma I_\delta} \sum_i \sum_j \left(\frac{\partial I_{\alpha\beta}}{\partial R_i}\right)_e (f^{-1})_{ij} \left(\frac{\partial I_{\gamma\delta}}{\partial R_j}\right)_e. \qquad (2.1.75)$$

The $I_{\alpha\beta}$ are instantaneous moments or products of inertia, and the R_i represent any suitable set of internal coordinates. The quantity $(f^{-1})_{ij}$ is an element of the inverse force constant matrix. The potential energy has been defined as

$$V = \tfrac{1}{2} \sum_i \sum_j f_{ij} R_i R_j.$$

In all cases the summation extends over the $3N - 6$ coordinates that are required to describe the internal vibrations.

The above expression differs somewhat from our previous formulation, which was carried out in terms of normal coordinates. Thus $\tau_{\alpha\beta\gamma\delta}$ contains derivatives $(\partial I_{\alpha\beta}/\partial R_i)$ rather than the $a_i^{\alpha\beta} = (\partial I_{\alpha\beta}/\partial Q_i)$ that appear in the H_2 term of the Hamiltonian of Eqs. (2.1.44). This is done because the derivatives with respect to the normal coordinates can become very complicated in larger molecules. The R_i may be chosen (say, as symmetry coordinates) to provide the most convenient relationship with the force field.

The first-order calculation of centrifugal distortion in asymmetric rotors requires the evaluation of the expectation value of the last term

[40] A. M. Mirri, *J. Chem. Phys.* **47**, 2823 (1967).
[41] D. Kivelson and E. B. Wilson, *J. Chem. Phys.* **20**, 1575 (1952); **21**, 1229 (1953).

of Eq. (2.1.74) in a rigid-rotor basis. Fortunately this term can be simplified considerably. By the use of commutation relations among the angular momentum components, Eq. (2.1.74) may be reduced to

$$H = A'P_a^2 + B'P_b^2 + C'P_c^2 + \tfrac{1}{4} \sum_{\alpha\alpha\beta\beta} \tau' P_\alpha^2 P_\beta^2, \qquad (2.1.76)$$

where

$$A' = A - \tfrac{1}{2}\tau_{abab} - \tfrac{1}{2}\tau_{caca} + \tfrac{3}{4}\tau_{bcbc},$$

$$B' = B - \tfrac{1}{2}\tau_{bcbc} - \tfrac{1}{2}\tau_{abab} + \tfrac{3}{4}\tau_{caca},$$

$$C' = C - \tfrac{1}{2}\tau_{bcbc} - \tfrac{1}{2}\tau_{caca} + \tfrac{3}{4}\tau_{abab},$$

$$\tau'_{\alpha\alpha\beta\beta} = \tau_{\alpha\alpha\beta\beta} + 2\tau_{\alpha\beta\alpha\beta} \qquad (\alpha \neq \beta). \qquad (2.1.77)$$

Thus there are six quartic terms left in the Hamiltonian, in addition to the three quadratic terms representing an effective rigid rotor, whose rotational constants are modified according to Eq. (2.1.77).

With further manipulation of the angular momenta, the first-order solution to the Hamiltonian of Eq. (2.1.76) may be written in the Kivelson–Wilson form

$$E = E_0 + A_1 E_0^2 + A_2 E_0 J(J+1) + A_3 J^2(J+1)^2$$
$$+ A_4 J(J+1)\langle P_z^2 \rangle + A_5 \langle P_z^4 \rangle + A_6 E_0 \langle P_z^2 \rangle. \qquad (2.1.78)$$

The coefficients $A_1, \ldots, A_6$ are related to the $\tau_{\alpha\beta\gamma\delta}$ through a complex set of equations given in the appendix of the first paper by Kivelson and Wilson.[41] E_0 is the rigid-rotor energy. The expectation values $\langle P_z^2 \rangle$ and $\langle P_z^4 \rangle$ are readily calculated from a computer solution of the rigid asymmetric-rotor problem. A useful table has been given by Schwendeman.[10] Therefore, the rotational transitions are expressible as functions of nine parameters, A', B', C', and $A_1, \ldots, A_6$.

It has been shown by Watson,[42] however, that Eq. (2.1.78) cannot in general be used for fitting an observed spectrum because of a further redundancy relation among the angular momentum components. This introduces a degree of arbitrariness into the formulation for the general asymmetric rotor. The form suggested by Watson leads to an energy expression

$$E = E_0 - d_J J^2(J+1)^2 - d_{JK} J(J+1)\langle P_z^2 \rangle - d_K \langle P_z^4 \rangle$$
$$- d_{WJ} E_0 J(J+1) - d_{WK} E_0 \langle P_z^2 \rangle. \qquad (2.1.79)$$

[42] J. K. G. Watson, *J. Chem. Phys.* **46**, 1935 (1967).

The five coefficients appearing in Eq. (2.1.79) are defined in terms of the $\tau_{\alpha\beta\gamma\delta}$ by Watson.[42] It should also be noted that the effective rotational constants are further modified in this formulation.

We see therefore that fitting an observed spectrum allows us to determine five centrifugal distortion constants, in addition to the three effective rigid-rotor constants. These five distortion constants are functions of the six independent $\tau_{\alpha\beta\gamma\delta}$, so that the latter constants cannot be determined uniquely (except in special cases). The six $\tau_{\alpha\beta\gamma\delta}$, in turn, are functions of a number of force constants which increases rapidly with the number of atoms in the molecule. In the simplest cases a measurement of the distortion constants allows the complete harmonic force field to be fixed (subject, of course, to the uncertainties introduced by the neglect of higher-order terms). When larger molecules are involved, the observed distortion constants provide constraints on the force field that are frequently useful in conjunction with data from the vibrational spectrum.

A certain simplification occurs in planar asymmetric rotors, as was first shown by Dowling.[43] Relationships among the $\tau_{\alpha\beta\gamma\delta}$, somewhat analogous to the planarity conditions on the moments of inertia, reduce the number of independent τ's from six to four.[42,43] Therefore, the τ's can be completely determined in this case.

The most detailed analysis of centrifugal distortion in asymmetric rotors has been done on bent XY_2 molecules. As pointed out above, four τ's can be derived from a fit of the spectrum, and these in principle determine the four harmonic force constants in such a molecule. A thorough study of F_2O was carried out by Pierce et al.[44] A good fit was obtained for transitions up to $J = 40$, although it was necessary to introduce higher-order (P^6) terms into the calculation. A very thorough study of the optimum way to handle centrifugal distortion in bent XY_2 molecules, with applications to SO_2, F_2O, and SiF_2, has been made by Kirchhoff.[45]

2.1.7. Inversion and Ring Puckering

The vibration–rotation interactions considered in Section 2.1.5 result from vibrational motions of small amplitude. Certain large-amplitude internal motions produce special effects that are not included in the formalism developed so far. Such motions fall into two categories,

[43] J. M. Dowling, J. Mol. Spectrosc. 6, 550 (1961).
[44] L. Pierce, N. di Cianni, and R. H. Jackson, J. Chem. Phys. 38, 730 (1963).
[45] W. H. Kirchhoff, J. Mol. Spectrosc. 41, 333 (1972).

internal rotations and large-amplitude oscillations, or "inversions." The latter type is considered in this section.

2.1.7.1. Inversion in Ammonia. The most widely studied microwave spectrum is the ammonia inversion spectrum. It is well known that the equilibrium configuration of NH_3 is pyramidal, with a relatively low potential barrier separating the two equivalent potential minima. The potential function of NH_3 is shown in Fig. 13. The eigenfunctions are alternately symmetric (s) and antisymmetric (a) with respect to a reflection through the plane defined when all four atoms are coplanar. It is seen that the first two levels are almost coincident, but the inversion doubling becomes larger as the levels approach the top of the barrier. Above the barrier, the levels approach a harmonic spacing appropriate to an out-of-plane bending vibration of a planar molecule.

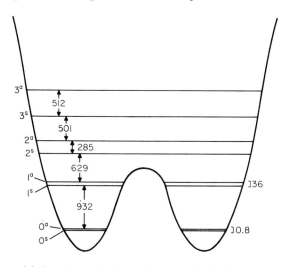

FIG. 13. Potential function and observed energy levels for the ν_2 mode of NH_3. The level separations are given in reciprocal centimeter units.

The separation of the 0^s and 0^a vibrational levels is about 0.79 cm^{-1} or 23,700 MHz in NH_3. Since the dipole selection rule is s $\leftrightarrow$ a, the pure rotational transitions $J, K \rightarrow J + 1, K$ appear as doublets; these fall in the far infrared region of the spectrum. In addition, $\Delta J = 0$ transitions between the 0^s and 0^a levels are allowod, and these are responsible for a very intense microwave spectrum.

In the zeroth approximation all of these $\Delta J = 0$ lines would be coincident. In actual fact, vibration–rotation interactions are quite large in NH_3, and the transitions extend over a wide range. The strongest lines, which are those from relatively low rotational states, fall near 24,000 MHz. A detailed discussion of the ammonia spectrum, including various functional representations for the transition frequencies, may be found in Townes and Schawlow,[1] Chapter 12.

A "pure inversion" spectrum of this type can exist only in a symmetric rotor molecule. No such spectra have been observed outside of NH_3 and its isotopic species. The inversion splitting is a very sensitive function of both the barrier height and the geometry. In analogs such as PH_3 and AsH_3 the splitting is evidently too small to be resolved by present techniques.

2.1.7.2. Inversion in Asymmetric Rotors. We now consider inversion-type motions in asymmetric rotors. In fact, any nonplanar molecule exists in two configurations that are related through an inversion of the internal coordinate system ($x \rightarrow -x, y \rightarrow -y, z \rightarrow -z$). If a physically meaningful path exists by which the molecule can tunnel between these configurations, a splitting of the inversion degeneracy may occur. The most important examples discovered so far are unsymmetric derivatives of ammonia and various ring compounds which undergo a puckering motion.

Let us consider a molecule NAB_2, where A and B may be atoms or groups. We assume a feasible path leading to an inversion, as shown in Fig. 14. The internal potential energy, expressed as a function of the "inversion coordinate," is assumed to have a symmetric, double-minimum form analogous to Fig. 13. If the barrier to inversion is relatively high (specifically, if the $0^s - 0^a$ interval is small compared to the $0^s - 1^a$ interval), each asymmetric-rotor level of the ground state may be regarded

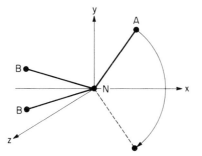

Fig. 14. Inversion motion in an NAB_2 type molecule. The B atoms lie in the xz plane, and the path of the A atom is in the xy plane.

as split into an (s) and (a) component. This will be a constant splitting if we neglect vibration–rotation interactions.

The selection rules for dipole transitions must take into account not only the rotational states but also the symmetry properties of the vibrational (inversion) states. It may be seen from Fig. 14 that the inversion operation $(x \to -x, y \to -y, z \to -z)$ is equivalent to a reflection through the xz plane followed by a $180°$ rotation about the y axis. This operation leads to a change of sign of the μ_y dipole component but leaves μ_x unchanged (μ_z is also unchanged, but in the special case considered here $\mu_z = 0$). Therefore, each rotational transition which is allowed by the μ_y component must connect levels of opposite inversion symmetry in order to ensure a nonvanishing dipole matrix element. Likewise, rotational transitions allowed by μ_x must connect levels of the same inversion symmetry. In summary, for a molecule in which the inversion coordinate is perpendicular to the xz plane, the selection rules are

$$\mu_x: \quad s \leftrightarrow s, \quad a \leftrightarrow a,$$

$$\mu_y: \quad s \leftrightarrow a.$$

The identification of x, y, z with the principal inertial axes a, b, c, depends, of course, on the particular molecule.

The transitions allowed by these selection rules are illustrated in Fig. 15. The left-hand side applies to the case of relatively high inversion barrier, where the inversion splitting is comparable to or smaller than the rotational constants. One sees that the μ_y-type transitions appear as doublets with a separation equal to twice the inversion splitting. The μ_x-type transitions are unresolved doublets in this approximation. However, vibration–rotation interactions may lead to an observable doublet splitting, although it will always be much less than the splitting of the μ_y-type transitions.

As the potential barrier is reduced (or, alternatively, as one observes vibrational levels approaching the top of the barrier), the situation on the right-hand side of Fig. 15 is approached. The μ_y transitions now occur at frequencies much higher than the normal microwave range. In the limit of very low barriers these μ_y transitions become ordinary vibrational transitions in the infrared; the inversion motion is now identified with the out-of-plane vibration of a planar NAB_2 molecule. The pair of μ_x transitions become rotational transitions in the ground and first excited states of the molecule. These will generally be well resolved because of the contribution of vibration–rotation interactions to the effective rotational constants.

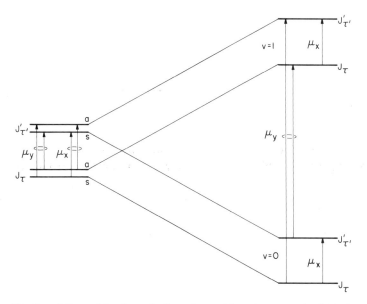

Fig. 15. Correlation of levels and allowed transitions between the high inversion barrier limit (left-hand side) and the zero or low barrier case (right-hand side). The asymmetric-rotor levels are designated by J_τ.

It will be recognized that the dipole components along the principal axes are not well-defined quantities in the situations just described. The μ_y component should be interpreted as a transition moment between the (s) and (a) states

$$(\mu_y)_{\text{sa}} = \int \Psi_s^* \mu_y \Psi_a \, d\tau,$$

where Ψ_s and Ψ_a are vibrational eigenfunctions. If the barrier is high, in the sense used above, this transition moment does not differ much from the average of the absolute value of the moment $|\mu_y|$ over the vibrational motion, and thus we can regard it approximately as a component of the permanent dipole moment. Likewise, μ_x can be interpreted as a similar average. In the low-barrier limit μ_y becomes a true vibrational transition moment of the planar molecule, and the permanent moment has a zero component along the y axis.

Inversion effects of the type described have been observed in several molecules. The deuterated ammonias,[46] NH_2D and ND_2H, are examples

[46] M. T. Weiss and M. W. P. Strandberg, *Phys. Rev.* **83**, 567 (1951).

of the high-barrier case, where each rotational transition appears as a recognizable doublet. Dimethylamine[47] is another example of this type. Formamide[48] and cyanamide[49] represent relatively low-barrier cases; transitions of the μ_x type appear in the microwave spectrum, but the μ_y transitions fall in the far infrared. Methylamine, CH_3NH_2, is a special case where the inversion motion is strongly coupled to the internal rotation of the CH_3 group.[50,51]

2.1.7.3. Ring Puckering. These same considerations apply to the puckering motion which can occur in ring compounds such as trimethyleneoxide, $(CH_2)_3O$. This compound can be represented by a diagram similar to Fig. 14 in which the puckering mode carries the skeleton of the molecule through a planar configuration to its inverted form. All of the above statements on selection rules hold for such ring compounds.

Ring-puckering effects have been studied in several four and five membered rings. In trimethylene-oxide[52] the barrier is very low (about 15 cm^{-1}), and the lowest level is slightly above the top of the barrier. In other ring compounds such as silocyclobutane[53] and trimethylenesulfide[54] the barrier is high enough that several levels fall below the top. In all such compounds a prominent sequence of vibrational satellites is observed, since the spacing of levels of the ring-puckering mode is generally much smaller than the frequencies of the other normal modes. Measurements of the rotational constants and relative intensities of these states provide valuable information on the shape of the potential function that governs the puckering motion. It should also be pointed out that transitions involving the higher levels are often observable in the far infrared. The combination of microwave and far infrared data[55] usually gives the best information on the potential function.

[47] J. E. Wollrab and V. W. Laurie, *J. Chem. Phys.* **48**, 5058 (1968).

[48] C. C. Costain and J. M. Dowling, *J. Chem. Phys.* **32**, 158 (1960).

[49] D. J. Millen, G. Topping, and D. R. Lide, *J. Mol. Spectrosc.* **8**, 153 (1962).

[50] T. Itoh, *J. Phys. Soc. Japan* **11**, 264 (1956).

[51] D. R. Lide, *J. Chem. Phys.* **27**, 343 (1957).

[52] S. I. Chan, J. Zinn, J. Fernandez, and W. D. Gwinn, *J. Chem. Phys.* **33**, 1643 (1960); **34**, 1319 (1961).

[53] W. P. Pringle, *J. Chem. Phys.* **54**, 4979 (1971).

[54] D. O. Harris, H. W. Harrington, A. C. Luntz, and W. D. Gwinn, *J. Chem. Phys.* **44**, 3467 (1966).

[55] S. I. Chan, T. R. Borgers, J. W. Russell, H. L. Strauss, and W. D. Gwinn, *J. Chem. Phys.* **44**, 1103 (1966).

2.1.7.4. Coriolis Interactions. Since inversion and ring-puckering motions can lead to a closely spaced pattern of vibrational energy levels, the chance of near-degeneracies is higher than normal. In particular, rotational levels of the ground state (0^s) may overlap those of the first excited state (0^a). Coriolis interactions (see Section 2.1.5.5) have been found in several cases of this type. The first example was in dideutero-cyanamide,[49] D_2NCN, where certain transitions in the first excited state 0^a were found to be shifted from the expected rigid-rotor position. This can be explained[56] as an interaction between a $K_{-1} = 1$ level of the 0^a state and a $K_{-1} = 2$ level of the 0^s state. The H_1 term in the Hamiltonian of Eqs. (2.1.44) is basically responsible for the interaction. However, since the motion is not of small amplitude, the detailed formulation is somewhat more complicated.[56]

The same type of perturbation occurs in several molecules that undergo ring-puckering motions.[52,53] The observation of such a perturbation is very important, since it permits the vibrational separation to be measured with high accuracy.

2.1.8. Internal Rotation

Another type of internal motion that can produce special effects on the microwave spectrum is internal rotation. The only internal rotations that are of practical importance are those about single bonds. When the rotation is restricted by a potential barrier that is not too high, tunneling between equivalent configurations may occur. This large-amplitude motion leads to effects that are not included in the framework developed for treating ordinary vibration–rotation interactions.

2.1.8.1. The Single-Top Problem. The most widely studied case is internal rotation in which one of the rotating groups is a symmetric top, such as a CH_3 group. There are three equivalent potential minima, and the internal potential function may be represented as a Fourier expansion in which the leading term is

$$V(\alpha) = \tfrac{1}{2}V_3(1 - \cos 3\alpha). \qquad (2.1.80)$$

Here α is the angle of internal rotation, measured from one of the potential minima, and V_3 is the height of the potential barrier (if higher terms in the Fourier expansion are neglected). The Hamiltonian for the pure

---

[56] D. R. Lide, *J. Mol. Spectrosc.* **8**, 142 (1962).

internal rotation problem is simply

$$H_\tau = Fp^2 + V(\alpha), \qquad\qquad (2.1.81)$$

where p is the angular momentum of the internal rotation and F is the reciprocal of a reduced moment of inertia which is explicitly defined below.

If the potential barrier V_3 is very high, the internal rotation has the character of a torsional oscillation, and this degree of freedom may therefore be approximated as an ordinary harmonic vibration. Approximating the potential energy by Eq. (2.1.80) and expanding the cosine function, one obtains a harmonic set of energy levels

$$E/hc = (v + \tfrac{1}{2})\omega_\tau,$$

where v is a torsional quantum number and ω_τ is given (in reciprocal centimeters units) by

$$\omega_\tau = 3(FV_3)^{1/2} = (9/2)Fs^{1/2}.$$

Here we have introduced a reduced barrier height s, given by

$$s = 4V_3/9F. \qquad\qquad (2.1.82)$$

Each level has an inherent threefold degeneracy because of the three equivalent potential minima.

This approximation breaks down as the barrier height is lowered. The exact solution to Eq. (2.1.81), which is a form of the Mathieu differential equation, shows that each torsional level splits into a totally symmetric component (species A) and a doubly degenerate component (species E). In physical terms this splitting is associated with the onset of tunneling through the potential barrier. A typical pattern of levels is shown in Fig. 16.

Since the dipole moment is independent of the torsional angle α, the selection rules for transitions between torsional levels are A $\leftrightarrow$ A and E $\leftrightarrow$ E. Thus the splittings considered up to now do not have a detectable effect on the microwave spectrum. When interactions between internal and overall rotation are taken into consideration, however, very pronounced effects are found. If we ignore all other internal vibrations, the Hamiltonian can be expressed as

$$H = H_r + H_{r\tau} + H_\tau.$$

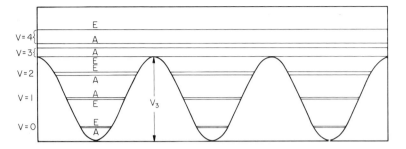

FIG. 16. Potential function and energy levels for internal rotation of a group with threefold symmetry. The levels are indicated for $F = 7\ \text{cm}^{-1}$ and $V_3 = 441\ \text{cm}^{-1}$ ($s = 28$). The abcissa is the torsional angle α.

Here H_τ is given by Eq. (2.1.81) and H_r is the appropriate rigid-rotor Hamiltonian (We note that H_r is independent of α, since the principal moments of inertia are unaffected by the internal rotation):

$$H_r = AP_a{}^2 + BP_b{}^2 + CP_c{}^2.$$

$H_{r\tau}$ is essentially the term H_1 from the general Hamiltonian of Eq. (2.1.44), i.e., the term that expresses the coupling between internal and overall angular momenta. In the present problem it takes the form

$$H_{r\tau} = -2F_p\mathscr{P}, \tag{2.1.83}$$

where

$$\mathscr{P} = (\lambda_a I_\alpha/I_a)P_a + (\lambda_b I_\alpha/I_b)P_b + (\lambda_c I_\alpha/I_c)P_c. \tag{2.1.84}$$

Here λ_a, λ_b, λ_c are direction cosines between the axis of internal rotation and the principal inertial axes of the molecule; I_α is the moment of the symmetric top about its axis and

$$F = h/8\pi^2 r I_\alpha, \tag{2.1.85}$$

where

$$r = 1 - [(\lambda_a{}^2 I_\alpha/I_a) + (\lambda_b{}^2 I_\alpha/I_b) + (\lambda_c{}^2 I_\alpha/I_c)].$$

Before considering the general problem, it is worthwhile to examine the case where both rotating groups (and hence the molecule as a whole) are symmetric tops.[57] If we label the symmetry axis as a, we find the

[57] J. S. Koehler and D. M. Dennison, *Phys. Rev.* **57**, 1006 (1940).

reduced moment rI_α becomes

$$rI_\alpha = I_\alpha(I_a - I_\alpha)/I_a.$$

The interaction term in the Hamiltonian is $H_{r\tau} = -2F(I_\alpha/I_a)pP_a$ $= -2F(I_\alpha/I_a)Kp$. Therefore, internal rotation in a symmetric rotor affects all levels of given K by the same amount; no detectable change in the microwave spectrum is produced because of the $\Delta K = 0$ selection rule. However, when interaction between internal rotation and the other vibrational modes is taken into account, the coupling through the $H_{r\tau}$ term becomes observable.[58,59]

There are two general methods for handling the $H_{r\tau}$ term in molecules consisting of a symmetric top attached to an asymmetric rotor, the principal axis method (PAM) and the intermediate axis method (IAM). In the former approach, which has been discussed fully by Herschbach,[60] this term is handled directly by perturbation theory. The matrix elements of p connecting the various torsional levels are removed by a high-order Van Vleck perturbation technique. In this way an effective Hamiltonian is obtained for each torsional state, in which the coupling term may be written

$$(H_{r\tau})_{v\sigma} = F \sum_n W_{v\sigma}^{(n)} \mathscr{P}^n. \tag{2.1.86}$$

The states are here labeled by the torsional quantum number v introduced above and a degeneracy index σ ($\sigma = 0$ corresponding to the A sublevel and $\sigma = \pm 1$ to the E sublevel). The perturbation coefficients $W_{v\sigma}^{(n)}$ are functions of only the reduced barrier height s, defined in Eq. (2.1.82). For levels of A symmetry the $W_{v\sigma}^{(n)}$ vanish when n is odd. A table of these coefficients has been given by Herschbach.[2,60] Since the coupling term of Eq. (2.1.86) differs for the A and E states, it will perturb the rigid-rotor spectra by different amounts and hence produce a splitting of each asymmetric-rotor transition. The components of the doublet are of equal intensity[61] in the case of a CH_3 group (or any other top in which the three off-axis atoms have nuclear spin $\frac{1}{2}$). The magnitude of the splitting is extremely sensitive to the barrier height and usually varies considerably from one rotational transition to another.

[58] D. R. Lide and D. K. Coles, *Phys. Rev.* **80**, 911 (1950).
[59] D. Kivelson, *J. Chem. Phys.* **22**, 1773 (1954); **23**, 2230 (1955).
[60] D. R. Herschbach, *J. Chem. Phys.* **31**, 91 (1959).
[61] R. W. Kilb, C. C. Lin, and E. B. Wilson, *J. Chem. Phys.* **26**, 1695 (1957).

The series in Eq. (2.1.86) converges fairly rapidly if the barrier is high. The $n = 2$ term, which is the leading term for the A levels, is quadratic in the overall angular momentum. Thus if higher terms can be neglected, the rotational spectrum of an A level will follow a rigid-rotor pattern, but with rotational constants which are modified by the effects of internal rotation. The rotational constants of the corresponding E level will be modified by a different amount. Thus the differences in effective rotational constants of A and E levels, ΔA, ΔB, ΔC, can be related to the barrier height through the coefficients $W_{v\sigma}^{(n)}$. Specifically, we can write, to a first approximation,

$$\Delta A = A_{vA} - A_{vE} = F(\lambda_a I_\alpha / I_a)^2 [W_{vA}^{(2)} - W_{vE}^{(2)}] \qquad (2.1.87)$$

and similarly for ΔB and ΔC. If the structure of the molecule is known well enough that I_α and the direction cosines λ_a, λ_b, λ_c, may be calculated, the barrier height V_3 can then be obtained from the differences in effective rotational constants.

An additional complication is present for the E levels because the odd-order perturbation terms do not vanish. This means that the effective Hamiltonian will contain linear terms in P_a, P_b, and P_c (as well as higher-order odd terms). These terms do not make a first-order contribution, since they have no diagonal elements in a rigid-rotor representation. However, their off-diagonal terms can become very important when there are near-degeneracies between the rigid-rotor levels. In such cases the E levels no longer follow a rigid-rotor pattern, and the rotational matrix must be diagonalized exactly.

The convergence of Eq. (2.1.86) becomes poorer as the barrier is reduced. Terms with $n > 2$ then become more significant, and the pseudo rigid-rotor character of the spectrum is lost. It may be noted that the $n = 4$ terms have the same formal effect as centrifugal distortion. In low-barrier cases many terms may be necessary to obtain a satisfactory fit of the spectrum. When the barrier is very low a different approach, described by Wilson et al.,[62] is more suitable. An important example of the low-barrier case is the CH_3BF_2-type molecule. Here the V_3 term in the potential energy vanishes because of symmetry considerations, and the next term, $V_6 \cos 6\alpha$, is found in practice to be very small.

When the torsional splittings are large, it is often desirable to treat the direction cosines as adjustable parameters in fitting the spectrum. This allows a better fit and, in addition, gives information of structural interest.

[62] E. B. Wilson, C. C. Lin, and D. R. Lide, *J. Chem. Phys.* **23**, 136 (1955).

It is also possible to include the next term in the potential energy expansion V_6 as a parameter. However, experience has shown that V_6 is usually much smaller than V_3 (unless V_3 vanishes identically because of symmetry), and it is very difficult to make a meaningful measurement of V_6 unless data are available for more than one value of the torsional quantum number v.

While the PAM approach has very general utility, the convergence tends to be poor in certain special cases. One such case occurs in a near-symmetric rotor in which the axis of internal rotation is almost parallel to the pseudo symmetry axis. Here the IAM method, which was developed by Hecht and Dennison,[63] can be used to advantage. In this method an axis system based on the axis of internal rotation is used, rather than the principal inertial axes. Furthermore, transformations are applied which minimize the coupling term between internal and overall rotation. In this manner the coupling can be easily handled as a small perturbation, but at the expense of dealing with a rotational Hamiltonian of rather complicated form.

A detailed comparison of the IAM and PAM formulations has been made by Lin and Swalen.[64] Specific IAM formulas applicable to the near-prolate rotor case have been given by Lide and Mann.[65]

2.1.8.2. Two-Top Molecules. The presence of more than one internal rotation leads to a somewhat more complicated problem. The simplest case, in which two equivalent tops are attached to a framework of C_{2v} symmetry, was first worked out by Pierce[66] and by Swalen and Costain.[67] The angular momentum of each top is coupled to the overall angular momentum, in an analogous manner to the single-top problem. In addition, there is a term in the kinetic energy that couples the tops to each other. The potential energy may be expressed in a Fourier series in which the leading terms are

$$V(\alpha_1, \alpha_2) = \tfrac{1}{2} V_3 (2 - \cos 3\alpha_1 - \cos 3\alpha_2) + V_3' \cos 3\alpha_1 \cos 3\alpha_2$$
$$+ V_3'' \sin 3\alpha_1 \sin 3\alpha_2. \tag{2.1.88}$$

Thus the potential energy also contains top–top coupling terms.

[63] K. T. Hecht and D. M. Dennison, *J. Chem. Phys.* **26**, 31 (1957).
[64] C. C. Lin and J. D. Swalen, *Rev. Mod. Phys.* **31**, 841 (1959).
[65] D. R. Lide and D. E. Mann, *J. Chem. Phys.* **27**, 868 (1957).
[66] L. Pierce, *J. Chem. Phys.* **34**, 498 (1961).
[67] J. D. Swalen and C. C. Costain, *J. Chem. Phys.* **31**, 1562 (1959).

Pierce[66] has described a method of treating both the kinetic and potential energy coupling terms by perturbation theory, which makes use of the perturbation sums for the single-top problem. In the ground state the various coupling terms split each rotational level into four sublevels, labeled A_1A_1, EE, A_1E, and EA_1. If the barrier is high, however, the splitting between A_1E and EA_1 is small compared to the other intervals. Thus it is characteristic to find triplets in the microwave spectra of two-top molecules, and these may resolve into quartets when the barrier is sufficiently low.

When the barrier is high it is often possible to fit the torsional splittings adequately in the ground state by neglecting the coupling terms V_3' and V_3''. In order to determine these coupling terms, data on excited torsional states or other isotopic species are usually required. Several methods of handling the coupling terms in excited torsional states have been described.[68]

2.1.8.3. Other Methods of Barrier Determination. The most sensitive method of measuring barriers to internal rotation is through the splittings produced by the coupling of internal and overall angular momenta. In cases where these splittings cannot be observed, other less accurate techniques are available. It has been mentioned that splittings predicted from the simple model used here cannot be directly observed in symmetric rotors, but do become detectable through interaction of the torsion with other vibrations. The theory of this effect has been developed by Kivelson,[59] and it has been applied to molecules such as CH_3SiH_3.

In molecules with extremely low threefold barriers the torsional levels can be described as the slightly perturbed levels of a free internal rotor. Here the barrier does not have a pronounced effect on the rotational spectrum. Furthermore, the spectra are so complex that it is difficult to identify many transitions. However, some information on the barrier can be extracted. Examples of such cases are provided by $H_3C—C{\equiv}C—CH_2Cl$ [69] and $H_3C—C{\equiv}C—SiH_3$.[70]

Finally, in molecules with rather high barriers (or heavy internal groups) it may be impossible to resolve the torsional splittings in the accessible transitions. If rotational transitions in the $v = 1$ level can be

[68] E. Hirota, C. Matsumura, and Y. Morino, *Bull. Chem. Soc. Japan* **40**, 1124 (1967); J. R. Hoyland, *J. Chem. Phys.* **49**, 1908 (1968); A. Trinkaus, H. Dreizler, and H. D. Rudolph, *Z. Naturforsch.* **23a**, 2123 (1968).

[69] V. W. Laurie and D. R. Lide, *J. Chem. Phys.* **31**, 939 (1959).

[70] W. H. Kirchhoff and D. R. Lide, *J. Chem. Phys.* **43**, 2203 (1965).

assigned, relative intensity measurements provide a value for the separation of $v = 0$ and $v = 1$ levels. The barrier height may be calculated from this value by using the eigenvalues of Eq. (2.1.81). Since accurate relative intensity measurements are difficult to make, barriers determined by this method are usually not reliable to better than 10%.

The location of the $v = 1$ torsional state can sometimes be determined very accurately by the observation of a Coriolis resonance with another vibrational mode. The barrier heights in C_2H_5CN [31] and PF_3BH_3 [32] have been measured in this way.

2.1.8.4. Rotational Isomerism. The effects of internal rotation are rather different when both of the rotating groups are asymmetric. Since the potential minima are not equivalent, the internal eigenfunctions describe configurations in which the molecule is approximately frozen in a particular geometry. This situation is referred to as *rotational isomerism*, i.e. we can think of an equilibrium mixture of isomers that interconvert rapidly, but with each isomer contributing its own characteristic rotational spectrum. The relative intensity of the spectra from the different isomers is a function of the differences in energy.

In the special case where each of the groups has a plane of symmetry, one often finds a *trans* isomer, with the angle of internal rotation $\alpha = 0$, and two equivalent *gauche* isomers with $\alpha \approx \pm 120°$. A typical example is n-propyl fluoride, $CH_3CH_2CH_2F$, which has been throughly studied by Hirota.[71] In such molecules each isomer usually exhibits a rigid-rotor spectrum to a good approximation. If one of the groups is light enough, however, tunneling between the two equivalent *gauche* minima introduces special effects.[72]

2.1.9. Determination of Molecular Structure

Since the predominant features of the microwave spectrum are determined by the moments of inertia of a molecule, the information that can be derived most directly from the spectrum concerns the molecular geometry. The procedures for extracting this structural information are discussed in the present section.

2.1.9.1. Gross Structural Features. A certain amount of information on molecular geometry is provided directly by the ground-state moments

---

[71] E. Hirota, *J. Chem. Phys.* **37**, 283 (1962).
[72] E. Hirota, to be published.

of inertia and by other observed features of the spectrum. In particular, the presence of symmetry elements can often be demonstrated on general grounds, without going into quantitative details of the molecular structure. Linear and symmetric rotors are usually identifiable immediately from the spectral pattern itself (Section 2.1.4.1). If a twofold symmetry axis is present in an asymmetric rotor, spin statistics lead to an intensity alternation (Section 2.1.3.3) that is detectable from careful measurements of relative intensities. Another method of showing the presence of a twofold axis is to measure the spectrum of a species that is isotopically substituted at a position on the symmetry axis. In the rigid-rotor approximation the moment about this axis will be unchanged, while the other two moments will change by the same amount. While these relations do not hold exactly for a real molecule, the deviations are usually extremely small (typically not more than 0.1–0.2%). Finally, a molecule with a twofold axis will have only one nonvanishing dipole component. Therefore, the failure to observe transitions allowed by other components, and, more critically, the ability to fit the Stark effect on the basis of a single component, provide strong evidence for a symmetry axis.

The planarity of a molecule can be demonstrated by measuring the inertial defect Δ, defined by

$$\Delta = I_c - I_a - I_b. \tag{2.1.89}$$

The inertial defect of a planar molecule vanishes identically in the rigid-rotor approximation; actual values of Δ, computed by using ground-state moments in Eq. (2.1.89), rarely exceed 0.2 amu Å². If we note that a pair of equivalent atoms of atomic mass 10 which deviate by 0.1 Å from the plane of the rest of the molecule will yield an apparent Δ of -0.4 amu Å² (see below), we see that the value of Δ provides a very critical test of planarity.

When there is sufficient information on the vibrational potential function, the inertial defect may be predicted and compared with the observed value to provide an even more sensitive test. The anharmonic contributions to the effective moments cancel in the expression for Δ, so that only the harmonic force field is required.[73,74] The general expression for the harmonic contribution to the inertial defect can be written

$$\Delta = \sum_s (v_s + \tfrac{1}{2}) \Delta_s,$$

[73] T. Oka and Y. Morino, J. Mol. Spectrosc. 6, 472 (1961); 8, 9 (1962).
[74] D. R. Herschbach and V. W. Laurie, J. Chem. Phys. 40, 3142 (1964).

where Δ_s is given by

$$\Delta_s = (h/8\pi^2 c) \sum_{s'} (8\omega_{s'}^2/\omega_s)(\omega_s^2 - \omega_{s'}^2)^{-1}$$
$$\times \{(\zeta_{ss'}^a)^2 + (\zeta_{ss'}^b)^2 - (\zeta_{ss'}^c)^2\} + (h/8\pi^2 c) \sum_t (12/\omega_t). \qquad (2.1.90)$$

In this expression the indices s and s' run over all the normal modes of the molecule, while t is restricted to the out-of-plane modes. The remainder of the notation is explained in Section 2.1.5. In addition, there are small contributions from the centrifugal distortion and from the mass of the electrons. However, these are generally only 1 or 2% of the contribution of Eq. (2.1.90).[73] If the force field has been accurately determined, the calculated value of Δ should agree to about 0.001 amu Å^2 with the value obtained from the observed moments.

We finally consider a nonplanar molecule which possesses a plane of symmetry. In the rigid-rotor approximation the moments are related by

$$I_x + I_y - I_z = 2 \sum_i m_i z_i^2, \qquad (2.1.91)$$

where the symmetry plane has been designated as xy. This equation may be refined by adding a term $-\Delta'$, referred to as a *pseudo-inertial defect*, to the right-hand side.[75] When the only nonplanar group is one whose structure varies little from one molecule to another, Eq. (2.1.91) can be used to test for a symmetry plane. A methyl group, for example, usually contributes 3.10–3.20 amu Å^2 to the right-hand side. Furthermore, an isotopic substitution in the symmetry plane does not affect the quantity $\Sigma m_i z_i^2$. Thus the invariance of $I_x + I_y - I_z$ to such a substitution is evidence for a symmetry plane.

2.1.9.2. r_e and r_0 Structures. In order to determine interatomic distances and angles quantitatively, it is necessary, except in a few simple cases, to have data on several isotopic species. If equilibrium moments of inertia are available for enough isotopic species, a set of simultaneous equations may, in principle, be solved for the structural parameters. This set of parameters is designated the "r_e structure."

Equilibrium moments require not only the ground-state rotational constants but also the vibration–rotation interaction constants α_s [see Eq. (2.1.46)]. The latter may be determined experimentally by measuring the spectrum in the first excited state of each normal mode. This is

[75] V. W. Laurie, *J. Chem. Phys.* **28**, 704 (1958).

difficult to do, however, when there are high-frequency modes (such as hydrogen stretching modes) in the molecule. An alternative is to calculate some of the α_s from the known (or estimated) potential constants.

Accurate r_e structures are available for many diatomics but have been obtained for only a small number of polyatomic molecules. In most cases one has only the effective ground-state moments

$$I_0^{(a)} = h/8\pi^2 A_0, \quad \text{etc.},$$

for all the isotopic species required for a structure determination. In this event the crudest approximation is to ignore the distinction between ground-state and equilibrium moments; the resulting parameters are designated the "r_0 structure." This is usually not a very satisfactory procedure, because the difference between I_0 and I_e (i.e., the contribution of zero-point vibrations to the effective moments) can change significantly on isotopic substitution. Furthermore, the structural parameters obtained depend very much on the particular combination of isotopic species chosen for the calculation. For this reason the r_0 structure is not well defined in an operational sense, and, of course, it is not simply related to the r_e structure.

A few examples of r_0 and r_e structures are given in Table IX. The range of r_0 parameters given is that found from various combinations of

TABLE IX. Comparison of r_0 and r_e Structures

Molecule	Parameter	r_0 [a]	r_e
OCS	$r(CO)$	1.155–1.165	1.154
	$r(CS)$	1.558–1.565	1.563
ClCN	$r(CCl)$	1.627–1.634	1.629
	$r(CN)$	1.157–1.166	1.160
HCN	$r(CH)$	1.058–1.069	1.066
	$r(CN)$	1.155–1.158	1.153
N_2O	$r(NN)$	1.12–1.14[b]	1.126
	$r(NO)$	1.18–1.20[b]	1.186

[a] The range of r_0 values represents the results obtained from various pairs of isotopic species.

[b] Certain combinations of isotopic species lead to imaginary distances.

input data. These results illustrate the extreme care which must be exercised in drawing any conclusions from r_0 structures.

2.1.9.3. r_s Structures. In an effort to avoid the deficiencies of the r_0 structure, Costain[76] introduced the concept of the substitution, or r_s structure. This idea is based upon a set of equations first given by Kraitchman.[77] If a single atom in a rigid rotor is isotopically substituted, the resulting changes in the moments of inertia fix the Cartesian coordinates of the substituted atom through a set of very simple relations. By substituting each atom in turn, a complete structure is obtained. If equilibrium moments are used, the results will be identical (except for a somewhat different weighting of experimental errors) with the structural parameters calculated in the conventional way. However, when the input data are ground-state moments, the use of isotopic changes in the moments ΔI_0, rather than the moments themselves, tends to give a calculation that is less sensitive to errors from zero-point effects. Furthermore, the computational procedure is well defined, and one avoids the arbitrary features of r_0 structures.

The general expression for calculating the r_s coordinates of a substituted atom is[76,77]

$$x^2 = \mu^{-1}\, \Delta P_x \{1 + [\Delta P_y/(I_x - I_y)]\}\{1 + [\Delta P_z/(I_x - I_z)]\}, \quad (2.1.92)$$

where

$$\Delta P_x = \tfrac{1}{2}(-\Delta I_x + \Delta I_y + \Delta I_z) \quad \text{and} \quad \mu = M\, \Delta m/(M + \Delta m).$$

In this expression ΔI_x, ΔI_y, ΔI_z are the isotope shifts in the principal moments of inertia produced by substitution of the indicated atom, M is the total mass of the original molecule, and Δm is the isotopic change in mass. The xyz frame is the principal axis system of the original molecule. The expressions for the other coordinates are obtained by cyclic permutation of xyz in Eq. (2.1.92).

When symmetry elements are present in the molecule, Eq. (2.1.92) simplifies considerably. Thus if z is an axis of symmetry, we have for an atom on the z axis

$$x^2 = y^2 = 0, \qquad z^2 = \mu^{-1}\, \Delta I_x = \mu^{-1}\, \Delta I_y. \quad (2.1.93)$$

[76] C. C. Costain, *J. Chem. Phys.* **29**, 864 (1958).

[77] J. Kraitchman, *Amer. J. Phys.* **21**, 17 (1953).

For substitution in a plane of symmetry (designated by xy) we have

$$x^2 = \mu^{-1} \Delta I_y [1 + \Delta I_x/(I_x - I_y)],$$

$$y^2 = \mu^{-1} \Delta I_x [1 + \Delta I_y/(I_y - I_x)], \qquad (2.1.94)$$

$$z^2 = 0.$$

It is seen from these equations that some redundancy exists in the calculation of r_s structures for molecules with symmetry elements. That is, different combinations of ΔI's may be used in calculating a given coordinate. In the absence of vibrational effects the results should be identical; the extent to which these calculations disagree gives a measure of the internal consistency of r_s structures. An example of such a comparison is given in Table X. It is seen that the r_s value for the C—C

TABLE X. Structural Parameters r_s of Propane Calculated from Various Combinations of Moments of Inertia

	I_x, I_y, I_z	I_x, I_y	I_y, I_z	I_x, I_z
r(CC)	1.5277 Å	1.5263	1.5252	1.5256
$\not\prec$CCC	112.24°	112.36°	112.50°	112.48°
CH$_2$ group				
r(CH)	1.0943	1.0952	1.0971	1.0963
$\not\prec$HCH	106.32°	106.16°	105.90°	106.04°
CH$_3$ group				
r(CH$_s$)	1.0863	1.0970	1.0799	1.0930
$\not\prec$CCH$_s$	111.90°	111.36°	112.81°	111.28°

distance in propane is internally consistent to 0.001 or 0.002 Å. This is usually the case for distances between heavy atoms, although some exceptions have been found. The CH distance in the methyl group, however, varies over a range of 0.018 Å, probably because of the large amplitude of the torsional oscillation.

In some cases it is difficult or impossible to carry out an isotopic substitution for every atom. Fluorine, for example, has no stable isotopes. A useful technique here is to apply the first moment, or center-of-mass, condition. Thus if the x coordinates of all atoms but one have been determined, the remaining coordinate can be fixed by requiring

$$\sum_i m_i x_i = 0.$$

Two other conditions are available, the second moment and product of inertia

$$\sum_i m_i(x_i^2 + y_i^2) = I_0^{(z)},$$

$$\sum_i m_i x_i y_i = 0.$$

However, these are generally less satisfactory than the first moment condition and should be used only when there is no alternative. The degree to which these conditions are satisfied by experimental r_s structures has been discussed by Costain.[76]

A serious problem arises when an atom is located very close to a principal axis. Whenever one of the ΔI's is very small, the contribution of vibrational effects (in absolute as well as percentage terms) becomes greatly magnified, and the error in this derived coordinate can be quite large. This problem has been discussed by Costain[76] and by Laurie and Herschbach.[20] In extreme cases the r_s coordinate may turn out to be imaginary. While it is difficult to give quantitative guidelines on when these errors become significant, they probably should be taken into consideration whenever a coordinate is less than about 0.20 Å. It is often desirable to use the first moment condition discussed above, rather than the direct r_s coordinate, to locate an atom with a very small coordinate. A double-substitution technique has also been successfully applied.[78]

One defect of the substitution structure is that its precise relationship to the equilibrium structure is obscure. In a diatomic molecule one can show that[76]

$$r_s = r_e(1 + \alpha_e/8B_e).$$

This indicates that r_s is larger than r_e by 0.05 to 0.2% for most diatomics (except for hydrides, where the difference may be as large as 0.5%). A general relation between r_s and r_e cannot be given for polyatomic molecules. However, where experimental comparisons can be made, the differences are found to be of the order of a few tenths of one percent, with the r_s distance generally larger than r_e.

[78] L. Pierce, *J. Mol. Spectrosc.* **3**, 575 (1959).

2.1.9.4. The Average Structure. Another method of accounting for vibrational effects is to determine the so-called *average structure*, designated as either $\langle r \rangle$ or r_z. It has been shown[79,80] that a correction for the harmonic contribution to the vibration–rotation interaction constants [i.e., the first term of Eq. (2.1.47)] leads to a set of moments that is identical to the moments of the hypothetical molecule in which each atom is frozen in its average position. If the harmonic force field is accurately known, one can convert the experimental ground-state moments I_0 into a set of moments I^* for the average configuration. The I^* for various isotopic species are then used to determine the structural parameters. Strictly speaking, $\langle r_{ij} \rangle$ is not the true average of the instantaneous distance between atoms i and j, but rather the average of the projection of this distance along the direction of the undisplaced bond.[20] This distinction is most important for hydrogen atoms, for which the amplitude of the bending modes can be quite high.

Explicit formulas for $I^* - I_0$ have been given by Laurie and Herschbach for several simple molecular types.[20] These expressions involve the harmonic vibrational frequencies and the Coriolis coupling constants. In more complicated molecules the calculation of I^* becomes much more difficult, since accurate knowledge of the normal coordinates is necessary.

Where data are available for diatomic and simple polyatomic molecules, $\langle r \rangle$ is generally found to be from 0.2 to 0.5% larger than r_e. In the case of hydrides, the differences tend to be in the 1–2% range, reflecting the much larger vibrational amplitudes for bonds involving hydrogen. As a rule, r_s is intermediate between r_e and $\langle r \rangle$.

The average structure has the important advantage of a well-defined physical meaning. However, there are practical difficulties in calculating it reliably, even if the harmonic force field is known. It has been found in ClCN, where data on four isotopic species are available,[33] that $\langle r \rangle$ parameters derived from the various pairs of isotopic molecules show rather poor internal consistency. This is undoubtably due to the fact that $\langle r \rangle$ is not isotopically invariant. Thus very small changes in the true $\langle r \rangle$ upon isotopic substitution can be magnified into much larger changes in the distances derived from the calculation.[20,33] In this respect $\langle r \rangle$ suffers from many of the drawbacks of the r_0 structure.

[79] D. R. Herschbach and V. W. Laurie, *J. Chem. Phys.* **37**, 1668 (1962).

[80] M. Toyama, T. Oka, and Y. Morino, *J. Mol. Spectrosc.* **13**, 193 (1964).

2.1.10. The Stark Effect

The shift or splitting of energy levels in the presence of an external field, commonly designated as the *Stark effect*, has played an important part in the development of microwave spectroscopy. Observation of the Stark effect is probably the most useful single tool in the assignment of rotational transitions. Quantitative Stark-effect measurements permit a very accurate determination of the electric dipole moment of a molecule. Finally, the Stark effect is the basis for the most widely used modulation method, as is discussed in Section 2.1.14.

If a molecule possesses a permanent dipole moment, the principal contribution to the Stark effect comes from the interactions of this moment with the external electric field. A much smaller contribution occurs from the polarization of the molecule by the external field and the subsequent interaction of this induced moment with the field. This polarization contribution is generally too small to detect unless the permanent moment vanishes. Since nonpolar molecules do not show a microwave spectrum, we shall restrict ourselves to the Stark effect produced by the permanent moment.

In the presence of an external electric field $\mathscr{E}$ an additional term appears in the Hamiltonian for a rotating molecule with permanent dipole moment $\boldsymbol{\mu}$

$$H_{\mathscr{E}} = -\boldsymbol{\mu} \cdot \mathscr{E}. \tag{2.1.95}$$

For typical laboratory fields this term is smaller than the average separation of rotational levels. We may therefore treat it by perturbation theory, using the rigid-rotor wave functions to evaluate its matrix elements. In the absence of degeneracies it may be shown on general grounds that the first-order contribution of $H_{\mathscr{E}}$ vanishes. There is always a contribution in second order, however, and degeneracies, both intrinsic and accidental, are sufficiently common that first-order Stark effects are frequently observed.

In order to carry out the perturbation calculation, we take $\mathscr{E}$ to be a uniform field of magnitude $\mathscr{E}$ directed along the space-fixed Z axis. The dipole moment $\boldsymbol{\mu}$ may be expressed in terms of its components μ_x, μ_y, μ_z along molecule-fixed axes. We can then write $\boldsymbol{\mu} \cdot \mathscr{E}$ in terms of the direction cosines discussed in Section 2.1.3.2, giving

$$H_{\mathscr{E}} = -(\mu_x \Phi_{Zx} + \mu_y \Phi_{Zy} + \mu_z \Phi_{Zz})\mathscr{E}. \tag{2.1.96}$$

This perturbation operator must be evaluated for the various types of rigid rotors.

2.1.10.1. Stark Effect of the Linear Rotor. The dipole moment of a linear rotor must necessarily lie along the rotor axis, so that $\mu_z = \mu$ and $\mu_x = \mu_y = 0$. The direction cosine matrix elements may be evaluated from Table III, remembering that $K = 0$. We obtain the nonvanishing matrix elements

$$
\begin{aligned}
(J, M \mid H_{\mathscr{E}} \mid J + 1, M) &= (J + 1, M \mid H_{\mathscr{E}} \mid J, M) \\
&= -\mu \mathscr{E}(J, M \mid \Phi_{Zz} \mid J + 1, M) \\
&= -\mu \mathscr{E}\left[\frac{(J + 1)^2 - M^2}{(2J + 1)(2J + 3)}\right]^{1/2}.
\end{aligned}
\tag{2.1.97}
$$

Each level therefore interacts only with its two neighboring levels. Applying second-order perturbation theory, with the energy levels given by Eq. (2.1.10), we have for the Stark energy

$$
W_{JM}^{(2)} = \frac{\mu^2 \mathscr{E}^2}{2B} \frac{J(J + 1) - 3M^2}{J(J + 1)(2J - 1)(2J + 3)}.
\tag{2.1.98}
$$

For the level $J = M = 0$ this becomes

$$
W_{00}^{(2)} = -\mu^2 \mathscr{E}^2/6B.
\tag{2.1.99}
$$

2.1.10.2. The Symmetric Rotor. The twofold degeneracy of all levels of the symmetric rotor with $K \neq 0$ (see Section 2.1.2.6) leads to a first-order Stark effect. Following the same procedure used for the linear rotor, we find an additional matrix element

$$
W_{JKM}^{(1)} = (JKM \mid H_{\mathscr{E}} \mid JKM) = -\mu \mathscr{E} KM/J(J + 1).
\tag{2.1.100}
$$

Since M can take positive or negative values, each level splits in first order into $2J + 1$ sublevels, symmetrically placed about the unshifted $(M = 0)$ level. The second-order Stark energy is found to be

$$
\begin{aligned}
W_{JKM}^{(2)} = \frac{\mu^2 \mathscr{E}^2}{2B} &\left\{ \frac{(J^2 - K^2)(J^2 - M^2)}{J^3(2J - 1)(2J + 1)} \right. \\
&\left. - \frac{[(J + 1)^2 - K^2][(J + 1)^2 - M^2]}{(J + 1)^3(2J + 1)(2J + 3)} \right\}.
\end{aligned}
\tag{2.1.101}
$$

This is much smaller than $W_{JKM}^{(1)}$ but must be taken into account in accurate work. In the case of $K = 0$, the first-order energy vanishes, and Eq. (2.1.101) reduces to the expression for a linear rotor, Eq. (2.1.98).

2.1.10.3. The Asymmetric Rotor. The Stark effect of the asymmetric rotor has been treated in detail by Golden and Wilson.[81] Since there are in general no restraints on the orientation of the dipole moment, we must calculate the contribution of each component μ_g (with $g = a, b, c$) to the Stark energy. In the simplest case where there are no near-degeneracies, the expression for the second-order energy is

$$
{}^gW_{J\tau M}^{(2)} = \mu_g^2 \mathscr{E}^2 \left\{ \frac{J^2 - M^2}{4J^2(4J^2 - 1)} \sum_{\tau'} \frac{(J\tau \mid \varPhi_{Zg} \mid J - 1, \tau')^2}{W_{J\tau}^0 - W_{J-1,\tau'}^0} \right.
$$

$$
+ \frac{M^2}{4J^2(J+1)^2} \sum_{\tau' \neq \tau} \frac{(J\tau \mid \varPhi_{Zg} \mid J\tau')^2}{W_{J\tau}^0 - W_{J\tau'}^0}
$$

$$
\left. + \frac{(J+1)^2 - M^2}{4(J+1)^2(2J+1)(2J+3)} \sum_{\tau'} \frac{(J\tau \mid \varPhi_{Zg} \mid J+1, \tau')^2}{W_{J\tau}^0 - W_{J+1,\tau'}^0} \right\}.
$$

$$\tag{2.1.102}$$

Here the asymmetric rotor levels are designated by J and τ (Section 2.1.2.7), and $W_{J\tau}^0$ is the unperturbed rigid-rotor energy. Note that the M-dependent part of the direction cosine elements (which is independent of the degree of asymmetry) has been written explicitly. In Eq. (2.1.102) the interacting levels have been separated into those with $J' = J - 1$, J, and $J + 1$. Within each group all levels that have significant matrix elements connecting the level $J\tau$ must be included. The number of such levels is restricted, of course, by the dipole selection rules discussed in Section 2.1.3.1.

In applying Eq. (2.1.102) the direction cosine matrix elements may be calculated conveniently from asymmetric-rotor line strengths through the relations[81]:

$$
(J\tau \mid \varPhi_{Zg} \mid J\tau')^2 = 4J(J+1)(2J+1)^{-1}S_{J\tau;J\tau'},
$$

$$
(J\tau \mid \varPhi_{Zg} \mid J+1, \tau')^2 = 4(J+1)S_{J\tau;J+1,\tau'}, \tag{2.1.103}
$$

$$
(J\tau \mid \varPhi_{Zg} \mid J-1, \tau')^2 = 4JS_{J\tau;J-1,\tau'}.
$$

It is seen that the second-order Stark energy can be expressed in the form

$$
{}^gW_{J\tau M}^{(2)} = \mu_g^2 \mathscr{E}^2 (A_{J\tau} + B_{J\tau} M^2). \tag{2.1.104}
$$

Thus each level splits into $J + 1$ sublevels, with $M = 0, 1, \ldots, J$. The

[81] S. Golden and E. B. Wilson, *J. Chem. Phys.* **16**, 669 (1948).

coefficients $A_{J\tau}$ and $B_{J\tau}$ have been tabulated in reduced form by Golden and Wilson[81] for certain low J levels. It is relatively simple to develop a computer program that will calculate the Stark coefficients along with other asymmetric-rotor properties.

The perturbation calculation breaks down if there are nearly degenerate levels that are connected by an allowed Stark interaction. Such near degeneracies frequently occur between levels of the same K_{-1} or K_1 (and the same J), which would be exactly degenerate in the prolate or oblate symmetric-rotor limit. Accidental degeneracies between levels whose J values differ by one unit must also be handled specially.

Degenerate perturbation theory, often referred to as the *Van Vleck perturbation technique*, may be used to treat such cases. That is, we calculate the second-order corrections by Eq. (2.1.102), but omitting the terms connecting degenerate levels, and then solve a secular equation involving the set of degenerate levels. If only two levels are involved with rigid-rotor energies W_j^0 and W_k^0, this equation takes the form

$$\begin{vmatrix} W_j^0 + W_j^{(2)} & \xi_{jk} \\ \xi_{jk} & W_k^0 + W_k^{(2)} \end{vmatrix} = 0, \qquad (2.1.105)$$

where

$$\xi_{jk} = (j \mid H_{\mathscr{E}} \mid k) + \mu_g^2 \mathscr{E}^2 \sum_m{}' (j \mid \varPhi_{Zg} \mid m)(m \mid \varPhi_{Zg} \mid k)$$

$$\times \left\{ \frac{1}{W_j^0 - W_m^0} + \frac{1}{W_k^0 - W_m^0} \right\}. \qquad (2.1.106)$$

The summation in Eq. (2.1.106) excludes the terms with $m = j$ and $m = k$.

The various types of degeneracy which can arise are discussed by Golden and Wilson.[81] We shall mention only the case of near K-type degeneracy in a slightly asymmetric rotor. If the second-order terms in Eq. (2.1.106) are neglected and use is made of Eq. (2.1.103), we find

$$\xi_{jk} = [S_{jk}/J(J + 1)(2J + 1)]^{1/2} \mu_g M \mathscr{E}. \qquad (2.1.107)$$

At very low field strengths, a normal quadratic Stark effect is found. As $\mathscr{E}$ increases and ξ_{jk} becomes comparable to $|W_j^0 - W_k^0|$, the two levels begin to repel each other more strongly, and an exact solution of Eq. (2.1.105) is required for an accurate representation of the Stark effect. At high fields the Stark shift becomes first order, and is given approximately by Eq. (2.1.107). Furthermore, if the line strength S_{jk}

is approximated by its value in the symmetric-rotor limit (see Table IV), namely

$$S_{jk} = K^2(2J + 1)/J(J + 1),$$

we find the Stark energy at high fields approaches

$$W^{(1)} = \mu_g KM\mathscr{E}/J(J + 1),$$

which agrees, as expected, with Eq. (2.1.100).

2.1.10.4. Selection Rules and Intensities of Stark Components. The selection rules for transitions in the presence of an electric field may be derived from the M-dependence of the dipole-moment matrix elements. If the external field direction is parallel to the electric vector of the microwave radiation, we are concerned with the matrix elements of Table III for which $F = Z$, and hence the Stark selection rule $\varDelta M = 0$ applies. For perpendicular orientation of the Stark and microwave electric vectors, the elements with $F = X$ and Y are pertinent, and the allowed transitions are those with $\varDelta M = \pm 1$. The relative intensities of Stark components for the two cases are easily calculated from the matrix elements. The results are given in Table XI.

TABLE XI. Relative Intensities of Stark Components

Stark field parallel to microwave field

$M \to M$

$J \to J - 1$	$I \propto J^2 - M^2$
$J \to J$	$I \propto M^2$
$J \to J + 1$	$I \propto (J + 1)^2 - M^2$

Stark field perpendicular to microwave field

$M \to M + 1$

$J \to J - 1$	$I \propto (J - M)(J - M - 1)$
$J \to J$	$I \propto (J - M)(J + M + 1)$
$J \to J + 1$	$I \propto (J + M + 1)(J + M + 2)$

$M \to M - 1$

$J \to J - 1$	$I \propto (J + M)(J + M - 1)$
$J \to J$	$I \propto (J + M)(J - M + 1)$
$J \to J + 1$	$I \propto (J - M + 1)(J - M + 2)$

2.1.10.5. Identification of Transitions from Stark Patterns. It should be clear from the foregoing description that the Stark pattern gives sensitive information on the nature of a transition. We shall consider various situations where information on the quantum numbers of the levels involved can be obtained from the Stark effect.

The normal quadratic Stark shifts in an asymmetric rotor follow the form [see Eq. (2.1.104)]

$$\Delta \nu = (A + BM^2)\mathscr{E}^2, \qquad (2.1.108)$$

where we take the more common case of a field configuration appropriate to a $\Delta M = 0$ selection rule. Typical patterns resulting from different ratios of A to B are illustrated in Fig. 17. Examination of these patterns shows that consideration of the relative intensities and frequency intervals usually allows a decision between a $\Delta J = 0$ and $\Delta J = \pm 1$ transition. The J value can then be determined from the number of components, namely, J for a $\Delta J = 0$ transition and $J + 1$ for a $J \rightarrow J + 1$ transition.

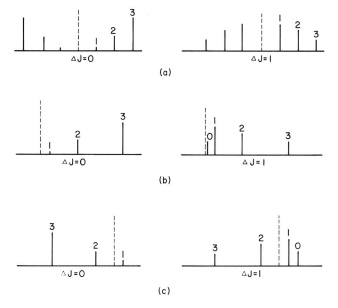

FIG. 17. Characteristic Stark patterns for $\Delta M = 0$ selection rules: (a) first order, (b) second order $(B/A = 5)$, (c) second order $(B/A = -0.5)$. The illustrations for $\Delta J = 0$ apply to $J = 3 \rightarrow 3$ transitions; those for $\Delta J = 1$ apply to $J = 3 \rightarrow 4$. The M values of the components are indicated. The dotted line gives the position of the unshifted transition.

In many situations one cannot be certain that all components have been observed, because of limitations in resolution, sensitivity, or available Stark fields. However, it is still possible to obtain useful information from the Stark effect. Golden and Wilson[81] have given formulas for extracting J values from the intervals observed in partially resolved Stark patterns. Another procedure is to assign an arbitrary value of M to one component and, after numbering the others sequentially, to plot the frequency shifts (at constant electric field) against M^2. When the correct value of M is chosen, for the initial assignment, a linear plot will be found according to Eq. (2.1.108). In applying any of these techniques, one must first ensure that the Stark shifts are quadratic in the electric field, since a deviation from quadratic field dependence necessarily implies that the dependence on M is not purely quadratic.

The observation of a first-order Stark effect always indicates a degeneracy, and therefore suggests either a symmetric-rotor or an asymmetric-rotor transition between levels that are close to the symmetric limit. In either case the J values involved may be determined by counting the Stark components. The relative intensities in the pattern show immediately whether the transition is of the $\Delta J = 0$ or $\Delta J = \pm 1$ type (see Fig. 17). In the former case the number of components is just $2J$, since the $M = 0 \to 0$ transition is not allowed when $\Delta J = 0$. The number of components for a transition $J \to J + 1$ is $2J + 1$. However, one of these, $M = 0 \to 0$, does not have a linear Stark effect, according to Eq. (2.1.100), so that only $2J$ components are observed at low fields. Degeneracies that are near, but not exact, lead to Stark effects in

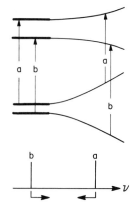

FIG. 18. Stark pattern for transitions involving nearly degenerate levels. The arrows in the bottom diagram indicate the predominant direction of the Stark shifts.

which neither the linear nor the quadratic contribution dominates. A very common situation is illustrated in Fig. 18. The nearly-degenerate levels repel each other in the presence of the field, leading to a mirror image pattern in which the Stark effect of the higher frequency line is predominately to low frequency and vice versa. A Stark pattern of this type is typical of parallel transitions in slightly asymmetric rotors (Section 2.1.4.3). Another example is the *l*-doublet patterns found in degenerate vibrational states of linear and symmetric rotors (Section 2.1.5.3).

2.1.11. Nuclear Quadrupole Hyperfine Structure

The interaction of nuclear spins with the rotational angular momentum can produce a hyperfine structure on the rigid-rotor spectrum. This effect complicates the microwave spectrum but at the same time provides much valuable information. The hyperfine structure is frequently helpful in assigning rotational transitions. Furthermore, the analysis of the hyperfine structure yields molecular parameters that are of great interest in themselves.

2.1.11.1. The Quadrupole Hamiltonian. Although there are magnetic interactions between nuclear spins and rotation, it is usually very difficult to resolve the resulting hyperfine structure in ordinary microwave spectra. A more easily observable coupling occurs through nuclear quadrupole interactions. Any nucleus of spin $I \geq 1$ may have a nonspherical charge distribution, leading to a finite electric quadrupole moment. This moment interacts with the electric-field gradient at the nucleus produced by all of the other charges in the molecule. The Hamiltonian for the interaction between a single nucleus of spin I and overall rotation is[1]

$$H_Q = \frac{eQq_J}{2J(2J-1)I(2I-1)} \left[3(\mathbf{I} \cdot \mathbf{J})^2 + \frac{3}{2} \mathbf{I} \cdot \mathbf{J} - \mathbf{I}^2\mathbf{J}^2 \right], \quad (2.1.109)$$

where e is the electronic charge, Q the nuclear quadrupole moment, and

$$q_J = \langle \partial^2 V / \partial Z^2 \rangle.$$

Here V is the electrostatic potential at the nucleus and Z is the space-fixed axis of quantization. The average is taken over the rotational wave-functions in the state $M_J = J$.

The Hamiltonian H_Q can usually be treated as a small perturbation to be added to the rigid-rotor Hamiltonian. In most, but not all, cases a first-order solution is adequate. We may think of the nuclear spin angular momentum $\mathbf{I}$ coupling with $\mathbf{J}$ to give a total angular momentum $\mathbf{F}$, so that

$$F = J + I, J + I - 1, \ldots, |J - I|.$$

Then the operator in brackets in Eq. (2.1.109) is easily shown to have the value

$$\tfrac{3}{4}C(C + 1) - I(I + 1) - J(J + 1),$$

where

$$C = F(F + 1) - I(I + 1) - J(J + 1).$$

The first-order quadrupole energy becomes

$$W_Q = eQq_J \frac{\tfrac{3}{4}C(C + 1) - I(I + 1)J(J + 1)}{2J(2J - 1)I(2I - 1)}. \qquad (2.1.110)$$

The problem remains of evaluating q_J. We may express the derivative with respect to the space-fixed Z axis in terms of molecule-fixed axes xyz by

$$\frac{\partial V}{\partial Z} = \Phi_{Zx} \frac{\partial V}{\partial x} + \Phi_{Zy} \frac{\partial V}{\partial y} + \Phi_{Zz} \frac{\partial V}{\partial z},$$

where the Φ_{F_g} are direction cosines between molecule-fixed and space-fixed axes (see Section 2.1.3.2). We then find

$$q_J = \langle \partial^2 V / \partial Z^2 \rangle = q_{xx} \langle \Phi_{Zx}^2 \rangle + q_{yy} \langle \Phi_{Zy}^2 \rangle + q_{zz} \langle \Phi_{Zz}^2 \rangle, \qquad (2.1.111)$$

where

$$q_{xx} = \partial^2 V / \partial x^2, \qquad \text{etc.}$$

The averages such as $\langle \Phi_{Zx} \Phi_{Zy} \rangle$ have vanished from symmetry considerations. The quantities q_{xx}, q_{yy}, and q_{zz} are molecular parameters, independent of rotational state. It is convenient to combine these with the constant factor eQ to give the so-called *quadrupole coupling constants*

$$\chi_{xx} = eQq_{xx}, \qquad \chi_{yy} = eQq_{yy}, \qquad \chi_{zz} = eQq_{zz}.$$

It may be shown from the fundamental properties of the electrostatic field that

$$\chi_{xx} + \chi_{yy} + \chi_{zz} = 0.$$

2.1.11.2. Linear and Symmetric Rotors. The computation of the quadrupole energy from Eqs. (2.1.110) and (2.1.111) requires the evaluation of the direction cosine averages for the level involved. This is readily done for a linear molecule by making use of the matrix elements of Table III and the following relation derived from the axial symmetry of the field

$$\chi_{xx} = \chi_{yy} = -\tfrac{1}{2}\chi_{zz}.$$

In linear molecules χ_{zz} is often written as eqQ. We find that

$$q_J = -q_{zz} J/(2J + 3)$$

so that

$$W_Q = -\chi_{zz} Y(J, I, F) = -eqQ\,Y(J, I, F). \qquad (2.1.112)$$

Here $Y(J, I, F)$ is Casimir's function,

$$Y(J, I, F) = \frac{\tfrac{3}{4}C(C + 1) - I(I + 1)J(J + 1)}{2J(2J - 1)(2J + 3)I(2I - 1)}, \qquad (2.1.113)$$

which is tabulated in many places.[1,4] It is seen from the form of $Y(J, I, F)$ that the hyperfine splittings become smaller with increasing J.

The hyperfine structure in the rotational spectrum of a linear molecule with one quadrupolar nucleus can therefore be fit with a single parameter, eqQ. The magnitude of the quadrupole splitting depends on the nuclear property Q and the molecular property q. The latter is sensitive to the type of chemical bonds emanating from the atom in question. In general, q is smallest for highly ionic bonds. Values of eqQ range from less than 1 MHz for deuterium compounds to more than 2000 MHz for compounds of heavy elements such as iodine.

The calculation is very similar for symmetric rotors, except for the additional quantum number K. The quadrupole energy is found to be

$$W_Q = -eqQ[1 - 3K^2/J(J + 1)]\,Y(J, I, F). \qquad (2.1.114)$$

The sensitivity of the quadrupole pattern to K should be noted. The near coincidence of lines of different K no longer holds for a symmetric rotor with hyperfine structure.

2.1.11.3. Asymmetric Rotors. The quadrupole energy for an asymmetric rotor cannot be expressed in closed form, except for a few levels of low J value. However, it is a relatively simple matter to calculate the direction cosine averages in Eq. (2.1.111) if a computer program for the asymmetric rotor is available. Also, they may be expressed in terms of tabulated line strengths by using the relations of Eq. (2.1.103). Another method, which expresses q_J in terms of the reduced asymmetric-rotor energy $E(\varkappa)$ and its derivative, has been given by Bragg and Golden.[82] Useful approximations for q_J can be developed for a slightly asymmetric rotor. In the near-prolate case we define a quadrupole asymmetry parameter η by

$$\eta = (q_{cc} - q_{bb})/q_{aa}.$$

Then, to second order in the Wang asymmetry parameter b (see Section 2.1.2.7), we have[1]*

$$K = 0: \quad q_J = \frac{q_{aa}}{(J+1)(2J+3)}\left\{-J(J+1) + \left(\frac{3}{2}b^2 - b\eta\right)f(J,1)\right\}$$

$$K = 1: \quad q_J = \frac{q_{aa}}{(J+1)(2J+3)}\left\{3 - J(J+1) \mp \frac{1}{2}\eta J(J+1)\right.$$
$$\left. + \left(\frac{3}{2}b^2 - b\eta\right)f(J,2)/4 \pm \frac{3}{128}b^2\eta J(J+1)f(J,2)\right\},$$

$$K = 2: \quad q_J = \frac{q_{aa}}{(J+1)(2J+3)}\left\{12 - J(J+1) + \left(\frac{3}{2}b^2 - b\eta\right)\right.$$
$$\left. \times\left[\frac{f(J,3)}{6} - \frac{f(J,1)}{2} \mp \frac{f(J,1)}{2}\right]\right\}, \qquad (2.1.115)$$

$$K = 3: \quad q_J = \frac{q_{aa}}{(J+1)(2J+3)}\left\{27 - J(J+1) + \left(\frac{3}{2}b^2 - b\eta\right)\right.$$
$$\left. \times\left[\frac{f(J,4)}{8} - \frac{f(J,2)}{4}\right] \mp \frac{3}{128}f(J,2)J(J+1)b^2\eta\right\},$$

$$K > 3: \quad q_J = \frac{q_{aa}}{(J+1)(2J+3)}\left\{3K^2 - J(J+1) + \left(\frac{3}{2}b^2 - b\eta\right)\right.$$
$$\left. \times\left[\frac{f(J,K+1)}{2(K+1)} - \frac{f(J,K-1)}{2(K-1)}\right]\right\}.$$

The function $f(J, n)$ is defined in Eq. (2.1.23). In these expressions the upper sign applies to the upper energy level of a K-type doublet. If the asymmetry is very small ($b \to 0$), we see that the hyperfine pattern reduces to that of a symmetric rotor, except for the $K = 1$ levels. Here

[82] J. K. Bragg and S. Golden, *Phys. Rev.* **75**, 735 (1949).

---

* A sign error in Ref. 1, Eq. (6-21), has been corrected.

the quadrupolar asymmetry η continues to make a large contribution even in the symmetric limit.

The same expressions apply to a near-oblate rotor if we define

$$\eta = (q_{bb} - q_{aa})/q_{cc}.$$

The $\pm$ signs must also be reversed.

If the first-order approximation is adequate, the hyperfine structure from a single quadrupolar nucleus in an asymmetric rotor may be fitted with two parameters, which may be taken as χ_{aa} (or χ_{cc}) and η. It should be emphasized that these parameters are referred to the principal inertial axis frame. Additional information is required to determine the quadrupole coupling tensor completely. This information can be supplied, for example, by measuring the change of χ_{aa} and η on isotopic substitution, since this usually produces a rotation of the inertial axis system relative to the principal axes of the quadrupole coupling tensor. Another way of obtaining additional information on the tensor components is through second-order quadrupole interactions; these may be particularly significant in cases of accidental near-degeneracy of rotational levels.[83]

2.1.12. Intensities and Line Widths

The observation of a microwave spectrum involves, at least in idealized form, an experiment in which energy is absorbed from the electromagnetic radiation field by molecules, which then dissipate this energy through the bulk gas by means of collisions. If the incident power level is small, the absorbed energy is rapidly dissipated and the normal thermal equilibrium is not disturbed. Since the energy levels are not perfectly sharp, radiation may be absorbed even though the incident microwave frequency (assumed monochromatic) does not coincide exactly with the absorption frequency predicted for an isolated molecule. Each allowed transition therefore appears as a "line" of finite width. We shall consider the various contributions to the line width and then proceed to the calculation of the absolute intensity of the absorption.

2.1.12.1. Line Widths. As discussed more fully in the next section, it is found that the line shape, at sufficiently high pressures, is determined primarily by intermolecular collisions. The line has a characteristic Lorentzian shape with a width that is directly proportional to the gas

[83] W. H. Flygare, and W. D. Gwinn, *J. Chem. Phys.* **36**, 787 (1962).

pressure. As the pressure is lowered, however, other sources of broadening become comparable in importance.

Doppler broadening results from the distribution of velocities of the molecules in the gas sample. If this is the dominant source of broadening, the line shape is Gaussian with a half-width-at-half-maximum intensity given by

$$\Delta \nu = (\nu/c)[(2kN_0T/M)\ln 2]^{1/2}, \qquad (2.1.116)$$

where k is the Boltzmann constant, N_0 is Avogadro's number, and M is the molecular weight. At $T = 300$ K, this expression becomes

$$\Delta\nu/\nu = 6.2 \times 10^{-6}M^{-1/2}.$$

For a molecular weight of 50 the Doppler width at $\nu = 30$ GHz is thus about 25 kHz. It should be noted that the Doppler width increases linearly with the frequency of the transition.

Broadening by collisions with the walls of the cell must be considered when the mean free path becomes comparable to the cell dimensions. As shown in Townes and Schawlow,[1] Chapter 13, this contribution to the half-width of the line may be approximated by

$$\Delta\nu \text{ (kHz)} \approx 10(A/V)M^{-1/2}, \qquad (2.1.117)$$

where A is the *surface* area in square centimeters and V is the volume of the cell in cubic centimeters, and the temperature has been taken as 300 K. For a molecular weight of 50 and a typical cell configuration (X-band Stark cell) this gives a $\Delta\nu$ of about 7 kHz. Thus the contribution of wall collisions is generally less than the Doppler width, except at very low frequencies.

Any modulation employed in a microwave spectrometer places a limitation on the minimum line width. The effect of Stark (or source) modulation on a collision-broadened line has been treated by Karplus.[84] If the modulation frequency is comparable to the other sources of line width, sidebands are produced which have the effect of broadening the line. While the details are complicated, this contribution to line width is of the order of magnitude of the modulation frequency.

Another instrumental broadening effect is power saturation, which occurs when the pressure is too low for the collisional dissipation of the absorbed energy. This is discussed in Section 2.1.13.1.

[84] R. Karplus, *Phys. Rev.* **73**, 1027 (1948).

From the foregoing we can conclude that the other sources of broadening are often negligible when the half-width from intermolecular collisional broadening is greater than about 100 kHz. This threshold is reached at pressures in the range of 5 to 50 mTorr (0.7–7 N m^{-2}) depending on the specific gas. Exceptions to this generalization occur at high microwave frequencies, when the modulation frequency is high, and when the incident power level is excessive.

2.1.12.2. Absorption Coefficient for Collision-Broadened Line. In discussing intensities of microwave lines we are interested in the absorption coefficient γ, which describes the fraction of power absorbed per unit path length as the radiation passes through the cell. If P_0 is the incident power, P the final power, and L the cell length, then γ is defined by

$$P = P_0 \exp(-\gamma L). \tag{2.1.118}$$

The function $\gamma(\nu)$ thus carries the full information on the line shape, width, and intensity.

Van Vleck and Weisskopf[85] derived an expression for the absorption coefficient under the following assumptions about the collisional process:

(1) Collisions are random with a mean collision time τ.

(2) The state of a molecule after a collision is essentially independent of its state before the collision ("hard" collisions).

(3) The duration of a collision is much shorter than the period of the radiation.

Another derivation, which is very convenient for extension to include power saturation, has been given by Javan.[86] If we restrict our attention to the pressure range suitable for microwave spectroscopy, the result is

$$\gamma(\nu) = \frac{4\pi}{3c\hbar} (n_1 - n_2) \mid \mu_{12} \mid^2 \frac{\nu \, \Delta\nu}{(\nu - \nu_0)^2 + (\Delta\nu)^2}, \tag{2.1.119}$$

where n_1, n_2 are the population densities (in molecules per cubic centimeter) of the two states involved in the transition, $\mid \mu_{12} \mid^2$ is the square of the dipole-moment matrix element connecting the states (see Section 2.1.3.2), and ν_0 is the center (peak) frequency of the transition. $\Delta\nu = 1/2\pi\tau$, where τ is the collision frequency. It is seen from this ex-

[85] J. H. Van Vleck and V. F. Weisskopf, *Rev. Mod. Phys.* **17**, 227 (1945).

[86] A. Javan, *Phys. Rev.* **107**, 1579 (1957).

pression that $\Delta\nu$ is the half-width at half-maximum intensity. Since τ is inversely proportional to pressure, the quantity $\Delta\nu/p$ is constant over a wide pressure range. This "collision-broadening parameter" is characteristic of the molecule responsible for the transition and the composition of the gas sample; it may also be a function of the rotational state. Typical values of $\Delta\nu/p$ range from 2 to 30 MHz/Torr.

When the pressure is high enough that other sources of broadening are negligible, Eq. (2.1.119) indicates that the peak absorption coefficient $\gamma_0 = \gamma(\nu_0)$ is independent of pressure. Thus γ_0 is a convenient measure of the absolute intensity of a transition. By expressing $n_1 - n_2$ in terms of a Boltzmann distribution, and noting that microwave frequencies are small compared to kT/h, we obtain for γ_0

$$\gamma_0 = \frac{8\pi^2}{3ckT} \; \frac{Nf_\nu\nu_0^2 \exp(-E_l/kT)}{Q_{\mathrm{r}} \, \Delta\nu} \; \mu_g^2 S, \qquad (2.1.120)$$

where N is the total number of molecules per cubic centimeter, f_v is the fraction of molecules in the particular vibrational state, E_l is the rotational energy of the lower state, Q_r is the rotational partition function, μ_g is the dipole moment component responsible for the transition, and S is the rotational line strength (Section 2.1.3.2). We have made use of the relation

$$(2J + 1) \, | \, \mu_{12} \, |^2 = S\mu_g^2.$$

For a linear molecule at $T = 300$ K, Eq. (2.1.120) reduces to

$$\gamma_0 = 1.092 \times 10^{-16} f_\mathrm{v} B \, \exp(-E_l/kT) S\mu^2\nu_0^2(\Delta\nu/p)^{-1} \quad \mathrm{cm}^{-1}$$
$$= 0.546 \times 10^{-16} f_\mathrm{v}\mu^2\nu_0^3 \exp(-E_l/kT)(\Delta\nu/p)^{-1} \quad \mathrm{cm}^{-1}. \qquad (2.1.121)$$

For the general asymmetric or symmetric rotor we have

$$\gamma_0 = 2.463 \; 10^{-20} f_\mathrm{v}(ABC)^{1/2} \exp(-E_l/kT) S\mu_g^2\nu_0^2(\Delta\nu/p)^{-1} \quad \mathrm{cm}^{-1}. \qquad (2.1.122)$$

In these expressions A, B, C, and ν_0 are measured in megahertz, μ in Debye units, and $(\Delta\nu/p)$ in megahertz per Torr. It may be noted that the numerical coefficient in Eq. (2.1.121) is proportional to T^{-3}, while that of Eq. (2.1.122) varies as $T^{-3.5}$.

Peak absorption coefficients rarely exceed 10^{-4} cm^{-1} at frequencies below 50 GHz. They may become much larger at higher frequencies because of the sensitive dependence on ν_0.

2.1.12.3. Integrated Intensity. It was shown in the last section that the peak absorption coefficient γ_0 is independent of pressure as long as the pressure is high enough that intermolecular-collision broadening is the dominant contribution to the line width. Under these conditions the integrated absorption coefficient is found to be

$$\gamma_{\text{int}} = \int \gamma(\nu) \, d\nu = \pi \gamma_0 \, \Delta\nu$$

$$= (8\pi^3/3ckTQ_r)Nf_\nu\nu_0{}^2 \exp(-E_l/kT)\mu_g{}^2 S. \qquad (2.1.123)$$

Thus the integrated absorption coefficient is proportional to N, the density of absorbing molecules, and independent of the relaxation time, as one expects intuitively.

The expressions given here for the absorption coefficient are applicable at pressures low enough that $\Delta\nu \ll \nu_0$. At higher pressures a second term in the Van Vleck–Weisskopf theory may become significant (see Townes and Schawlow,[1] Chapter 13). This term may be safely neglected, however, for all normal microwave spectroscopy.

2.1.13. Saturation and Double Resonance

The discussion of intensities and line widths in Section 2.1.12 assumed that the microwave power level is low enough that the absorbed energy is dissipated without any disturbance of thermal equilibrium in the gas sample. Under actual experimental conditions this assumption is frequently invalid. The resulting *power saturation* affects both the width and peak intensity of a line and also provides a basis for the important technique of double resonance.

2.1.13.1. Power Saturation. The effects of power saturation have been treated in several places.[1,87] A very convenient formulation is contained in the treatment of the three-level maser by Javan.[86] A detailed consideration of the two processes, absorption (plus stimulated emission) of microwave power and dissipation of this energy through collisions, shows that the usual Boltzmann distribution is disturbed in the following way.

$$\bar{n}_1 - \bar{n}_2 = (n_1 - n_2) \frac{1 + 4\pi^2(\nu - \nu_0)^2\tau^2}{1 + 4\pi^2(\nu - \nu_0)^2\tau^2 + 4y^2\tau^2}. \qquad (2.1.124)$$

Here the symbols have the same meaning as in Eq. (2.1.119); n_1, n_2

[87] R. Karplus and J. Schwinger, *Phys. Rev.* **73**, 1020 (1948).

are the population densities appropriate to a Boltzmann distribution and $\bar{n}_1$, $\bar{n}_2$ are the densities in the presence of the microwave radiation field. The quantity y measures the strength of the radiation field

$$y = \mu_{12}E_0/2h,$$

where E_0 is the amplitude of the microwave field. In practical units y is given by

$$y^2 = 629 \mid \mu_{12} \mid^2 (P_0/A), \qquad (2.1.125)$$

where y is in megahertz, μ_{12} is in Debye units, P_0 is in watts, and A is the cross-sectional area of the cell in centimeters squared.

The absorption coefficient under conditions of power saturation is obtained by replacing $n_1 - n_2$ in Eq. (2.1.119) with $\bar{n}_1 - \bar{n}_2$ from Eq. (2.1.124). The result is simply to replace the Lorentz shape factor

$$\frac{\nu \, \Delta\nu}{(\nu - \nu_0)^2 + (\Delta\nu)^2}$$

with

$$\frac{\nu \, \Delta\nu}{(\nu - \nu_0)^2 + (\Delta\nu)^2 + y^2/\pi^2}. \qquad (2.1.126)$$

Therefore, when the power level is high enough that y/π cannot be neglected in comparison to $\Delta\nu = 1/2\pi\tau$, the effective line width is increased and the peak height reduced. Insertion of typical numbers into the above equations shows that this is usually a significant effect unless the power level is kept extremely low.

2.1.13.2. The Γ Absorption Coefficient. Harrington[88] has introduced a somewhat different absorption coefficient which explicitly recognizes the general presence of power-saturation effects. Noting that $\gamma L \ll 1$ in any practical laboratory measurement of a microwave spectrum (except, perhaps, at very high frequencies), we see that Eq. (2.1.118) is adequately approximated by

$$\gamma = P/LP_0.$$

The Γ coefficient proposed by Harrington is defined by

$$\Gamma = P/LP_0^{1/2} = \gamma P_0^{1/2}. \qquad (2.1.127)$$

[88] H. W. Harrington, *J. Chem. Phys.* **46**, 3698 (1967); **49**, 3023 (1968).

It may be shown[88] that the actual output intensity measure of most types of microwave spectrometers (e.g., a chart-recorder deflection) is directly proportional to Γ. Restricting our attention to peak absorption coefficients, we may write, by making use of Eqs. (2.1.119) and (2.1.126)

$$\Gamma_0 = \gamma_0^0 P_0^{1/2}/(1 + KP_0), \qquad (2.1.128)$$

where γ_0^0 is the peak absorption coefficient in the limit of zero power, as given by Eq. (2.1.120). Here K is an abbreviation for

$$K = 4y^2\tau^2/P_0 = 629 \mid \mu_{12} \mid^2/\pi^2 A(\Delta\nu)^2, \qquad (2.1.129)$$

where use has been made of Eq. (2.1.125).

We may rewrite Eq. (2.1.128) as the product of two factors

$$\Gamma_0 = (\gamma_0^0/K^{1/2})[(KP_0)^{1/2}/(1 + KP_0)]. \qquad (2.1.130)$$

The second factor is dimensionless, and has a maximum value of $\frac{1}{2}$ (reached when $KP_0 = 1$). Therefore, the first factor, $\gamma_0^0/K^{1/2}$, can be determined experimentally by noting the maximum value of Γ_0 (i.e., the maximum output signal) as the power level is varied. This has a considerable experimental advantage over determination of γ_0^0 by extrapolation to zero power, since one then must deal with very weak output signals.

The quantity $\gamma_0^0/K^{1/2}$ determined in this way has as important property. As may be seen from Eqs. (2.1.120) and (2.1.129), it is independent of $\Delta\nu$ (i.e., of the relaxation time τ). In fact, $\gamma_0^0/K^{1/2}$ is proportional to the integrated absorption coefficient defined in Eq. (2.1.123). Thus $\gamma_0^0/K^{1/2}$ provides a direct measure of the number density of absorbing molecules in the cell, independent of the other constituents of the gas. It is a much more useful quantity for purposes of quantitative analysis than is γ_0^0 itself, since the latter depends upon the relaxation time, which is a function of the composition of the gas mixture.

The advantage of this formulation is that it permits a convenient separation of the intensity measure into two parts, one involving the density of absorbing molecules and the other the relaxation time. Suitable experiments can be designed to study each of these separately. The equations given here are applicable to the case of uniform spacial distribution of the microwave power; the modifications required for actual waveguide absorption cells are discussed by Harrington.[88]

2.1.13.3. Double Resonance. Since power saturation disturbs the Boltzmann distribution, we can expect that saturation of one microwave transition will affect the intensity of every other transition that has a level in common with the first. This is the basis of the technique of double resonance. In its simplest form, a double-resonance experiment involves the application of a high-power microwave signal at or near a molecular absorption frequency ν_{13} concurrently with a low-power monitoring signal at a frequency ν_{12} (see Fig. 19). Early experiments of this type were carried out by Gozzini and co-workers[89] and by Shimoda.[90] The basic theory was formulated by Javan;[86] other discussions have been given by Di Giacomo[91] and Yajima.[92]

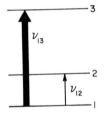

Fig. 19. Double-resonance in a three-level system. The pumping transition is ν_{13} and the monitoring transition is ν_{12}.

The detailed theory of the three-level system shows that the line shape of the monitoring transition is affected not only by the change in population of the states but also by more specific quantum-mechanical effects produced by the saturating field. We shall consider a simple case where the high-power (pumping) signal has a frequency exactly equal to the peak absorption frequency ν_{13}. Then the peak absorption coefficient γ_0 of the monitoring transition ν_{12} is related to its value in the absence of the pumping signal $\gamma_0{}^0$ by[86]

$$\gamma_0 = \gamma_0{}^0 \frac{(1 + y^2\tau^2) - 3y^2\tau^2[(\nu_{13}/\nu_{12}) - 1]}{(1 + y^2\tau^2)(1 + 4y^2\tau^2)}. \qquad (2.1.131)$$

The second term in the numerator of Eq. (2.1.131) is negative and therefore represents emission. At some value of $y^2\tau^2$ the peak absorption at

[89] A. Battaglia, A. Gozzini, and E. Polacco, *Nuovo Cimento* **14**, 1076 (1959).
[90] K. Shimoda, *J. Phys. Soc. Japan* **14**, 954 (1959).
[91] A. Di Giacomo, *Nuovo Cimento* **14**, 1082 (1959).
[92] T. Yajima, *J. Phys. Soc. Japan* **16**, 1594 (1961).

ν_{12} will thus become zero, and higher pumping power will produce a net emission of microwave power at ν_{12}.

Javan[86] has given expressions for the general case where both the pumping and monitoring signals may be displaced from the peak transition frequencies. Figure 20 shows a typical line shape of the monitoring transition when the pumping signal is on-resonance (exactly equal to ν_{13}). When the pumping signal is off-resonance, subsidiary peaks (two-quantum transitions) appear. Examples of the various situations that are possible have been given by Cox et al.,[93] Favero et al.,[94] and Oka.[95]

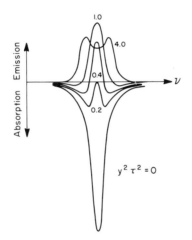

FIG. 20. Line shape of the monitoring transition for various values of $y^2\tau^2$ when the pumping signal is exactly on-resonance. This figure applies to the level pattern of Fig. 19 with $\nu_{13}/\nu_{12} = 3$. [After A. Javan, *Phys. Rev.* **107**, 1579 (1957).]

Double resonance is an extremely valuable tool in spectral analysis, since it allows one to identify two transitions that share a common energy level. It is particularly useful in very complex spectra where Stark effects cannot be studied. By scanning the pumping frequency over a wide range and looking for effects on a given monitoring transition, a set of closely related transitions can be selected from a dense, overlapping spectrum. In some cases this has proved to be the only practical way of

[93] A. P. Cox, G. W. Flynn, and E. B. Wilson, *J. Chem. Phys.* **42**, 3094 (1965).

[94] P. G. Favero, F. Scappini, and A. M. Mirri, *Boll. Sci. della Facoltá di Chim. Indust.* (*Bologna*) **24**, 93 (1966).

[95] T. Oka, *Can. J. Phys.* **47**, 2343 (1969).

analyzing a complicated spectrum.[96] Double resonance can also be used as a method of modulation.[97]

2.1.13.4. Measurement of Rotational Relaxation Times.

Many of the expressions given in this and the preceding section have involved the molecular relaxation time τ. It may be worthwhile to summarize the various methods that are available for determining this quantity from microwave experiments.*

In the Van Vleck–Weisskopf theory for the unsaturated absorption coefficient, the quantity $\Delta\nu = 1/2\pi\tau$ is the half-width of the line at half-maximum intensity [Eq. (2.1.119)]. Thus a direct measurement of the line width yields a value for τ. The measurement should be done over a range of pressures to insure that other sources of broadening are negigible. Line widths are somewhat difficult to measure accurately, especially with spectrometers employing Stark modulation. Details of measurement techniques and some representative results have been reported by Rinehart et al.[98] and by Rusk.[99]

The observation of saturation effects provides a means of determining τ from intensity, rather than line-width data, as may be seen from Eq. (2.1.126). The formulation in terms of the Γ coefficient (Section 2.1.13.2) is rather convenient for this purpose. A determination of K from a measurement of Γ_0 as a function of microwave power, according to Eq. (2.1.130), leads to a value of τ. However, absolute power measurements are required, which is a disadvantage.

Double-resonance experiments provide another means of obtaining τ. According to Eq. (2.1.131) one can determine τ from the value of y (i.e., of the power) that makes γ_0 vanish. Another technique, in which the

---

[96] O. L. Stiefvater, H. Jones, and J. Sheridan, *Spectrochim. Acta* **26A**, 825 (1970).

[97] R. C. Woods, A. M. Ronn, and E. B. Wilson, *Rev. Sci. Instrum.* **37**, 927 (1966).

[98] E. A. Rinehart, R. H. Kleen, and C. C. Lin, *J. Mol. Spectrosc.* **5**, 458 (1960); E. A. Rinehart and C. C. Lin, *Rev. Sci. Instrum.* **32**, 562 (1961); R. L. Legan, J. A. Roberts, E. A. Rinehart, and C. C. Lin, *J. Chem. Phys.* **43**, 4337 (1965).

[99] J. R. Rusk, *J. Chem. Phys.* **42**, 493 (1965).

---

* We have assumed that there is no distinction between collisions which merely interrupt the absorption or emission of radiation and those which change the rotational energy. Strictly speaking, the τ that appears in the Van Vleck–Weisskopf theory refers to the mean time between collisions of the former type. A different relaxation time, which might in principle be longer, should be used in the discussion of power saturation and disturbance of thermal equilibrium. However, there is no experimental evidence that these two relaxation times differ by a measurable amount.

pumping signal is suddenly removed and the relaxation of the intensity of the monitoring signal is measured directly, has been used by Unland and Flygare[100] to determine rotational relaxation times in OCS.

Finally, mention should be made of the experiments of Oka[101] which involve an extension of double-resonance techniques to four-level systems. Oka has found that saturation of one microwave transition sometimes affects the intensity of other transitions which do not share a common level with the pumped transition. This is interpreted as a selective transfer of the disturbance of the Boltzmann distribution to certain other rotational levels. In other words, collision-induced transitions between rotational levels appear to follow selection rules, rather than being entirely random in character. These experiments provide a measure of the relative probabilities of rotational relaxation through different channels.

2.1.14. Experimental Techniques

The microwave technology that has been developed for communications and military applications provides most of the components required for constructing a microwave spectrometer. Since many options are available, and since the technology is still moving rapidly, no attempt will be made to discuss specific configurations. However, the general techniques will be described briefly. Further details may be found in the works of Townes and Schawlow[1] and Wollrab.[2]

2.1.14.1. Basic Spectrometer Design. As in other types of absorption spectroscopy, the essential components of a microwave spectrometer are the radiation source, absorption cell, and detector. The source is usually a vacuum-tube oscillator whose frequency may be varied smoothly and continuously in some convenient manner. For many years reflex klystrons received the greatest use in microwave spectroscopy. More recently, backward wave oscillators (BWO's) have begun to replace klystrons, and the development of solid state microwave sources has now reached the point where they are becoming competitive in certain frequency regions. In each of these sources the frequency stability is quite high, even for a free-running source, as long as the voltages are adequately stabilized and the source is properly insulated from thermal or mechanical fluctuations. For high-resolution work and for measurements carried out

---

[100] M. L. Unland and W. H. Flygare, *J. Chem. Phys.* **45**, 2421 (1966).

[101] T. Oka, *J. Chem. Phys.* **45**, 754 (1966); **47**, 13, 4852 (1967); **48**, 4919 (1968); **51**, 3027 (1969).

with long time constants, greater stability may be necessary. In such cases, techniques are available for phase-locking of the microwave source to a crystal-controlled frequency standard. By using commercially available components it is therefore possible to construct a microwave source with adequate monochromaticity for normal microwave spectroscopy.

The simplest type of absorption cell is a length of standard rectangular waveguide. An absorption cell is easily made by sealing thin mica windows to the ends of the waveguide. Other materials such as teflon and quartz have also been used. The window should be as thin as possible in order to minimize the reflection of microwave power at the entrance and exit of the cell. A small hole or slot in the waveguide wall placed along the no-current line provides a connection to a vacuum manifold from which the sample is admitted. In the centimeter wavelength region (20–40 GHz), absorption cells are typically from 1 to 5 m in length. Much longer cells have been used in some cases and shorter cells are satisfactory at higher frequencies.

The most commonly used detectors in microwave spectroscopy are crystal diodes, of which a wide variety are available. The crystal is mounted in a suitable waveguide holder which is attached to the end of the absorption cell.

In the simplest type of a microwave spectrometer, the video spectrometer, the output from the crystal detector is passed through a suitable amplifier and displayed on an oscilloscope or recorder. In the former case the oscillator frequency is swept by a saw-tooth voltage over the range of a few megahertz and the same saw-tooth is used to drive the horizontal plates of the oscilloscope. Thus one obtains a direct display of transmitted microwave power versus frequency. An absorption line appears as a small, sharp dip in this trace. In recorder display the oscillator is usually swept either mechanically or electronically at a very slow rate.

2.1.14.2. Stark Modulation. The very low absorption coefficients in the microwave region make such a video spectrometer of limited use. The basic problem is the difficulty in detecting a very small absorption (often in the few parts per million level) in the presence of the very large low-frequency noise which is characteristic of most crystal diodes. The sensitivity of a spectrometer can be greatly increased by operating the detection system at a higher frequency. Superheterodyne systems have been used in some cases, but these are not very convenient in a spectrometer that is to be used over a broad frequency region. Another technique that has been tried is source modulation. However, this has the disadvantage that all of the reflections and power variations in the absorption cell show up on the output, and these are generally many orders of magnitude larger than the absorption signal which one wishes to detect. A far more convenient technique is Stark modulation,[102] in which one modulates the frequency of a molecular absorption line rather than the source of microwave power. As discussed in Section 2.1.10, the Stark shifts of rotational transitions can usually be observed with convenient laboratory electric fields. In an absorption cell made from standard rectangular waveguide the Stark field may be applied through a septum placed in the waveguide and insulated from the walls. A modulation field of square-wave shape is almost universally used because it simplifies the pattern that is observed. Modulation frequencies may range from about 5–100 khz. Lower frequencies have a minimum effect on the microwave line widths, but higher frequencies permit somewhat greater sensitivity. In a Stark-modulation spectrometer the crystal diode acts as a mixer; the output of this diode at the modulation frequency is passed through a preamplifier and then into a phase-sensitive detector. The final output may be displayed either on an oscilloscope or a chart recorder.

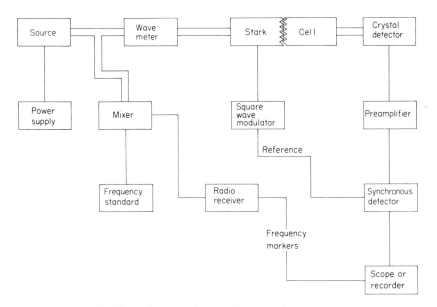

FIG. 21. Block diagram of a Stark-modulation spectrometer.

[102] K. B. McAfee, R. H. Hughes, and E. B. Wilson, *Rev. Sci. Instrum.* **20**, 821 (1949).

A block diagram of a Stark-modulation spectrometer is shown in Fig. 21. A single waveguide cell can be used through most of the common microwave region (excluding very low or very high frequencies); however, the oscillator, detector, and other waveguide components must be changed according to the particular frequency band.

2.1.14.3. Frequency and Intensity Measurements. Frequency measurements in the microwave region are normally made by reference to a crystal-controlled frequency standard. Commercial frequency standards of satisfactory accuracy and stability are readily available with outputs in the lower microwave region. This output is mixed with a small part of the source power in a crystal diode mixer. The diode acts both as a multiplier, generating harmonics of the standard frequency in the neighborhood of the source frequency, and also as a mixer of these harmonics with the source frequency. The beat frequency between the source and frequency-standard harmonic can then be measured with a communications receiver. The particular harmonic can be identified through a rough frequency measurement carried out with a cavity wavemeter. In this way standard frequency markers are generated which may be displayed along with the absorption lines on the oscilloscope or chart recorder. An alternative scheme is to lock the microwave source to harmonics of a stable frequency standard whose frequency may be varied continuously and smoothly. The frequency standard may be monitored with a conventional frequency counter.

The accuracy with which frequencies of microwave absorption lines can be measured depends upon a number of factors. In a normal spectrometer the limiting line width, expressed as the half-width at half-maximum intensity, is usually in the range of 50 to 100 kHz (see Section 2.1.12.1). It is therefore relatively easy to measure the peak frequency of an isolated absorption line to an accuracy of about 10 kHz. With adequate care, greater accuracy can be achieved. However, attention must be given to the possible distortion of the line shape by the amplifying system. When absorption lines are weak, it may not be possible to determine the peak frequency to this accuracy; likewise, the accuracy is degraded in complex spectra with extensive overlapping of lines.

The measurement of intensities in a microwave spectrum, even on a relative basis, presents many problems. Relative intensity measurement of two absorption lines is subject to a number of rather obvious sources of error, such as nonlinearities in the detector and amplifiers. Different degrees of modulation can also introduce errors. Even if these sources are eliminated, however, there remains a difficulty in determining the effective path length of the absorption cell at any given frequency. If the matching of the absorption cell to the rest of the microwave transmission line is not perfect, the reflections of microwave power will lead to an effective path length of undetermined magnitude. Since these reflections are very frequency sensitive, the effective path lengths may differ by a significant amount even for two lines relatively close in frequency. For this reason, intensity measurements of better than 5–10% accuracy require extreme care.

2.1.14.4. Cavity Cells. As an alternative to using a waveguide absorption cell a microwave resonant cavity may be employed. Since high sensitivity can be achieved in a cavity absorption cell with a length of the order of few centimeters, cavities have an advantage over long waveguides. However, cavity cells suffer from two great disadvantages: first, the microwave power density in a cavity is high, which intensifies problems of power saturation of the absorption lines, and secondly, a cavity by its nature is narrow banded. In fact, an efficient cavity will have a resonance that is not much broader than a typical absorption line. In order to scan the spectrum it is necessary to adjust the resonant frequency of the cavity in synchronism with the source frequency. While techniques for doing this are available, they require a rather complex control system. Because of these drawbacks cavities have not been used to any extent in spectrometers intended for searching a broad spectral range. However, they have found application in studying details of individual lines, such as collision broadening and power-saturation effects. In the higher-frequency microwave region the Fabry–Perot type of cavity cell has been used to advantage.[103]

2.1.14.5. Complete Spectrometer System. During the first 20 years of the development of microwave spectroscopy it was necessary to construct one's own spectrometer from commercially available components. However, complete spectrometer systems are now available from the Hewlett–Packard Company for the range 8000–40,000 MHz. Those spectrometers employ BWO sources whose frequency is indicated continuously on a digital meter. There are provisions for manual or automatic frequency scanning over any desired region at various scanning rates. It is possible to program search routines and to store and manipulate the resulting spectral data by interfacing with a small computer.

---

[103] M. Lichtenstein, J. J. Gallagher, and R. E. Cupp, *Rev. Sci. Instrum.* **34**, 843 (1963).

In one mode of operation a microwave bridge is employed, which permits accurate intensity measurements to be made. A calibrated modulator may be inserted in the transmission line to simulate the signal from an absorption line. In this way a measure of the absolute absorption coefficient of the line may be obtained. Further details on this type of spectrometer may be found in the work of Harrington.[88]

2.1.14.6. High Frequency Spectroscopy. Microwave spectroscopy in the millimeter and submillimeter range requires rather special techniques. There are only a few microwave sources available at frequencies above 100 GHz which are suitable for spectroscopy. Therefore, most of the work in the higher frequency range has been done with harmonic generators. Specially constructed frequency multipliers driven by sources in the 20 to 60 GHz range provide small amounts of power even at very high harmonics. The transmission and detection of such small power levels requires very special care. As a compensating factor, however, absorption coefficients of microwave transitions are much larger at higher frequencies. Since Stark modulation is usually not very suitable in this region, simple video spectrometers are generally used.

2.1.14.7. Compilations of Microwave Data. There are a number of valuable compilations of experimental data obtained from microwave spectra. A comprehensive list of measured frequencies has been published in the National Bureau of Standards Monograph 70 series.[104] Volume V of this work contains a list of all reported lines ordered by frequency. This is very useful for identifying suspected impurity lines in a spectrum. A revision of these tables is in progress at the Microwave Data Center at the National Bureau of Standards.

A general compilation of derived data (rotational constants, quadrupole coupling constants, barriers to internal rotation, etc.) has been prepared by Starck.[105] Selective compilations of such data may be found in the literature.[1-4] A critical compilation of dipole moments determined by microwave spectroscopy and other methods has been made by Nelson, Lide and Maryott.[106] Barriers to internal rotation determined by various

---

[104] Microwave Spectral Tables, Nat. Bur. Std. (U.S.) Monograph 70, Vol. I (1964); Vol. III (1969); Vol. IV (1968); Vol. V (1968).

[105] B. Starck, "Molecular Constants from Microwave Spectroscopy," Landolt-Börnstein (N.S.) Group II, Vol. 4. Springer, Berlin, 1967.

[106] R. D. Nelson, D. R. Lide, and A. A. Maryott, Selected Values of Electric Dipole Moments for Molecules in the Gas Phase. *Nat. St. Ref. Data Ser., Nat. Bur. Std.* (U.S.) **10** (1967).

techniques have been listed by Lowe,[107] and quadrupole coupling constants may be found in the book by Lucken.[108] No comprehensive compilation of structural parameters has been published since the tables of Sutton.[109]

Annotated bibliographies of the microwave literature appear in the works of Townes and Schawlow[1] and Wollrab.[2] Comprehensive bibliographies have also been prepared by Favero[110] and Starck.[111] Recent research in the field has been reviewed by Lide,[112] Flygare,[113] Morino and Hirota,[114] and Rudolph.[115]

[107] J. P. Lowe, *Progr. Phys. Org. Chem.* **6**, 1 (1968).

[108] E. A. C. Lucken, "Nuclear Quadrupole Coupling Constants." Academic Press, New York, 1969.

[109] L. E. Sutton, "Tables of Interatomic Distances and Configuration in Molecules and Ions," Special Publ. 11. Chem. Soc., London, 1958; Supplement, 1960.

[110] P. G. Favero, Microwave Gas Spectroscopy Bibliography. Laboratory of Radio-frequency Spectroscopy, University of Bologna, Italy (1963, 1966, 1969).

[111] B. Starck, Bibliography of Microwave Spectroscopic Investigations of Molecules. Structure Documentation Section, University of Ulm, Germany (1962, 1965, 1970).

[112] D. R. Lide, *Ann. Rev. Phys. Chem.* **15**, 225 (1964).

[113] W. H. Flygare, *Ann. Rev. Phys. Chem.* **18**, 325 (1967).

[114] Y. Morino and E. Hirota, *Ann. Rev. Phys. Chem.* **20**, 139 (1969).

[115] H. D. Rudolph, *Ann. Rev. Phys. Chem.* **21**, 73 (1970).

2.2. Infrared*

2.2.1. Introduction

The subject of infrared spectroscopy has its origin in the very earliest investigations into the nature of light. The first demonstration that white light could be dispersed into a spectrum was performed by Isaac Newton in 1666. Roughly a century later, in 1752, Thomas Melvill observed the first emission from a sodium flame, and in the *Philosophical Transactions* of 1800 Sir William Herschel published the first investigation of the distribution of radiant heat from the sun. By placing thermometers in the solar spectrum he discovered that the heating effect was most intense beyond the red end of the visible spectrum. Since these very early experiments the names of Fraunhofer, Sir John Herschel (the son of Sir William), Kirchhoff, Bunsen, Tyndall, Langley, Rubens, Nichols, Coblentz, Wood, Pfund, Randall, and others have been connected in an important manner with the growth of the science. For example, in 1840 Sir John Herschel demonstrated selectivity of absorption by showing that a black paper soaked in alcohol dried more readily when exposed to certain spectral regions than to others. In a series of papers entitled "Investigations of Infrared Spectra," Coblentz,[1] in 1905, published prism spectra of a large number of substances in the gaseous, liquid, and solid states, and details of new techniques developed largely by himself.

It is not the purpose of this section to recount the complete history of infrared spectroscopy, but rather to lay a little foundation for the more recent happenings theoretically and experimentally in an effort to show whence they came and how they are important to modern-day science. By 1814, 700 dark absorption lines had been observed by Fraunhofer and in 1835 Talbot wrote in the *Philosophical Magazine*[†]

---

[1] W. W. Coblentz, "Investigations of Infrared Spectra," Parts I-VII. Carnegie Inst. of Washington, Washington, D.C., 1905, 1906, and 1908.

[†] W. H. F. Talbot, *Philosophical Magazine* **7**, 113 (1835).

* Chapter 2.2 is by W. E. Blass and A. H. Nielsen.

I conclude that light when traversing a transparent medium is able to excite motion among its particles. This being admitted, let us suppose iodine vapor so constituted that its molecules are disposed to vibrate with a rapidity not altogether dissimilar to that of light. Now, if the differently colored rays differ also (as is probable) in rapidity of vibration, some of them will vibrate in accordance and others in discordance with the vibrations of the iodine gas. And these accordances and discordances will succeed each other in regular order from the red end of the spectrum to the violet end; each discordance being marked by a dark line or deficiency in the spectrum because the corresponding ray is not able to vibrate through the medium but is arrested by it and absorbed.

Although the truth of the matter is just the reverse of the idea expressed by Talbot, i.e., the light absorbed is that for which the frequencies coincide with the medium, his words express the fundamental notion involved in the study of molecular structure.

Progress in infrared spectroscopy has been rapid since about 1942 with the development by Barnes, Wright, V. Z. Williams, Strong, and others of automatically recording prism spectrometers. This was made possible by the results of intense wartime research on infrared detectors and infrared transparent materials. Automatic recording immediately made possible intense effort in the analysis of inorganic and organic molecules, and the current catalog of well-known infrared spectra of pure compounds runs into the thousands.

The electromagnetic energy spectrum extends from the extremely short-wave gamma rays of, say, 10^{-10} cm, representing millions of electron volts, up through the visible region, and on into the radio frequencies with wavelengths of many meters and representing millionths of electron volts. Within this vast range of wavelengths lies a narrow band from, let us say, the ultraviolet of 0.4 to 1000 μm comprising the range of energy of most of the motions of molecules. If the electronic arrangement of the molecule is disturbed, the wavelength of the absorbed or emitted light is in the ultraviolet or visible spectral region representing energy of the order of electron volts. If the molecular motion is simply that of vibration of the nuclei, the wavelengths are in the near infrared from about 1–25 μm corresponding to tenths of electron volts: and if the molecule just rotates, the wavelengths are, in general, longer than 25 μm and the energies are measured in hundredths of electron volts, or less.

Molecules are generally thought of as being composed of atoms, or groups of atoms, bound together in some geometric framework by forces related to the electronic structure of the atoms composing the molecule. Molecules can be separated again into neutral atoms or ions depending upon the nature of the characteristic binding forces. To a good approximation such a framework can further be thought of as mass points bound by forces which, regardless of their origins, behave like Hooke's law forces. Such a framework composed of N particles has $3N$ degrees of freedom of which, in general, six may be separated as being used up in translation and rotation. There should, therefore, remain $3N - 6$ degrees of freedom representing the vibrations of the framework ($3N - 5$ for linear frameworks). These vibrations, depending on the masses and forces involved, have frequencies distributed throughout the infrared spectrum corresponding to a wavelength of about 3.0–50 μm. If there is associated with any one of these vibrations a change in the electric moment of the molecule, energy may be absorbed or emitted. It is primarily with the problem of discovering these $3N - 6$ (or $3N - 5$) fundamental frequencies that infrared prism spectroscopy is concerned.

To a surprisingly good first approximation a potential function composed only of quadratic terms in the nuclear displacements may be made to fit the vibrational spectrum of most molecules. Furthermore, the potential energy constants for isotopic molecules are nearly the same as in the normal molecule, depending presumably on the electronic structure rather than on the nuclear structure of the isotope. The latter has been abundantly demonstrated by the agreement that exists between the observed frequencies for the deuterium halides and the frequencies predicted from the hydrogen halides upon substitution of the appropriate reduced masses. The identity of the potential constants is a fortunate circumstance because when a potential function is written in its most general form to include quadratic cross-product terms, i.e., bond–bond or bond–angle, there are usually more potential constants than there are frequencies, and it becomes impossible to calculate the constants uniquely from vibrational data for a single molecule. If, however, the frequencies for the isotopic molecule are available, the number of independent relations available is increased and it frequently becomes possible to determine the potential constants. Detailed studies of force constants are important since these studies lead to a clearer understanding of molecular binding.

Since the infrared spectrum of a molecule is completely unique, it has found considerable use as a key physical characteristic of a molecule. This fact, coupled with the additional fact that infrared spectra of compounds in multicomponent mixtures are nearly additive, has made it practicable to record and catalog the infrared spectra of thousands of extremely pure compounds as reference spectra for analysis and to use the data and techniques of infrared spectroscopy as analytical tools in the chemical industries. That the forces between atoms in molecules are not strictly Hooke's law forces is evident from the fact that overtone frequencies of the fundamentals and sum and difference frequencies occur. The existence and intensity of such frequencies is indicative of the amount of anharmonicity of the motion. The appropriate potential function should, therefore, include cubic and quartic terms. The inclusion of such terms in polyatomic molecules enormously complicates the problem of obtaining the harmonic frequencies since the number of frequencies required is enormously increased.

In addition to the vibrational degrees of freedom, the molecule also has, in general, three degrees of rotational freedom. If it is now assumed that the molecular framework is a rigid structure, the various possible molecular configurations fall into the categories; spherical tops, symmetrical tops, and asymmetrical tops in which the principal moments of inertia are, respectively, all equal, two equal and one different, and all unequal. Symmetrical tops may further be classified as prolate and oblate. Quantum mechanically there exist discrete rotational energy levels, the energies of which depend upon the quantum number of total angular momentum, and the quantum number of angular momentum about the unique axis. The spectrum of the simple, rigid, nonvibrating, symmetric rotator would be composed of equally spaced lines arising from transitions between these levels according to well-known selection rules. The line separation is inversely proportional to the moments of inertia. Thus if this separation is observable, information about the internuclear distances can be obtained.

2.2.2. General Theory of Infrared Spectra[2-7]

2.2.2.1. Introduction. In the present review we shall confine the discussion to those motions which involve energies that produce spectra in the infrared region, i.e., vibration and rotation, to the types of spectra

[2] H. H. Nielsen, in "Handbuch der Physik" (S. Flugge, ed.), Vol. XXXVII/1, p. 173. Springer-Verlag, Berlin, 1959.

[3] G. Herzberg, "Spectra of Diatomic Molecules," 2nd ed. Van Nostrand–Reinhold, Princeton, New Jersey, 1950.

[4] G. Herzberg, "Infrared and Raman Spectra of Polyatomic Molecules." Van Nostrand–Reinhold, Princeton, New Jersey, 1945.

[5] H. C. Allen, Jr., and P. C. Cross, "Molecular Vib-Rotors." Wiley, New York, 1966.

[6] J. Charette, "Theory of Molecular Structure." Reinhold, New York, 1966.

[7] H. H. Nielsen, Rev. Mod. Phys. 23, 90 (1951).

that occur, and to the explanation of their appearance. We shall consider molecules as being of two classes, develop the diatomic case in some detail, and generalize the ideas presented to the more complex polyatomic case.

2.2.2.2. The Diatomic Molecule[2,3,8,9]

2.2.2.2.1. THE VIBRATIONAL PROBLEM.[10]   In the formation of a stable molecule, it is generally supposed that when the atoms are a distance r apart, the mutual force between them is the algebraic sum of an attractive force and a repulsive force. Both of these forces vary rapidly with r and are balanced at some internuclear distance r_e, the equilibrium distance of the molecule. When the internuclear distance is less than r_e the repulsive force is the greater: when r is somewhat larger than r_e the attractive force is the greater. A potential energy function which is a function of r alone may then be written. This function has a minimum at r_e, goes to infinity when $r = 0$, and levels off to some fixed value for $r = \infty$. The energy difference between $r = r_e$ and $r = \infty$ represents the work or heat of dissociation of the molecule.

When the amplitude of oscillation is small, to a good first approximation the resultant forces between the atoms of molecules may be considered as of the Hooke's law type. The force may be taken as proportional to the displacement coordinate $x = r - r_e$ and opposite in sign. The potential energy function is, therefore, a parabola and may be expressed as $V = Kx^2/2$, where K is the force constant, and is related to the frequency of vibration along the internuclear axis by the relation

$$\omega_e \ [\text{cm}^{-1}] = (\tfrac{1}{2}\pi c)K^{1/2}/\mu^{1/2}, \tag{2.2.1}$$

where ω_e is the frequency for small amplitudes and μ is the reduced mass equal to $m_1 m_2/(m_1 + m_2)$.

It may easily be seen at this point how substitution of one isotope for another would change the frequency of vibration of the molecule.[3] It is a well-known and reasonable notion that as isotopes differ only in nuclear structure and not in their electronic structure and as molecular forces depend upon the extranuclear properties of the molecule, the potential constant K should be the same in isotopic molecules. If the above formula

[8] L. D. Landau and E. M. Lifshitz, "Quantum Mechanics," p. 261. Pergamon, Oxford, 1958.

[9] P. R. Bunker, *J. Mol. Spectrosc.* **35**, 172 (1970).

[10] L. Pauling and E. B. Wilson, Jr., "Introduction to Quantum Mechanics," p. 259. McGraw-Hill, New York, 1935.

were applied to HCl and DCl, the isotopic shift would depend on the ratio of the reduced masses, $\sim\sqrt{2}$, with ω_e for DCl about $0.7\omega_e$ for HCl. If this isotopic relation were applied to HCl^{35} and HCl^{37}, the effect would be much smaller and depend on the ratio 1.0015.

Although the simple harmonic approximation is quite good, it is evident from observed spectra that it is not good enough. To allow for these observations the theory is modified such that the potential energy is expressed as a Taylor series expansion of the true but unknown potential function V. The resulting potential energy may be written as

$$V = Kx^2/2 + K_{111}x^3 + K_{1111}x^4 + \cdots, \qquad (2.2.2)$$

where $K \gg K_{111} \gg K_{1111}$, etc.

The quantum-mechanical solution to the vibrational problem of the simple harmonic oscillator is quite straightforward and gives the energy in terms of the vibrational quantum number v as

$$E_v/hc \quad [\text{cm}^{-1}] = \omega_e(v + \tfrac{1}{2}), \qquad (2.2.3)$$

where ω_e is the frequency for infinitesimal amplitudes and v may take integral values $0, 1, 2, \ldots$. It may be seen that the energy levels occur at $\omega_e/2, 3\omega_e/2, \ldots$, etc., and are equally spaced. The selection rule for electric dipole transition is $\Delta v = \pm 1$, and this predicts a single line occurring at ω_e cm^{-1} in the spectrum. The introduction of higher-order terms in V causes the wave equation not to have an exact solution and a perturbation theory calculation must be carried out. This results in additional terms involving quadratic and cubic, etc., terms in the vibrational quantum number

$$E_v/hc \quad [\text{cm}^{-1}] = \omega_e(v + \tfrac{1}{2}) - x_e\omega_e(v + \tfrac{1}{2})^2 + y_e\omega_e(v + \tfrac{1}{2})^3$$
$$+ z_e\omega_e(v + \tfrac{1}{2})^4 + \cdots, \qquad (2.2.4)$$

where the $x_e\omega_e$ and $y_e\omega_e$, etc., are the anharmonic constants and are related to the K_{111} and K_{1111}, etc., in the potential energy. The selection rule now becomes $\Delta v = $ any integer. Thus overtones are accounted for by the theory when anharmonic terms are included in the potential energy function.

These remarks are illustrated in Fig. 1, which shows the potential energy curves of the harmonic and anharmonic oscillator for a hydrogen fluoride-like molecule. The theoretically predicted spectrum for the harmonic case is shown below the harmonic potential curve in Fig. 1a.

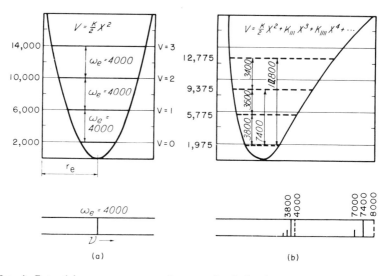

FIG. 1. Potential energy curves and energy levels in the nonrotating case for (a) the harmonic oscillator and (b) the anharmonic oscillator showing allowed transitions with $\omega_e = 4000$ cm^{-1} and $x_e\omega_e = 100$ cm^{-1}.

The spectrum consists of a single transition at 4000 cm^{-1}. The effect of the anharmonic terms in the potential energy is illustrated in Fig. 1b. The potential energy curve is unsymmetrical—steeper as $r \to 0$ and less steep as r increases—thus the nuclei spend more time at $r > r_e$ than for $r < r_e$ in each cycle. From the energy relations, it is clear that the energy levels are no longer equally spaced because of the higher terms in v. For illustrative purposes in the present case, $x_e\omega_e$ has been chosen 100 cm^{-1}, which is about right for HF, and $y_e\omega_e$, $z_e\omega_e$, etc., have been set equal to zero. Thus the zero-point energy level comes at 1975 cm^{-1}, and the level for $v = 1$ comes at 5775 cm^{-1}, and so forth. It is then clear that the transition $v'' = 0$ to $v' = 1$ falls at 3800 cm^{-1} in the spectrum instead of at 4000 cm^{-1} in the harmonic case. Furthermore, the transition $v'' = 1$ to $v' = 2$ no longer coincides with $v'' = 0$ to $v' = 1$, but falls at a still lower frequency. With regard to the overtone, it may be seen that the transition $v'' = 0$ to $v' = 2$ falls at 7400 cm^{-1} instead of at 8000 cm^{-1}.

The intensity of absorption I_v depends essentially upon the number of molecules per cubic centimeter in the initial state, and upon the transition probability from the initial to the final state. The number of molecules in a vibrational state is proportional to the Boltzmann factor $e^{-E/kT}$. The distribution is shown in Fig. 2 for the case of iodine vapor I_2 at

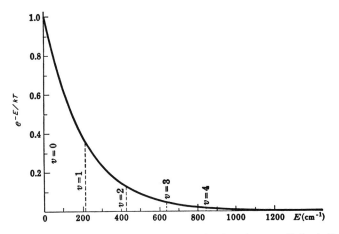

FIG. 2. Thermal distribution of vibrational levels plotted versus E for iodine vapor I_2 at $T = 300$ K (after G. Herzberg, "Spectra of Diatomic Molecules," 2nd. ed. Van Nostrand-Reinhold, Princeton, N. J., 1950). I_2 vapor has no infrared spectrum.

$T = 300$ K with $e^{-E/kT}$ plotted versus E, and showing the levels $v = 0, 1, 2, \ldots$. Thus as the number of molecules in state $v = 1$ is much less than in $v = 0$, the transitions from $v'' = 1$ to $v' = 2$ are fewer than $v'' = 0$ to $v' = 1$. This effect is illustrated in the spectrum at the bottom of Fig. 1b.

As has been remarked, overtones are permitted for the anharmonic case through the selection rule $\Delta v =$ any integer. However, deviations from the harmonic oscillator are usually small and as a result the transition probabilities for the overtones are much smaller than for the fundamental. Thus the first overtone is of the order of 50 times weaker than the fundamental. Figure 1b does not illustrate this disparity in intensities.

2.2.2.2.2. THE PURE ROTATIONAL PROBLEM.[3,11-13] In addition to the vibrational degree of freedom, the molecule may also rotate about an axis perpendicular to the internuclear axis. We shall first treat the molecule as a simple nonvibrating rigid rotator with reduced mass μ and equilibrium distance r_e. The classical kinetic energy of such a rotator is $E_R = P^2/2I_e$, where P is the total angular momentum and $I_e = \mu r_e^2$

[11] C. H. Townes and A. L. Schawlow, "Microwave Spectroscopy." McGraw-Hill, New York, 1955.

[12] J. E. Wollrab, "Rotational Spectra and Molecular Structure." Academic Press, New York, 1967.

[13] I. Kovacs, "Rotational Structure in the Spectra of Diatomic Molecules." American Elsevier, New York, 1969.

is the moment of inertia of the molecule. If the molecule has a permanent electric dipole moment, there is a large probability that it will emit or absorb electromagnetic radiation, thereby changing its rotational energy state. The rotational energy may be written in quantum-mechanical form by substitution of $J(J + 1)h^2/4\pi^2$ for P^2, where J is the quantum number of the total angular momentum and takes values $0, 1, 2, \ldots$. The rotational energy is generally written in the form

$$E_R/hc \quad [\text{cm}^{-1}] = J(J + 1)B_e \qquad (2.2.5)$$

where B_e is the equilibrium rotational constant and is equal to $h/8\pi^2 I_e c$. From this relation it may be seen that a succession of energy levels spaced $0, 2B_e, 6B_e, 12B_e$, etc., ensues. This is shown in Fig. 3a with the levels actually drawn with $B_e = 20$ cm^{-1}. The level spacings are then 40, 80, 120, etc., cm^{-1}. It can be shown that the electric dipole selection rule gives $\Delta J = \pm 1$, so that the transitions $J = 0 \rightarrow 1$, $1 \rightarrow 2$, etc., as shown are permitted ($\Delta J = -1$ corresponds, for pure rotation transitions, to emission). Thus the first line in the pure rotation absorption spectrum should fall at 40 cm^{-1} or 250 μm, the second line should occur at 80 cm^{-1} or 125 μm, the third at 120 cm^{-1} or 85 μm, and so forth.

Actually, the molecule is not a rigid rotator, and the energy is affected both by the vibration and the centrifugal forces. As the vibrational energy increases, i.e., as v increases, the average distance between the nuclei r increases because of the asymmetry of the potential energy function. This has the effect of increasing the moment of inertia and reducing B, the rotational constant. At the same time, the centrifugal forces also increase the average internuclear distance so that B depends upon the angular momentum symbolized by the rotational quantum number J. This gives rise to a term in the energy proportional to $J^2(J + 1)^2$.

The rotational energy (in the $v = 0$ vibrational state) may be written

$$E_R/hc \quad [\text{cm}^{-1}] = [B_e - \tfrac{1}{2}\alpha - (D_e + \tfrac{1}{2}\beta)J(J+1)]J(J+1) \qquad (2.2.6)$$

or

$$E_R/hc \quad [\text{cm}^{-1}] = B_0 J(J + 1) - D_0 J^2(J + 1)^2, \qquad (2.2.7)$$

where B_0 is the ground vibrational state B_v and D_0 is the centrifugal distortion constant in the ground state corresponding to D_v, i.e.,

$$B_v = B_e - \alpha(v + \tfrac{1}{2}), \qquad (2.2.8)$$

$$D_v = D_e + \beta(v + \tfrac{1}{2}). \qquad (2.2.9)$$

Thus if we consider the rotational energy levels of a diatomic molecule in some vibrational state given by v, the energy levels become

$$E_R/hc = B_v J(J+1) - D_v J^2(J+1)^2. \qquad (2.2.10)$$

In some cases, HF being an example, it is necessary also to include a term proportional to $J^3(J+1)^3$. The effect of the increase in moment of inertia due to vibration and rotation is a convergence of the lines of the pure rotation spectrum.

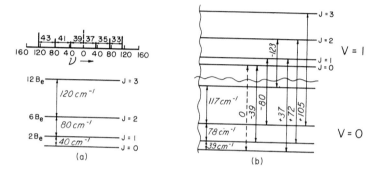

Fig. 3. Energy levels and allowed transitions for (a) the rigid and (b) the nonrigid diatomic rotator showing the appearance of the spectrum, with $B_e = 20$ cm^{-1} and $\alpha = 1$ cm^{-1}.

The thermal distribution of rotational levels is not simply expressed by the Boltzmann factor but depends also on the fact that the rotational levels are $(2J+1)$-fold degenerate, i.e., a rotational level with quantum number J is composed of $2J+1$ coincident levels in zero field. This number is also called its *statistical weight*. The number of molecules N_J in the J level in the lowest vibrational state at temperature T is then

$$N_J \propto (2J+1)e^{-BJ(J+1)hc/kT}. \qquad (2.2.11)$$

$2J+1$ increases linearly with J, while the exponential term decreases in the manner shown in Fig. 2. Therefore, the distribution goes through a maximum. For HF the maximum comes at about $J = 2$. The behavior of this relation plotted versus J is shown in Fig. 4. The broken line ordinates show the relative populations of the rotational levels for the case of HCl, $T = 300$ K, $B = 10.44$ cm^{-1}. Figure 5 shows a record of the pure rotational emission spectrum of HF taken with a Perkin–Elmer model 12C. The lines shown are the transitions $J = 10 \rightarrow 11$ up to

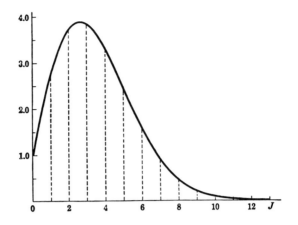

FIG. 4. Intensity relations of vibration–rotation lines for the diatomic molecule HCl, $T = 300$ K (after G. Herzberg, "Spectra of Diatomic Molecules," 2nd. ed. Van Nostrand-Reinhold, Princeton, N.J., 1950).

$J = 15 \rightarrow 16$ which fall in a convenient region because of the extremely large value of B for HF.

2.2.2.2.3. THE VIBRATING ROTATOR. The next step in the development is to combine the anharmonic vibrator with the nonrigid rotator. As has already been shown, if the molecule is not rigid then the rotational energy levels converge and this has an interesting effect on the vibrational spectrum. Classically, one can see that the absorbing frequencies for such a system are $\omega_e \pm \omega_R$. In Fig. 3b, the energy levels for the vibrational state $v = 0$ and $v = 1$ are shown with reasonable values assumed for B_e and for α, the difference between the B values for $v = 0$ and $v = 1$.

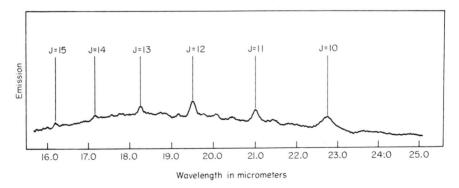

FIG. 5. Pure rotational spectrum of HF in emission [after G. A. Kuipers, D. Smith, and A. H. Nielsen, *J. Chem. Phys.* **25**, 275 (1956)].

The selection rule $\Delta J = \pm 1$, but not 0, shows how the vibration–rotation lines occur. For example, on the high wave number side, or R branch, the transition $J'' = 0$ to $J' = 1$ is 37 cm^{-1} higher than the forbidden $J'' = 0$ to $J' = 0$ or pure vibrational band center. The next line $J'' = 1 \rightarrow J' = 2$ is 35 cm^{-1} higher, the next 33, and so forth—showing a convergence that, if it went far enough, would bring this branch to a head. On the low wave number side, or P branch, the first line $J'' = 1 \rightarrow J' = 0$, however, is 39 cm^{-1} lower than the pure vibration frequency. In this branch the lines diverge steadily. In the upper part of Fig. 3a may be seen a diagram of the predicted spectrum of HF.

Fig. 6. Vibration–rotation spectrum of HF.

In Fig. 6 is shown an actual record of the fundamental band $v'' = 0 \rightarrow v' = 1$ for HF recorded on a vacuum grating spectrograph equipped with a 6000 lines/cm grating and a lead sulfide detector. It is to be noted that there is a gap at the center—the so-called "missing line." This is due to the fact that $\Delta J = 0$ transitions are forbidden for a diatomic

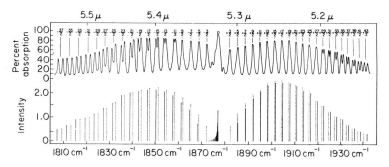

Fig. 7. Vibration–rotation spectrum of the NO fundamental showing the Q branch [after A. H. Nielsen and W. Gordy, *Phys. Rev.* **56**, 781 (1939)].

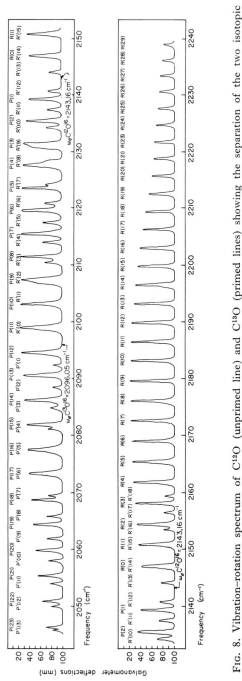
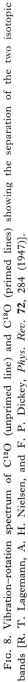

FIG. 8. Vibration–rotation spectrum of $C^{12}O$ (unprimed line) and $C^{13}O$ (primed lines) showing the separation of the two isotopic bands [R. T. Lagemann, A. H. Nielsen, and F. P. Dickey, *Phys. Rev.* **72**, 284 (1947)].

molecule except in certain unusual cases and also because $J = 0 \to 0$
transition are strictly forbidden electric dipole transitions. There is,
however, at least one case of a diatomic molecule having a central line or
Q branch. This molecule is nitric oxide and its spectrum is shown in
Fig. 7. The situation in NO is that the "extra" electron has angular
momentum about the internuclear axis. Another effect already discussed
may be seen in Fig. 8, which is a record of the spectrum of $C^{12}O$ and $C^{13}O$
showing the vibrational frequency separation of the two species produced
by the difference in mass of the two carbon isotopes, as well as the
convergence of the rotation lines. The isotope effect is also shown in
Fig. 9, which pictures the fundamental band for HCl^{35} and HCl^{37}.
The vibrational band center does not suffer nearly such a large shift as
may readily be seen in the closeness of the rotation lines. Evidence of
the 3 to 1 abundance ratio for the chlorines is also clearly shown.

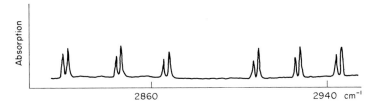

FIG. 9. Central lines of the vibration–rotation spectrum of HCl showing the separation of the HCl^{35} and HCl^{37} isotopic bands (recorded by W. F. Herget, The University
of Tennessee).

Drawing together the separate aspects of the spectrum of a diatomic
molecule, the energy of vibration–rotation in a vibrational state characterized by the vibrational quantum number v and the total angular
momentum quantum number J may be written as

$$E_{VR}(v, J) = \omega_e(v + \tfrac{1}{2}) - x_e\omega_e(v + \tfrac{1}{2})^2 + y_e\omega_e(v + \tfrac{1}{2})^3 + \cdots$$
$$+ B_v J(J+1) - D_v J^2(J+1)^2 + H_v J^3(J+1)^3, \quad (2.2.12)$$

where

$$B_v = B_e - \alpha(v + \tfrac{1}{2}) + \gamma(v + \tfrac{1}{2})^2 = B_0 - \alpha v + \gamma(v^2 + v), \quad (2.2.13)$$

$$D_v = D_e + \beta(v + \tfrac{1}{2}) = D_0 + \beta v, \quad (2.2.14)$$

$$H_v = H_e + \delta(v + \tfrac{1}{2}) = H_0 + \delta v. \quad (2.2.15)$$

Transitions in a vibration–rotation absorption spectrum may be characterized by the ground-state values of v and J (often described by v'', J'' but herein simply called v and J) and their changes in a transition to an excited state, Δv and ΔJ. Thus Δv and ΔJ are simply the selection rule values for the changes in v and J, i.e., the upper states v and J (often denoted by v' and J') are given by $v + \Delta v$ and $J + \Delta J$. A generalized transition frequency may be written as $E_{VR}(v', J') - E_{VR}(v'', J'')$ or more explicitly as $E_{VR}(v + \Delta v, J + \Delta J) - E_{VR}(v, J)$. For transition out of the ground vibrational state ($v = 0$), the transition frequency can be written as

$$
\begin{aligned}
E_{VR}(\Delta v, J + \Delta J) - E_{VR}(0, J) ={} & \omega_e \, \Delta v - x_e \omega_e (\Delta v + 1) \, \Delta v \\
& + y_e \omega_e (\Delta v^2 + (3/2) \, \Delta v + 3/4) \, \Delta v \\
& + \cdots + B_0 (2J + 1 + \Delta J) \, \Delta J \\
& - \alpha \, \Delta v (J + \Delta J)(J + 1 + \Delta J) \\
& + \gamma (\Delta v + 1) \, \Delta v (J + \Delta J)(J + 1 + \Delta J) \\
& - D_0 [(J + \Delta J)^2 (J + 1 + \Delta J)^2 \\
& - J^2 (J + 1)^2] \\
& + \beta \, \Delta v (J + \Delta J)^2 (J + 1 + \Delta J)^2 \\
& + H_0 [(J + \Delta J)^3 (J + 1 + \Delta J)^3 \\
& - J^3 (J + 1)^3],
\end{aligned}
\tag{2.2.16}
$$

where the contribution of δ to H has been neglected. By substituting $\Delta J = +1$ into Eq. (2.2.16) the frequency of $R(J)$ is obtained and if $\Delta J = -1$ is substituted, $P(J)$ results. Although Eq. (2.2.16) looks complicated, the spectrum of a diatomic molecule is generally quite simple. However, to extract the most meaningful results from experimental data, all observations of a single diatomic molecule could be analyzed using Eq. (2.2.16) properly restated for least-squares analysis.[14,15] From an analysis of the vibration–rotation bands the constants ω_e, $x_e \omega_e$, $y_e \omega_e$, and so forth, may be determined, and these in turn make it possible to determine the potential energy constants K_i, and in some cases the dissociation energy. When it is possible to resolve the rotational structure, the moment of inertia in the various vibrational energy levels may be determined because the spacing between the rotational lines

[14] D. B. Keck and C. D. Hause, *J. Mol. Spectrosc.* **26**, 163 (1968).

[15] A. W. Mantz, E. R. Nichols, B. D. Alpert, and K. N. Rao, *J. Mol. Spectrosc.* **35**, 325 (1970).

TABLE I. Molecular Constants for HF and DF $(cm^{-1})^a$

Observed vibrational data	HF^b	DF^c
$v'' = 0 - v' = 1$	3961.42	2906.84
$v'' = 0 - v' = 2$	7750.81	5722.27
$v'' = 0 - v' = 3$	11372.87	
$v'' = 0 - v' = 4$	14831.71	

Observed rotational data	HF	DF
B_0	20.555	10.857
B_1	19.795	10.567
B_2	19.035	10.263
B_3	18.303	—
B_4	17.573	—
D_0	0.0022	—
D_1	0.001_9	—
D_2	0.003_2	0.0006
D_3	0.002_2	—
D_4	0.001_6	—

Calculated molecular constants (vib.)	HF^d	DF^e
$v'' = 0 - v' = 1$		2906.86
$v'' = 0 - v' = 2$		5722.17
ω_e	4138.52	3000.36
$x_e\omega_e$	90.07	47.34
$y_e\omega_e$	0.980	0.373
$z_e\omega_e$	0.025	0.007

Calculated molecular constants (rot.)	HF	DF
B_e	20.939	11.005
α_e	0.771	0.294
γ_e	0.005	0.001
D_e	0.002_2	0.0006_5
I_e	$1.336_3 \times 10^{-40}$ g cm^2	—
r_e	0.9170 Å	—

[a] R. M. Talley, H. M. Kaylor, and A. H. Nielsen, *Phys. Rev.* **77**, 529 (1950).
[b] Observed band center values.
[c] Observed values.
[d] Calculated from observed HF bands.
[e] Calculated from HF data.

is approximately $2B$. From this datum and the masses of the atoms, the equilibrium internuclear distance r_e may be calculated. These features provide unique information about the molecular forces and geometry. Table I gives a list of such constants as obtained by Talley and co-workers[16] for HF and DF. The following things are noteworthy in the light of the previous discussion: the rough $2^{1/2}$ to 1 relation between the HF and DF vibration wave numbers; the rough 2 to 1 relation between the B values; the decrease of B_v with increasing v; the agreement between observed band center of DF and that calculated from HF; and the relative sizes of ω_e, $x_e\omega_e$, and $y_e\omega_e$.

2.2.2.3. Polyatomic Molecules[2,4-8]

2.2.2.3.1. INTRODUCTION. The basic features of the vibration and rotation of a diatomic molecule are also characteristic of polyatomic molecules (number of nuclei $N \geq 3$). Because of the larger number of nuclei both rotation and vibration become more complex. The major source of the complexity is the dynamic coupling between vibration and rotation and the coupling between different vibrational modes. An adequate dynamic model of a polyatomic molecule is a collection of N mass points (nuclei) bound together by a collection of "springs" that nearly obey Hooke's law (the intramolecular force field). The energy of such a molecular model may be expressed as a function of three angles (to describe rotation) and $3N - 6$ vibrational coordinates. As in the case of a diatomic molecule, the energy of the model depends upon parameters which specify molecular geometry and the intramolecular force field. Both the classical Hamiltonian and the quantum-mechanical counterpart that characterize the energy of the molecular model are extremely complex as one might expect in the case of an N-particle system.

In order to understand and analyze observed absorption spectra, the Hamiltonian is usually simplified to the greatest extent possible consistent with the apparent resolution of the observed spectrum and the precision with which data are obtained. The usual approximations used can be cataloged as shown in the accompanying tabulation. Approximation A comprises the vibrational problem that, when treated separately from the rotational problem, may include anharmonicity. When dealing with low-resolution spectra, the vibrational problem alone is generally of prime interest if for no other reason than the rotational substructure is not resolved. Of course, solid and liquid spectra generally only exhibit

---

[16] R. M. Talley, H. M. Kaylor, and A. H. Nielsen, *Phys. Rev.* **77**, 529 (1950).

Model component	Approximation				
	A	B	C	D	E
Harmonic oscillator	x	x	x	x	x
Anharmonic correction			x	x	x
Rigid rotor		x	x	x	x
Nonrigid correction				x	x
Coriolis interaction			x	x	x
Accidental and essential resonance					x

vibrational transitions. In cases where rotational structure is even partially resolved, adequate characterization of an observed spectrum requires the application of approximation B or C and as resolution increases D or E become mandatory for useful characterization of the features of the spectrum.

At times the presentation in the literature of one or the other of these approximations tends to mask a somewhat subtle but really quite simple point. The molecule in actuality is a complex system in which any given energy state is affected to some degree by numerous other energy states of the molecule. In other words, even though the harmonic oscillator–rigid rotor approximation adequately characterizes a particular observed spectrum, the molecule does not vibrate harmonically nor rotate as a rigid rotor, a fact that the very same spectrum observed under conditions of higher resolution would always indicate.

2.2.2.3.2. THE VIBRATION PROBLEM.[2,4,17] An N-atomic nonlinear molecule has $3N - 6$ degrees of freedom associated with its vibrational motion. The standard approach used is to assume an harmonic intramolecular force field and perform a normal coordinate analysis of the vibrational motion. Numerous normal coordinate treatments of nearly all simple molecules can be found in the literature.[4] From these treatments one can readily find a pictorial characterization of the motion of the nuclei in each normal mode. Since these treatments are usually carried out using the observed fundamental vibration frequencies to fix the values of the force constants, the principal use of such treatments (from a spectroscopist's point of view) is to provide a starting point for a perturbation

---

[17] E. B. Wilson, Jr., J. C. Decius, and P. C. Cross, "Molecular Vibrations." McGraw-Hill, New York, 1955.

treatment in order to include anharmonic effects or as a basis for a non-quadratic force field calculation. It is beyond the province of this treatment to discuss this problem in detail. However, considerable work is being carried out on nonquadratic force fields.[18-22] The data available for input to the problem (e.g., anharmonic constants, corrections to moments of inertia in excited vibrational states, etc.) have apparently reached a "critical mass" making it possible for several molecules at least to achieve meaningful results from a nonquadratic force field calculation.

2.2.2.3.2.1. *Normal Modes of Linear Molecules.* An extremely complete catalog of normal modes of simple molecules is available in the literature. However, several examples are presented in the following to indicate the type of results available.

Linear molecules are an interesting special case, since these molecules possess $3N - 5$ vibrational and two rotational degrees of freedom as opposed to $3N - 6$ and three degrees of freedom, respectively, for nonlinear molecules. This results from the dynamic specification of the problem as a *linear* molecule problem wherein rotation about the cylinder axis of the molecule is not defined. Nevertheless, some characterization of rotation about this axis arises out of the vibrational problem. Since CO_2 is a triatomic linear molecule, it thus possesses four vibrational degrees of freedom. The four normal modes are shown in Fig. 10. The mode labeled v_2 is degenerate and gives rise to a rotational angular momentum, a picture of which may be had by visualizing two out-of-phase components of v_2 providing a rotation-like motion. The phase difference between the vibrations of the two degenerate components gives rise to an effective rotation about the cylinder axis of the molecule.

In Fig. 10, v_1 is inactive, and only v_2 and v_3 will be observed in the infrared. In order for a molecule to change its vibrational state by means of an electric dipole-type transition the electric dipole moment of the molecule must have a nonzero time-varying component in the excited state. For CO_2, it is apparent that a center of charge and the center of mass of the molecule maintain their relative positions in the v_1 mode and that therefore no electric dipole transition is possible from the ground

[18] S. Reichman and J. Overend, *J. Chem. Phys.* **48**, 3095 (1968).

[19] K. Machida, *J. Chem. Phys.* **44**, 4186 (1966).

[20] M. A. Pariseau, I. Suzuki, and J. Overend, *J. Chem. Phys.* **42**, 2335 (1965).

[21] Y. Morino and T. Nakagawa, *J. Mol. Spectrosc.* **26**, 496 (1968).

[22] I. Suzuki, *J. Mol. Spectrosc.* **25**, 479 (1968).

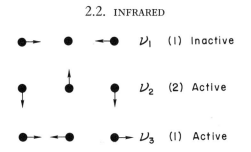

FIG. 10. Normal vibrations of the linear triatomic molecule.

vibrational state to the ν_1 state. In the ν_2 state it is apparent that the electric dipole moment of CO_2 has a time-varying component perpendicular to the cylinder axis of the molecule and in ν_3 the time-varying component is parallel to the cylinder axis. Both ν_2 and ν_3 are infrared active in the sense discussed above. If a linear molecule such as HCN is considered, one finds that the ν_1 mode is infrared active. For such a nonsymmetrical linear molecule, the centers of mass and charge do change their relative positions in the ν_1 mode and thereby the molecule possesses the requisite time-varying dipole moment.

2.2.2.3.2.2. *Nonlinear Molecules.·* The next most simple vibrational problem is the nonlinear XY_2 of which water vapor is an example. This molecule has a permanent dipole moment and all the modes are active in the infrared. The vibrations may be represented as shown in Fig. 11.

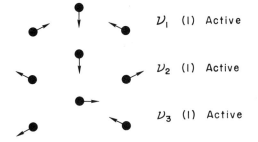

FIG. 11. Normal vibrations of the nonlinear triatomic molecule.

These are all nondegenerate because this is a space figure and the number of vibrational modes is $3N - 6 = 3$. Such a molecule is, except accidentally, an asymmetrical top with modes ν_1 and ν_2 vibrating along the same axis and ν_3 vibrating perpendicularly to this axis. For H_2O, the observed band centers ν_1 and ν_3 are close together 3651.7 cm^{-1} and 3755.8 cm^{-1}, and ν_2 is much lower 1595.0 cm^{-1} as given in Table II.

A group of molecules of considerable interest and importance is the ZYX_3 of which the methyl halides and substituted methanes are examples. The normal modes for this type of molecule, typified by CH_3Cl, number $3N - 6 = 9$, of which three are doubly degenerate and all are infrared active: ν_1, ν_2, and ν_3 are vibrations parallel to the axis of geometrical symmetry, which is also the unique axis of the symmetric top: ν_4, ν_5, and ν_6 are doubly degenerate and produce angular momentum of vibration about the figure axis. The three parallel vibrations are, respectively, the C—H stretch, the flattening of the CH_3 pyramid, and the C—Cl bond stretch, while the three perpendicular ones represent the various tipping motions of the pyramid against the Cl while maintaining the center of gravity position fixed, and linear momentum and angular momentum equal to zero.

TABLE II. Table of Fundamental Frequencies for Bent Triatomic Molecules[a]

	H_2O	D_2O	H_2S	D_2S	F_2O	Cl_2O
ν_1:	3651.7	2666	2610.8	1891.6	830	680
ν_2:	1595.0	1178.7	1290	934	490	330
ν_3:	3755.8	2789	2684	1999	1110	973

[a] Values are in units of reciprocal centimeters.

The last type of molecule to be treated in this discussion is the XY_4 type, of which CH_4 is the common example. In this molecule the X atom is at the center and the Y atoms are situated tetrahedrally about the center giving the molecule, to first approximation, spherical symmetry. The number of modes is 9 as in the previous case. Because of the high symmetry, they give rise to only four distinct frequencies of which ν_1 is the inactive singly degenerate breathing frequency, ν_2 is the inactive doubly degenerate bending of the Y—X—Y angle, ν_3 is the active triply degenerate bending vibration in which the bending is out of phase, and ν_4 is the active triply degenerate motion of the four Y atoms against the X atom.

2.2.2.3.2.3. *Isotope Effects.*[4]   To illustrate the effect of substitution of isotopes or other atoms from the same group, the following examples are cited.

In Table II the effect of the deuterium substitution is clearly seen for water and hydrogen sulfide in the first four columns, while the effect of substituting chlorine for fluorine may be seen in the last two columns. In Table III the same effects may be seen for the methyl halide group. Consideration of the normal modes of CH_3X (X is a halogen) will show, for example, that in ν_1 the halogen does not enter into the vibration very much and thus this frequency remains substantially the same, while in ν_3 the frequency changes greatly from the fluorine to the iodine case. A similar situation is to be noted in ν_4 as compared with ν_6, which is halogen-dependent.

TABLE III. Fundamental Frequencies for the Methyl Halides

	ν_1	ν_2	ν_3	ν_4	ν_5	ν_6
CH_3F:	2964.5	1460.0	1048.6	3009.1	1475.5	1183.2
CH_3Cl:	2967.8	1354.9	731.5	3054.0	1452.1	1017.3
CH_3Br:	2973.0	1305.9	610.5	3056.6	1442.7	954.7
CH_3I:	2953.2	1250.7	533.2	3063.2	1438.0	882.4

2.2.2.3.2.4. Group Frequencies.[23-26] It should finally be pointed out that when a particular bond or chemical group appears in a molecule, certain frequencies which are representative of the bond or group appear as absorption bands in the spectrum with a frequency which depends somewhat upon the environment, but which appears to be characteristic of the bond or group. This is, of course, an extremely important point because of its implications for analytical use of infrared spectroscopy. The few examples shown in Table IV will suffice as illustrations, since this point is extensively treated in connection with analytical problems in the literature.

[23] R. B. Barnes, R. C. Gore, U. Liddel, and V. Z. Williams, "Infrared Spectroscopy." Van Nostrand–Reinhold, Princeton, New Jersey, 1944.

[24] H. M. Randall, R. G. Fowler, N. Fuson, and J. R. Dangl, "Infrared Determination of Organic Structures." Van Nostrand–Reinhold, Princeton, New Jersey, 1949.

[25] L. J. Bellamy, "The Infrared Spectra of Complex Molecules." Wiley, New York, 1954.

[26] R. T. Conley, "Infrared Spectroscopy." Allyn and Bacon, Boston, Massachusetts, 1966.

TABLE IV. A Few Examples
of Characteristic Group Frequencies

$\equiv$C—H	3300 cm^{-1}
—O—H	3680 cm^{-1}
—S—H	2570 cm^{-1}
$\diagdown$C$=$O	1700 cm^{-1}

2.2.2.3.3. THE ROTATIONAL PROBLEM.[2,4,5,11–13] As in the diatomic molecule, rotation may also occur, but here the problem is more complicated. The pure rotation case, while of interest in the infrared, has been adequately treated in microwave spectroscopy discussions. The following treatment of the vibration–rotation problem includes the pure rotation problem as a special case.

Rotational properties are used as a convenient method of classifying molecules. Molecules for which the principal moments of inertia I_x, I_y, and I_z are equal are called *spherically symmetric* (often *spherical tops*). If $I_x = I_y \neq I_z$, the molecule is *axially symmetric* (*symmetric top*) whereas if $I_x \neq I_y \neq I_z$, the molecule is *asymmetric* (*asymmetric top*). For symmetric tops if $I_z < I_x = I_y$, the molecule is called a *prolate top*: if $I_z > I_x = I_y$, the molecule is called an *oblate top*. Linear molecules have $I_x = I_y$, and I_z is actually undefined. In the following we specialize to symmetric tops observed in the gas phase, although much of the general outline is true for the other cases. A similar attack on the asymmetric molecule is available[2,4,5] whereas for the spherical molecule, the original literature seems to be the best source.[2,4,27,28]

2.2.2.3.4. THE VIBRATION–ROTATION PROBLEM.[2,29]

2.2.2.3.4.1. *Summary of the Development of the Rotation–Vibration Hamiltonian.* Although the general problem of rotation and vibration of a molecule is quite complex, the development of the problem is straightforward. From a rather simple dynamical model it is possible to derive a classical Hamiltonian and ultimately a quantum-mechanical Hamil-

[27] K. T. Hecht, *J. Mol. Spectrosc.* 5, 355 (1960).
[28] K. T. Hecht, *J. Mol. Spectrosc.* 5, 390 (1960).
[29] G. Amat, H. H. Nielsen, and G. Tarrago, "Higher Order Rotation–Vibration Energies of Polyatomic Molecules." Dekker, New York, 1971.

tonian.[30,31] In the following it is assumed that the reader has in fact become familiar with such a development or is willing to assume the results of such a treatment as given.

Briefly, one begins with a classical Hamiltonian and converts it into a quantum-mechanical Hamiltonian. This Hamiltonian is then expanded in a power series in the normal coordinates and the resulting terms are grouped according to the estimated order of magnitude of the contribution to the energy of rotation–vibration. The expanded Hamiltonian is diagonal with respect to J and M but may contain terms off-diagonal with respect to v_s, l_s, m_s, and K.* At this point a contact transformation is made on the Hamiltonian[32-37]

$$H = H_0 + H_1 + H_2 + \cdots \tag{2.2.17}$$

such that the resulting Hamiltonian

$$h' = H_0 + h_1' + h_2' + h_3' + \cdots \tag{2.2.18}$$

is diagonal, through h_1', with respect to all quantum numbers for axially symmetric molecules. (This is valid only in the absence of accidental resonances. In the event of such resonances a special transformation must be performed on H to preserve the validity of the contact transformation.[38,39]) For axially symmetric molecules, since H_0 and h_1' are diagonal,

[30] E. B. Wilson, Jr. and J. B. Howard, *J. Chem. Phys.* **4**, 262 (1936).

[31] B. T. Darling and D. M. Dennison, *Phys. Rev.* **57**, 128 (1940).

[32] S. Maes and G. Amat, *Cah. Phys.* **11**, 277 (1957).

[33] M. Goldsmith, G. Amat, and H. H. Nielsen, *J. Chem. Phys.* **24**, 1178 (1957).

[34] G. Amat, M. Goldsmith, and H. H. Nielsen, *J. Chem. Phys.* **27**, 838 (1957).

[35] G. Amat and H. H. Nielsen, *J. Chem. Phys.* **29**, 665 (1958); **34**, 339 (1961); **36**, 1859 (1962).

[36] M. L. Grennier-Besson, G. Amat, and H. H. Nielsen, *J. Chem. Phys.* **36**, 3454 (1962).

[37] J. K. G. Watson, *Mol. Phys.* **15**, 479 (1968).

[38] H. H. Nielsen, *Phys. Rev.* **68**, 181 (1945).

[39] S. Maes, *J. Phys.* **27**, 37 (1966).

---

* One defines the following symbols: J is the quantum number associated with the total angular momentum, M is the quantum number of the projection of the total angular momentum on a space-fixed axis, K is the quantum number of the projection of the total angular momentum on the body-fixed symmetry axis of the molecule, v_s is the quantum number of the s normal vibration, l_s is the second quantum number of doubly degenerate normal vibrations (associated with internal angular momentum), and m_s is the third quantum number of threefold degenerate normal vibrations.

and since the first nondiagonal terms occur in h_2', the energy of any particular level is given through third order by the diagonal elements of $H_0 + h_1' + h_2' + h_3'$, i.e., in the absence of accidental resonances, off-diagonal terms of h_2' will not contribute to the energy before fourth order.

Amat and Nielsen[35] performed a second contact transformation which diagonalized h' through second order with respect to the vibrational quantum number v_s. For axially symmetric molecules, the twice-transformed Hamiltonian

$$h^+ = H_0 + h_1' + h_2^+ + h_3^+ + h_4^+ + \cdots \qquad (2.2.19)$$

is diagonal with respect to the vibrational quantum numbers v_s through h_2^+. However, h_2^+ has off-diagonal terms with respect to K and l_s. In addition $h_3^+, h_4^+, \ldots$ have terms off-diagonal with respect to v_s, l_s, and K.

For symmetric tops, i.e., axially symmetric molecules, the twice transformed Hamiltonian h^+ yields energies good through third order since the terms in h_2^+ that are off-diagonal in K and l_s do not in general contribute before the fourth order. Since h^+ is diagonal with respect to v_s through h_2^+, terms off-diagonal with respect to v_s in h_3^+ will not contribute to the energy before sixth order.[40,41]

2.2.2.3.4.2. *Rotation–Vibration Energy through Third Order.* The expansion of the Hamiltonian is effected with an assumed expansion parameter on the order of $1/30$.* It is assumed that a zero-order contribution to the energy is approximately 1000 cm^{-1}. One should realize that order-of-magnitude considerations are approximations and are not to be taken too literally. They should therefore serve as a helpful guide rather than as restrictions in the analysis of spectra.

The twice transformed Hamiltonian

$$h^+ = H_0 + h_1' + h_2^+ + h_3^+ + h_4^+ + \cdots \qquad (2.2.20)$$

may be used to obtain the rotation–vibration energy of axially symmetric

[40] S. Maes, *J. Mol. Spectrosc.* **9**, 204 (1962).
[41] G. Tarrago, *Cah. Phys.* **176–177**, 150 (1965).

---

* The expansion parameter can be shown to be $(B_e/\omega)^{1/2}$. The value $1/30$ is characteristic of a number of molecules.[2,32]

molecules by solving the secular equation

$$\det[\langle JMK \cdots v_s, l_s \cdots | H_0 + h_1' + h_2^+ + h_3^+ + h_4^+$$
$$+ \cdots | JMK \cdots v_{s'}, l_{s'} \cdots \rangle$$
$$- \delta_{KK'} \cdots \delta_{v_s v_{s'}} \cdots \delta_{l_s l_{s'}} \cdots E_{\mathrm{VR}}] = 0, \qquad (2.2.21)$$

where $\delta_{ii'}$ is the Kronecker symbol and E_{VR} is the energy of rotation–vibration which we wish to obtain.

For energies through third order, in the absence of accidental resonances the secular equation which must be solved is simple enough since h^+ is diagonal and the energy of rotation–vibration is given by the diagonal elements of $H_0 + h_1' + h_2^+ + h_3^+$. Table V presents the diagonal terms in the Hamiltonian and the resulting contribution to E_{VR}. Table VI gives a list of symbols found in Table V with appropriate definitions. In the absence of accidental resonances, the energy of rotation–vibration to any desired order through the third may be obtained from Table V. In addition, a partial fourth-order correction may be obtained but special caution is required since there are, to that order of approximation, additional contributions from terms in h_2^+ that are off-diagonal in K and l_t.[41]

In Table V a simplified notation has been used for the Hamiltonian term, which follows the convention that operators in the Hamiltonian having the forms $q_a q_b$, $p_a p_b$, or $q_a p_b$ be called r^2, etc., that $P_\alpha P_\beta P_\gamma$ be called P^3, etc., and that the sub- and superscripts on the Y's and Z's be dropped.

The order of magnitude expansion of the Hamiltonian is valid for $J \sim K \sim 30$. For other values of J and K, a term may be demoted or promoted from its present formal order to a different order which would then be called the "true" order of magnitude of its contribution.

A relatively simple approach to the problem leads one to discuss the order of magnitude of a given *constant* appearing in h_m^+. Consider the generic term $_{(m)}Z r^n P^s$, which is a term (from the mth order transformed Hamiltonian) with the constant coefficient $_{(m)}Z$, a product of n vibrational operators and s rotational operators. The order of magnitude of the *constant* $_{(m)}Z$ is $m + s$. For example, D_e^J is of order 6 since $m = 2$ and $s = 4$. Table V lists the formal order of magnitude of the molecular parameters occurring in each term and estimated magnitude based on an expansion parameter of $1/30$ and a zero-order contribution of 1000 cm^{-1}, i.e., parameter magnitude ~ 1000 cm$^{-1}/(30)^{m+s}$.

TABLE V. Symmetric Top Energy[a]

h_m^+	Operator	Constant	Order of constant	Approximate magnitude of constant (cm^{-1})	Axial molecule energy
H_0	$_{(0)}XP^2$	A_e, B_e	2	1	$B_e J(J+1) + (A_e - B_e)K^2$
	$_{(0)}Xr^2$	ω_s	0	1000	$\sum_s \omega_s(v_s + \frac{1}{2}g_s)$
h_1'	$_{(1)}Yr^2P$	$2A_e \sum_t \zeta_t^z l_t$	2	1	$-2A_e \sum_t \zeta_t^z l_t K$
h_{21}^+	$_{(2)}ZP^4$	$D_e^J, D_e^{JK}, D_e K$	6	1×10^{-6}	$-D_e^J J^2(J+1)^2 - D_e^{JK}K^2J(J+1) - D_e^K K^4$
h_{22}^+	$_{(2)}Zr^2P^2$	α_s^A, α_s^B	4	1×10^{-3}	$-\sum_s \alpha_s^B(v_s + \frac{1}{2}g_{ss})J(J+1) - \sum_s (\alpha_s^A - \alpha_s^B)(v_s + \frac{1}{2}g_s)K^2$
h_{23}^+	$_{(2)}Zr^4$	$x_{ss'}, x_{l_t l_{t'}}$	2	1	$\sum_{\substack{ss' \\ s\leq s'}} x_{ss'}(v_s + \frac{1}{2}g_s)(v_{s'} + \frac{1}{2}g_{s'}) + \sum_{\substack{tt' \\ t\leq t'}} x_{l_t l_{t'}} l_t l_{t'}$
h_3^+	$_{(3)}Zr^2P^3$	$\eta_{l_t}^J, \eta_{l_t}^K$	6	1×10^{-6}	$\sum_t l_t \eta_{l_t,J} J(J+1)K + \sum_t l_t \eta_{l_t,K}K^3$
	$_{(3)}Zr^4P$	$\eta_{l_t}, \eta_{l_t s}$	4	1×10^{-3}	$+\sum_t l_t \left[\eta_{l_t} + \sum_s \eta_{l_t,s}(v_s + \frac{1}{2}g_s)\right]K$

h_{41}^+	$_{(4)}Zr^2P^4$	$\beta_s^J, \beta_s^{JK}, \beta_s^K$	8	1×10^{-9}	$\sum_s \beta_s^J[v_s + \frac{1}{2}g_s]J^2[J+1]^2 + \sum_s \beta_s^{JK}[v_s + \frac{1}{2}g_s]J[J+1]K^2$ $+ \sum_s \beta_s^K K^4[v_s + \frac{1}{2}g_s]$
	$_{(4)}Zr^4P^2$	$\gamma_{ss'}^A, \gamma_{ss'}^B, \gamma_{t_l t'}^A, \gamma_{t_l t'}^B$	6	1×10^{-6}	$\sum_{\substack{ss' \\ s\le s'}} \{\gamma_{ss'}^B J(J+1) + [\gamma_{ss'}^A - \gamma_{ss'}^B, K^2]\}\{v_s + \frac{1}{2}g_s\}\{v_{s'} + \frac{1}{2}g_{s'}\}$ $+ \sum_{\substack{tt' \\ t\le t'}} \{\gamma_{t_l t'}^B J(J+1) + [\gamma_{t_l t'}^A - \gamma_{t_l t'}^B]K^2\}l_t l_{t'}$
	$_{(4)}ZP^2$	$\Delta A_e, \Delta B_e$	6	1×10^{-6}	$\Delta B_e[J(J+1) - K^2] + \Delta A_e K^2$
	$_{(4)}Zr^6$	$Y_{ss'}'', Y_{s_l t_l t'}$	4	1×10^{-3}	$\sum_{\substack{ss's'' \\ s\le s'\le s''}} Y_{ss's''}[v_s + \frac{1}{2}g_s][v_{s'} + \frac{1}{2}g_{s'}][v_{s''} + \frac{1}{2}g_{s''}]$ $+ \sum_{\substack{st t' \\ t\le t'}} Y_{s_l t_l t'}l_t l_{t'}[v_s + \frac{1}{2}g_s]$
h_{44}^+	$_{(4)}Zr^2$	$\Delta\omega_s$	4	1×10^{-3}	$\sum_s \Delta\omega_s[v_s + \frac{1}{2}g_s]$
h_{43}^+	$_{(4)}ZP^6$	H^J, H^{JK}, H^{KJ}, H^K	10	1×10^{-12}	$H^J J^3(J+1)^3 + H^{JK} J^2(J+1)^2 K^2 + H^{KJ} J(J+1)K^4$ $+ H^K K^6$

[a] See Table VI for a list of symbols.

TABLE VI. List of Symbols from Table V

Symbol	Description
α	x, y, z (likewise for β and γ) (body-fixed)
s	index for the normal vibration
n	index for specifically nondegenerate normal vibration
t	index for specifically twofold degenerate vibration
Y, Z	constant coefficients (generally complicated) of the transformed Hamiltonians
P_α	rotational angular momentum operator
q_s	normal coordinate operator
p_s	momentum operator conjugate to q_s
I_α	principal equilibrium moment of inertia about the α axis
B_e	$= h/(8\pi^2 c I_x)$, $I_x = I_y$
A_e	$= h/(8\pi^2 c I_Z)$, I_Z is unique $I_Z \neq I_x = I_y$
ω_s	harmonic frequency (cm^{-1}) of s normal mode
λ_s	$= (2\pi c \omega_s)^2$
g_s	degeneracy of the s normal mode
ζ_t^z	Coriolis coupling constant—a function of molecular geometry
D_e^m	$m = J, K, JK$—the equilibrium centrifugal distortion constant
α_s^B	correction to B_e due to vibration
α_s^A	correction to A_e due to vibration
$X_{ss'}, X_{l_t l_{t'}},$	
$y_{ss's''}, y_{sl_t l_t}$	anharmonic vibrational constants
$\eta_{t,m}, \eta_t^m$	$m = J, K, S,$ or nothing: third-order constants
β_s^m	$m = J, K, JK$—the correction to D_e^m due to vibration
$\gamma_{ss'}^B$	correction to B_e due to vibration
γ_{ss}^A	correction to A_e due to vibration
$\Delta B_e, \Delta A_e, \Delta \omega_s$	corrections to A_e, B_e, ω_s
H^m	$m = J, K, JK, KJ$—the corrections to distortion constants due to rotation

The computation of the "true" order of magnitude of the contribution of a *term* in E_{VR} for a particular J and K may be carried out by substituting the values of J and K in question into the contribution of the term of E_{VR} using the approximate magnitude of the constant from Table V. For example, D_e^J is of order 6: however, the contribution to the energy from the term containing D_e^J is of order 2 for $J \sim 20$ and $K \sim 0$ or

$K \sim 30$. A more meaningful order of magnitude of the contribution of a term to E_{VR} may be computed if one has available the actual value of the constant involved.

2.2.2.3.4.3. *The Transition Frequency for Symmetric Tops.* In this section the development of a single expression that simultaneously represents all* allowed rotation–vibration transitions of a particular molecule is reviewed. The generalized transition frequency expression is of special value to the spectroscopist since it enables him to perform a simultaneous least-squares analysis of all unperturbed transitions or any subset of transitions of a particular molecule. This then results in a consistent set of molecular constants and accordingly, as more and more transitions are analyzed simultaneously, the statistics involved become more and more favorable.

Consider the energy levels of an axially symmetric molecule with a negligible inversion probability. These energy levels are given by the energy expressions presented previously and are specified by the quantum numbers $v_n, v_{n+1}, \ldots, v_t, l_t, \ldots, J, K, M$.[†]
Consider the transition

$$v_n + \Delta v_n, \ldots, v_t + \Delta v_t, l_t + \Delta l_t, \ldots, J + \Delta J, K + \Delta K$$
$$\leftarrow v_n, \ldots, v_t, l_t, \ldots, J, K \qquad (2.2.22)$$

which is a transition[‡] between the lower state

$$E''_{VR}(v_n, v_{n+1}, \ldots, v_t, l_t, v_{t+1}, \ldots, J, K) \qquad (2.2.23)$$

and the upper state

$$E'_{VR}(v_n + \Delta v_n, \ldots, v_t + \Delta v_t, l_t + \Delta l_t, \ldots, J + \Delta J, K + \Delta K). \qquad (2.2.24)$$

A generalized transition frequency expression is obtained in a straightforward manner when the upper and lower levels are written as in Eqs. (2.2.24) and (2.2.23), respectively. Alternative forms of the transition

---

* It should be pointed out that the resulting single polynomial expression also represents pure rotation and Raman transitions as well as infrared allowed vibration-rotation transitions.

[†] In the absence of external electric or magnetic fields, M shall be omitted since the energy E_{VR} is then independent of M.

[‡] For each transition, the set $\Delta v_n, \Delta v_{n+1}, \ldots, \Delta v_t, \Delta l_t, \ldots, \Delta K, \Delta J$ are determined by the selection rules for that particular transition.

frequency corresponding to the transition given by Eq. (2.2.22) are

$$(v_n, v_{n+1}, \ldots, v_t, l_t, \ldots, \Delta v_n, \ldots, \Delta v_t, \Delta l_t, \ldots, \Delta K, \Delta J, K, J)$$
$$= (v_n, v_{n+1}, \ldots, v_t, \Delta_t, \ldots, \Delta v_n, \ldots, \Delta v_t, \Delta l_t, \ldots)^{\Delta K} \Delta J_K(J)$$
$$= E'_{VR}(v_n + \Delta v_n, \ldots, v_t + \Delta v_t, l_t + \Delta l_t, \ldots, J + \Delta J, K + \Delta K)$$
$$- E''_{VR}(v_n, \ldots, v_t, l_t, \ldots, JK). \tag{2.2.25}$$

Table VII presents the generalized transition frequency expression* defined by Eq. (2.2.25) for the case $\sum_s v_s = 0$ in E''_{VR} complete (in the absence of accidental resonances) through the contributions of $h_3{}^+$. Table VII also contains a partial contribution from $h_4{}^+$.

The definitions given in Eqs. (2.2.26)–(2.2.31) identify some of the constants appearing in Table VII with those found in the literature and in papers on analysis of spectra.

$$B_0 = B'' = B_e - \sum_s (g_s/2)\alpha_s{}^B + \sum_s \sum_{\substack{s' \\ s \leq s'}} (g_{s'}g_s/4)\gamma_{ss'}^B + \Delta B_e, \tag{2.2.26}$$

$$B_v = B' = B_0 - \sum_s v_s\alpha_s{}^B + \sum_s \sum_{\substack{s' \\ s \leq s'}} (v_s v_{s'} + v_s g_{s'}/2 + v_{s'}g_s/2)\gamma_{ss'}^B$$
$$+ \sum_t \sum_{\substack{t' \\ t \leq t'}} \gamma_{l_t l_{t'}}^B l_t l_{t'}, \tag{2.2.27}$$

$$A_0 = A'' = A_e - \sum_s (g_s/2)\alpha_s{}^A + \sum_s \sum_{\substack{s' \\ s \leq s'}} (g_{s'}g_s/4)\gamma_{ss'}^A + \Delta A_e, \tag{2.2.28}$$

$$A_v = A' = A_0 - \sum_s v_s\alpha_s{}^A + \sum_s \sum_{\substack{s' \\ s \leq s'}} (v_s v_s + v_s g_{s'}/2 + v_{s'}g_s/2)\gamma_{ss'}^A$$
$$+ \sum_t \sum_{\substack{t' \\ t \leq t'}} \gamma_{l_t l_{t'}}^A l_t l_{t'}, \tag{2.2.29}$$

$$D_0{}^m = D_m'' = D_e{}^m + \sum_s (g_s/2)\beta_s{}^m, \qquad m = J, K, JK, \tag{2.2.30}$$

$$D_v{}^m = D_m' = D_0{}^m + \sum_s v_s\beta_s{}^m, \qquad m = J, K, JK. \tag{2.2.31}$$

Table VIII contains descriptions of symbols found in Table VII but not previously defined.

---

* The terms in Table VII are grouped according to the various molecular constants. However, special considerations must be taken into account before using the expressions given in this chapter in the numerical analysis of spectra in order to avoid linear dependence or a poorly conditioned set of normal equations.

TABLE VII. Generalized Transition Frequency Expression[a]

$$(v_n, v_{n+1}, \ldots, v_t, l_t, v_{t+1}, l_{t+1}, \ldots)^{\Delta K} \Delta J_K(J)$$

$$= \sum_s v_s(\omega_s + \Delta\omega_s) + \sum_{\substack{s\,s'\\s\leq s'}} x_{ss'}\left[\left(v_s + \frac{g_s}{2}\right)\left(v_{s'} + \frac{g_{s'}}{2}\right) - \frac{g_s g_{s'}}{4}\right] + \sum_{\substack{t\,t'\\t\leq t'}} x_{l_t l_{t'}} l_t l_{t'}$$

$$+ \sum_{\substack{s\,s'\,s''\\s\leq s'\leq s''}} y_{ss's''}\left[\left(v_s + \frac{g_s}{2}\right)\left(v_{s'} + \frac{g_{s'}}{2}\right)\left(v_{s''} + \frac{g_{s''}}{2}\right) - \frac{g_s g_{s'} g_{s''}}{8}\right]$$

$$+ A_0[(K + \Delta K)^2 - K^2] + B_0[(J + \Delta J)(J + 1 + \Delta J) - J(J + 1) - (K + \Delta K)^2 + K^2]$$

$$+ \left[-2A_e \sum_t \zeta_t^z l_t + \sum_t\left\{\eta_t + \sum_s \eta_{t,s}\left(v_s + \frac{g_s}{2}\right)\right\} l_t\right][K + \Delta K] - D_0^J[(J + \Delta J)^2(J + 1 + \Delta J)^2 - J^2(J + 1)^2]$$

$$- D_0^{JK}[(K + \Delta K)^2(J + \Delta J)(J + 1 + \Delta J) - K^2 J(J + 1)] - D_0^K[(K + \Delta K)^4 - K^4]$$

$$+ \left[-\sum_s \alpha_s^A v_s + \sum_{ss'} \gamma_{ss'}^A\left(v_s v_{s'} + \frac{v_s' g_{s'}}{2} + \frac{v_s' g_s}{2}\right) + \sum_{\substack{tt'\\t\leq t'}} \gamma_{l_t l_{t'}}^A l_t l_{t'}\right][(K + \Delta K)^2]$$

$$+ \left[-\sum_s \alpha_s^B v_s + \sum_{ss'} \gamma_{ss'}^B\left(v_s v_{s'} + \frac{v_s' g_{s'}}{2} + \frac{v_s' g_s}{2}\right) + \sum_{\substack{tt'\\t\leq t'}} \gamma_{l_t l_{t'}}^B l_t l_{t'}\right][(J + \Delta J)(J + 1 + \Delta J) - (K + \Delta K)^2]$$

$$+ \sum_t \eta_t^J l_t(K + \Delta K)(J + \Delta J)(J + 1 + \Delta J) + \sum_t \eta_t^K l_t(K + \Delta K)^3 + \sum_s \beta_s^J v_s[(J + \Delta J)^2(J + 1 + \Delta J)^2]$$

$$+ \sum_s \beta_s^{JK} v_s[(K + \Delta K)^2(J + \Delta J)(J + 1 + \Delta J)] + \sum_s \beta_s^K v_s(K + \Delta K)^4 + H^J[(J + \Delta J)^3(J + 1 + \Delta J)^3 - J^3(J + 1)^3]$$

$$+ H^{JK}[(J + \Delta J)^2(J + 1 + \Delta J)^2(K + \Delta K)^2 - J^2(J + 1)^2K^2] + H^{KJ}[(J + \Delta J)(J + 1 + \Delta J)(K + \Delta K)^4 - J(J + 1)K^4]$$

$$+ H^K[(K + \Delta K)^6 - K^6]$$

[a] See Table VIII for a list of symbols.

TABLE VIII. List of Symbols from Table VII

Symbol	Description
B_0, A_0	ground state inverse moment of inertia corresponding to B_e and A_e
α_s^A, α_s^B	coefficients of the correction to A_e and B_e, respectively, due to vibration
D_0^J, D_0^{JK}, D_0^K	ground state centrifugal distortion constants
s	refers generically to any vibrational mode
n	refers specifically to nondegenerate vibrations
t	refers specifically to twofold degenerate vibrations
v_s	vibrational quantum number for the s mode, e.g., if three quanta of ν_4 are excited, $v_4 = 3$
l_t	internal angular momentum quantum number associated with twofold degenerate vibrations where l may take on values $\pm v_t$, $\pm(v_t - 2)$, ..., 0 or ± 1
J	total angular momentum quantum number
K	quantum number associated with the projection of J on the symmetry axis of the molecule
ΔJ	change in J in the transition being considered, i.e., if $J_2 \leftarrow J_1$, then $\Delta J = J_2 - J_1$
ΔK	similarily, for $K_2 \leftarrow K_1$, $\Delta K = K_2 - K_1$

There is an important restriction to be observed regarding the expression in Table VII. The quantum number K may properly take the values $\pm |K|$. However, in Table VII, one need only take $+ |K|$ values into account, and in fact, if one desires to retain simplicity in selection rules, only $+ |K|$ should be used.

Returning to the significance of the general transition frequency expression of Table VII, it is important to note that simply by specifying for a given transition, $v_1, v_2, \ldots, v_n, v_t, v_{t+1}, \ldots, l_t, l_{t+1}, \Delta v_1, \Delta v_2, \ldots, \Delta v_n, \Delta v_t, \Delta l_t, \Delta v_{t+1}, \Delta l_{t+1}, \ldots, J, \Delta J, K$, and ΔK as independent variables, the transition frequency stands, in relation to these, as the dependent variable. Thus a simultaneous least-squares analysis of all unperturbed transitions of a given molecule may be obtained.[42-45]

[42] H. Kurlat, M. Kurlat, and W. E. Blass, *J. Mol. Spectrosc.* **38**, 197 (1971).
[43] T. L. Barnett and T. H. Edwards, *J. Mol. Spectrosc.* **20**, 347 (1966).
[44] T. L. Barnett and T. H. Edwards, *J. Mol. Spectrosc.* **20**, 352 (1966).
[45] T. L. Barnett and T. H. Edwards, *J. Mol. Spectrosc.* **23**, 302 (1967).

Quite often this is impractical, especially if any of the levels involved are perturbed, but the fact remains that it is otherwise possible. Of course, any subset of transitions believed to be unperturbed may be analyzed using an appropriate modification of the generalized expression of Table VII.

For example, the sum of the purely vibrational contributions in Table VII may be called ν_0, and all the spectral lines of a given rotation–vibration band may be simultaneously analyzed. There are now a considerable number of examples of this approach in the literature.[46–49] Another consideration in the analysis of spectra involves the expected order of magnitude of the contribution of any particular term found in Table VII to the vibration–rotation energy. However, since the transition frequency results from the difference of E'_{VR} and E''_{VR}, certain of the terms in the transition frequency expression are demoted to a higher order of magnitude than they had in the energy expression. Furthermore, as mentioned previously the formal order of magnitude of some contributions to the energy are "correct" only for $J \approx K \approx 30$. For $J < 30$ or $K < 30$ various terms are demoted or promoted from their formal order of magnitude to a higher order or lower order.

Thus the spectral data with which one deals will often determine which terms in the transition frequency are of primary importance, and more important, which terms are likely to be significant in representing observed transitions. For example, if one observes transitions for $K = 0$ to 4 and $J = 0$ to 40 it can easily be seen that $D_0{}^K$ is of little significance in many cases while $D_0{}^J$ takes on added significance because of the high values of J involved.

2.2.2.3.4.4. *Selection Rules.*[2,4] For an electric dipole absorption transition between two states a and b to be observable it is necessary that at least one of three integrals

$$\int \psi^*_{\text{exc}} \hat{M}_i \psi_{\text{grd}} \, d\tau, \qquad (2.2.32)$$

where M_i are the components of the dipole moment along three space-fixed axes, be different from zero (exc is excited and grd is ground). Using the symmetry of the zero-order wave functions, the well-known

[46] W. E. Blass and T. H. Edwards, *J. Mol. Spectrosc.* **25**, 440 (1968).

[47] M. Morillon-Chapey and G. Graner, *J. Mol. Spectrosc.* **31**, 155 (1969).

[48] W. E. Blass and T. H. Edwards, *J. Mol. Spectrosc.* **24**, 111 (1967).

[49] S. G. W. Ginn, D. Johansen, and J. Overend, *J. Mol. Spectrosc.* **36**, 448 (1970).

selection rules

$$\Delta K = 0, \quad \Delta J = 0, \pm 1 \quad \text{if } K \neq 0$$
$$\Delta J = \pm 1, \quad \text{if } K = 0 \qquad \text{parallel bands} \qquad (2.2.33)$$

and

$$\Delta K = \pm 1, \quad \Delta J = 0, \pm 1 \qquad \text{perpendicular bands} \qquad (2.2.34)$$

are obtained.

However, when all the splittings introduced by higher-order terms in the Hamiltonian are considered, a more complete set of selection rules is obtained. Tarrago[41] has studied the problem in detail for axially symmetric molecules. The following is specialized to the case of C_{3v} symmetry.

In order to determine the complete set of selection rules, one has to define symmetry-adapted wave functions. These new functions are given by

$$| + \rangle \equiv | \{v_s\}, \{l_s\}, K, J \rangle + | \{v_s\}, \{-l_s\}, -K, J \rangle, \qquad (2.2.35)$$

$$| - \rangle \equiv | \{v_s\}, \{l_s\}, K, J \rangle - | \{v_s\}, \{-l_s\}, -K, J \rangle. \qquad (2.2.36)$$

The symmetry properties of these wave functions, under the full point group C_{3v}, are summarized in Table IX. The phase of the wave functions is an important consideration in arriving at Table IX.[50]

Making use of the symmetry-adapted wave functions, the symmetry selection rules given in Table X are obtained.[41]

TABLE IX. Symmetry Properties of the Wavefunctions

	$K - \sum_t l_t = 3p$		$K - \sum_t l_t \neq 3p$
	$\tau = J + K + \sum_t l_t + \sum_{n(A_2)} v_n$		
	τ even	τ odd	
$\lvert + \rangle$	A_1	A_2	
			E
$\lvert - \rangle$	A_2	A_1	

[50] W. E. Blass, *J. Mol. Spectrosc.* **31**, 196 (1969).

TABLE X. Symmetry Selection Rules for C_{3v} Molecules

		$\sum\limits_{t} \Delta l_t + \sum\limits_{n(A_2)} \Delta v_n$	
		Even	Odd
Parallel bands $\sum\limits_{t} \Delta l_t = 3p$	Q branches $\Delta J = 0$	$\|+\rangle \leftrightarrow \|-\rangle$	$\|+\rangle \leftrightarrow \|+\rangle$ $\|-\rangle \leftrightarrow \|-\rangle$
$\Delta K = 0$	R and P branches $\Delta J = \pm 1$	$\|+\rangle \leftrightarrow \|+\rangle$ $\|-\rangle \leftrightarrow \|-\rangle$	$\|+\rangle \leftrightarrow \|-\rangle$
Perpendicular bands $\sum\limits_{t} \Delta l_t - \Delta K = 3p$	Q branches $\Delta J = 0$	$\|+\rangle \leftrightarrow \|+\rangle$ $\|-\rangle \leftrightarrow \|-\rangle$	$\|+\rangle \leftrightarrow \|-\rangle$
$\Delta K = \pm 1$	R and P branches $\Delta J = \pm 1$	$\|+\rangle \leftrightarrow \|-\rangle$	$\|+\rangle \leftrightarrow \|+\rangle$ $\|-\rangle \leftrightarrow \|-\rangle$
	$A_1 \leftrightarrow A_2$;	$E \leftrightarrow E$	

2.2.2.3.4.5. *Intensities of the Transitions.*[2,4] For the allowed transitions, that is, for those transitions for which the transition moment integral does not vanish identically, it is possible to calculate this integral and thus determine a relative intensity of the transitions observed in absorption spectra. If $F(K, J, \Delta K, \Delta J)$ is the transition moment, then the intensity of a line is proportional to

$$\nu G F(K, J, \Delta K, \Delta J) \exp[-(E_0(K, J)/kT)], \qquad (2.2.37)$$

where G is a statistical factor which depends on the spin of the nuclei and the degeneracy of the ground state, k is the Boltzmann constant, T is the absolute temperature, $E_0(K, J)$ is the ground state energy, and ν is the frequency of the transitions.

The transition moments have been calculated[4] and are given by

$$F(K, J, \Delta K, \Delta J) = \begin{cases} (J+2+K\,\Delta K)(J+1+K\,\Delta K)/(J+1)(2J+1) \\ \qquad\qquad\qquad \text{if } \Delta J = +1, \quad (2.2.38) \\ (J+1+K\,\Delta K)(J-K\,\Delta K)/J(J+1) \\ \qquad\qquad\qquad \text{if } \Delta J = 0, \quad (2.2.39) \\ (J-1-K\,\Delta K)(J-K\,\Delta K)/2(2J+1) \\ \qquad\qquad\qquad \text{if } \Delta J = -1, \quad (2.2.40) \end{cases}$$

for the case $\Delta K = \pm 1$, while for $\Delta K = 0$

$$
F(K, J, 0, \Delta J) =
\begin{cases}
[(J+1)^2 - K^2]/(J+1)(2J+1) & \\
\qquad\qquad \text{if } \Delta J = +1, & (2.2.41) \\
K^2/J(J+1) & \text{if } \Delta J = 0, & (2.2.42) \\
(J^2 - K^2)/2(2J+1) & \text{if } \Delta J = -1. & (2.2.43)
\end{cases}
$$

For molecules belonging to the C_{3v} group having three identical nuclei, the factor G can be written as

$$
G = (2 - \delta_{K,0})(2J+1)\left(1 + \frac{3}{4I(I+1)}\delta_{K,3n}\right), \qquad (2.2.44)
$$

where $\delta_{i,j}$ is the Kronecker delta, n is an integer, I is the spin of the identical nuclei, and K is the ground state K value.

2.2.2.3.4.6. *Systematics of the Band Structure.*[4]　The formulas of the preceding sections allow one to determine the overall appearance of vibration–rotation bands. Before entering into the discussion, recall the usual convention for naming transitions. It is customary to associate the letters P, Q, and R with the values -1, 0, and $+1$ of ΔK and ΔJ. Furthermore, a transition is labeled as $^{\Delta K}\Delta J_K(J)$. Thus a label $^R Q_K(J)$ indicates a transition from the K, J level of the ground vibrational state to the $K+1$, J level of an excited vibrational state.

Consider a fixed value of K and observe the behavior of the transitions as ΔK, ΔJ, and J change. One must first recall that for a transition to be possible, both the lower and upper states must exist. This imposes the restriction that for both states J must be greater than or equal to K. Thus $^R R_K(J)$, $^Q R_K(J)$, and $^Q Q_K(J)$ are allowed only if $J \geq K$; $^R Q_K(J)$ and $^Q P_K(J)$ are allowed only if $J \geq K+1$; $^R P_K(J)$ is allowed only if $J \geq K+2$; and $^P R_K(J)$, $^P Q_K(J)$, and $^P P_K(J)$ are allowed only if $J \geq K$ and $K \geq 1$. It must be noted, however, as is readily seen from the expression for $F(K, J, \Delta K, \Delta J)$, that transitions of the type $^Q Q_0(J)$ have zero intensity.

Inspection of Eqs. (2.2.37)–(2.2.43) shows that, except for transitions of the type $^Q Q_K(J)$, for a fixed value of K the intensity is the product of a linearly increasing function of J and an exponentially decreasing function of J. This means that the intensity as a function of J increases initially, goes through a maximum, and then decreases. Furthermore, the maximum moves toward $J = K$ as K increases. Equations (2.2.37) and (2.2.42) show that for $^QQ_K(J)$ transitions the $K = J$ transition is always the strongest.

Let us now examine the problem of the position of the lines. Considering a fixed value of K it is easily seen that for a given value of J the separation of adjacent lines in the "K subband" is given by

$$\Delta \nu = 2[B_0 \, \Delta J - \Sigma a_s v_s (J + \Delta J + 1)] - 4D_0^{\ J}[(J + \Delta J + 1)^3 - (J + 1)^3]$$
$$-2D_0^{JK}[(K + \Delta K)^2(J + \Delta J + 1) - K^2(J + 1)]. \qquad (2.2.45)$$

One can see that, in first approximation, the separation between lines varies linearly with J. The resulting subband structure is shown in the second line of Fig. 12. In addition, Fig. 12 shows how the different K

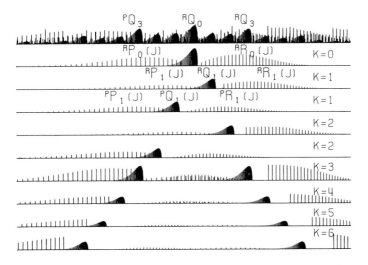

FIG. 12. Computer-generated perpendicular band spectrum of a prolate axially symmetric molecule (typical of $\nu_3 + \nu_4$ of CH_3F). The K subband structure is shown below the composite spectrum.

subbands combine to form a perpendicular band of a prolate molecule like CH_3F, $\nu_3 + \nu_4$.[46] The half-width of the Gaussian response function (instrument function convoluted with line shape) in the calculated spectrum of Fig. 12 is 0.02 cm^{-1}. Figure 13 shows the same results for a parallel band $\nu_1 + \nu_3$ of CH_3F. In order to see how the structure changes for oblate molecules, a perpendicular band,[51] ν_{11} of C_3H_6 is shown in Fig. 14. A special case wherein all the "Q branches" pile up in

[51] G. J. Cartwright and I. M. Mills, *J. Mol. Spectrosc.* **34**, 361 (1970).

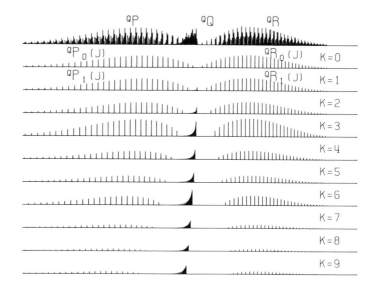

FIG. 13. Computer-generated parallel band spectrum of a prolate axially symmetric molecule (typical of $\nu_1 + \nu_3$ of CH_3F). The K subband structure is shown below the composite spectrum.

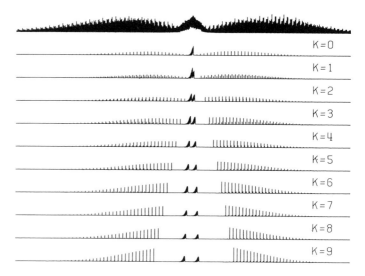

FIG. 14. Computer-generated perpendicular band spectrum of an oblate symmetric top molecule (typical of ν_{11} of C_3H_6). The K subband structure is shown below the composite spectrum.

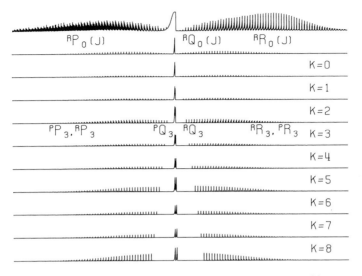

FIG. 15. Computer-generated perpendicular band spectrum of an oblate symmetric top molecule (typical of ν_4 of NF_3). The K subband structure is shown below the composite spectrum.

FIG. 16. Computer-generated perpendicular band spectrum of an oblate symmetric top molecule (typical of ν_3 of NF_3). The K subband structure is shown below the composite spectrum.

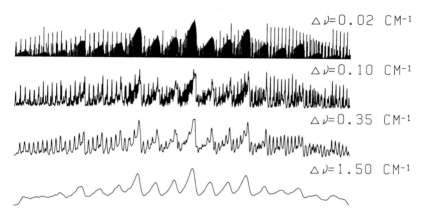

FIG. 17. Computer-generated spectra showing the effect of instrumental resolution $\Delta\nu$ on the spectrum shown in Fig. 12.

one place[52] is characteristic of numerous heavy oblate molecule perpendicular bands and this is represented by ν_4 of NF_3 in Fig. 15. Another type of "structure" for an oblate perpendicular band is represented in Fig. 16, ν_3 of NF_3. For all of these calculated spectra the effective resolution is 0.02 cm^{-1} in order that the structure of the bands might be readily seen.

Because most infrared spectra are obtained at somewhat lower resolution, Figs. 17–20 indicate the effect of degraded resolution on the spectra

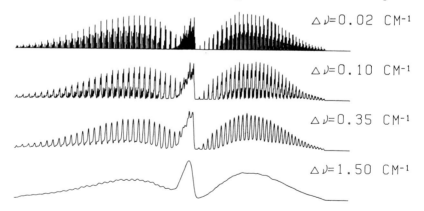

FIG. 18. Computer-generated spectra showing the effect of instrumental resolution $\Delta\nu$ on the spectrum shown in Fig. 13.

[52] R. J. L. Popplewell, F. N. Masri, and H. W. Thompson, *Spectrochim. Acta* **23A**, 2797 (1967).

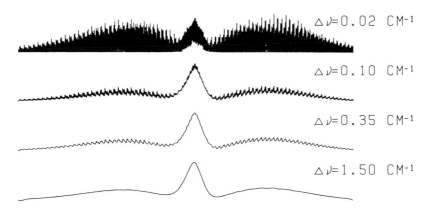

FIG. 19. Computer generated spectra showing the effect of instrumental resolution Δv on the spectrum shown in Fig. 14.

shown in Figs. 12–14 and 16. The figures are self-explanatory. A corresponding representative for Fig. 15 is not shown, since the result of degraded resolution is similar to that shown for Fig. 13.

2.2.2.3.4.7. Analysis of Vibration–Rotation Bands. In the previous sections, the results of the theory of vibration–rotation spectra of symmetric molecules have been reviewed. These results, however, are of small value if the spectroscopist is unable to proceed from the observed data to the determination of the different parameters of the theory.

The study of molecular vibration–rotation spectra can be conveniently separated into three distinct but highly interrelated phases. First, the spectrum has to be observed, the transition frequencies and intensities

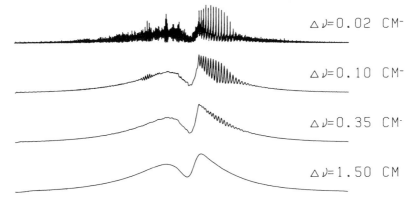

FIG. 20. Computer-generated spectra showing the effect of instrumental resolution Δv on the spectrum shown in Fig. 16.

measured, and all the information recorded to form the working data set. Next, the observed absorption lines have to be assigned to transitions between particular energy levels. Finally, the data have to be reduced and numerical values for the molecular parameters obtained. It is desirable, especially in the case of complicated spectra, to let high-speed digital computers perform many of the operations required for the study of the data. In order to make maximum use of the power of digital computers, a clear understanding of the problem and a systematic approach are mandatory. The systematic approach reveals that many of the functions usually performed by a researcher can be successfully programmed for a computer. However, not all the load can be removed from the researcher, since the study of molecular spectra involves a large number of decisions based on experience. These cannot readily be programmed.

The data acquisition phase determines the form of the data that is the working data set. A spectrometer is used to generate a visual display of the spectrum. From this display, the observer determines the position of the absorption lines relative to some arbitrary point in the display. This can be done either by visual inspection or with the aid of a computer procedure. Next, the position of several recorded absorption standards are used to determine the frequency of the observed lines.[53] Finally, the intensities are measured and all the information recorded, both in graphical and tabular form.

Once the experimentally observed data have been recorded, the assignment phase can begin. Initially, some of the outstanding features predicted by the theory are used to establish a set of tentative assignments. From this point on the computer can take much of the load from the researcher if a model representing the observed data can be constructed. A model is some mathematical representation of the observed data. Within this context, two kinds of model can be defined: (a) an ad hoc or prediction model, and (b) a physical model. While the latter type gives a representation that can be interpreted in terms of physical principles, the first type is a simple representation of the data that can be derived from the data itself.

Prediction models can be used to obtain assignments for all the absorption lines in a spectrum. Such a model can be generated by means of a least-squares procedure from an initial set of assignments. The

[53] K. N. Rao, C. J. Humphreys, and D. H. Rank, "Wavelength Standards in the Infrared." Academic Press, New York, 1966.

model can then be used to predict the position in the spectrum of un-assigned transitions. The observed data can then be scanned for lines in the predicted positions. This procedure can be done very efficiently by a computer. Human intervention is needed for a final evaluation of the predictions.

The final and most rewarding phase of the study of molecular spectra is the correlation of the observations with a physical model. This is the interpretation of the observations in terms of a molecular model derived from first principles. In the analysis phase a least-squares procedure is used to obtain numerical values for the parameters of the physical model from the assigned transitions. In many cases, anomalies are detected at this stage. Such anomalies are indicative of perturbations that cause deviations from the patterns predicted by the theory. When perturbations are observed a suitable scheme has to be developed to treat them. This probably constitutes the most interesting aspect of the study of molecular spectra.

The final analysis of the data can be performed by a computer.[42–46,54] It is even possible to devise procedures to detect some of the possible anomalies. Furthermore, in some cases corrective actions can be taken automatically so that the parameters obtained are free from the effects introduced by the perturbations. In fact, it has become apparent that digital computers can be used to perform rapidly the mechanical functions of the study of molecular spectra. This has the advantage of allowing the spectroscopist to proceed rapidly and without excessive effort to the more delicate and interesting aspects of the problem.

2.2.2.3.4.8. Anomalies in Vibration–Rotation Spectra. As has been shown, an expression may be developed from quantum-mechanical considerations for the vibration–rotation energies of molecules. Unfortunately, very few molecules have spectra that are consistent with these results in all details. This is now known to be due to resonance.[2,55] i.e., to interactions that become large because certain combinations of frequencies are equal—or nearly so—and cannot be treated by the usual quantum-mechanical methods. Such resonances may be of any order—first, second, third order, etc. They may, moreover, be divided roughly into two classes, vibrational resonances and rotational resonances.[56]

Most important of the vibrational resonances[2] is the Fermi–Dennison type of resonance which was discovered in the spectrum of CO_2, and

[54] R. M. Lees, *J. Mol. Spectrosc.* **33**, 124 (1970).
[55] G. Amat, *Cah. Phys.* **77**, 26 (1957).
[56] G. Amat and H. H. Nielsen, *J. Mol. Spectrosc.* **23**, 359 (1967).

which is a first-order resonance. It arises from a term in the potential energy of the form $k_{abb}q_a q_b{}^2$ and produces resonance if $2\omega_b \simeq \omega_a$. A more general first-order resonance of this type arises from a term $k_{abc}q_a q_b q_c$ and produces resonance if $\omega_a \simeq \omega_b + \omega_c$.

Of the second-order anharmonic resonances, the Darling–Dennison resonance is the most important. It arises from a term in the potential energy of the form $k_{aabb}q_a{}^2 q_b{}^2$ and similar terms, and when $2\omega_a \simeq 2\omega_b$, resonance occurs. The resonance does not affect the fundamentals ω_a and ω_b.

A third-order anharmonic resonance was discovered in the case of HCN by Douglas and Sharma. This resonance was first discussed theoretically by Hansen and Nielsen and arises from terms of the type $k_{aaabb}q_a{}^3 q_b{}^2$ in the potential energy.

The principal effect of anharmonic resonance is to displace the center of the bands in question from their predicted or unperturbed positions by amounts of the order of the magnitude of the term. They do, however, also have an effect upon the values of the rotational constants B_v and D that is significant, and cannot in general be neglected.[57,58]

In the case of CO_2, the totally symmetric Raman active (infrared-inactive) frequency ν_1 lies very close to the infrared-active bending frequency $2\nu_2$. Instead of finding a single Raman line with this frequency, two lines separated by about 100 cm^{-1} occur. These are explained on the basis that the Fermi–Dennison resonance is ν_1 displaced, and $2\nu_2$ is both enhanced and displaced by about 50 cm^{-1} on each side of the position where ν_1 would be found without resonance. This type of resonance has been observed in many molecules.

The most important rotational resonance is the Coriolis resonance,[2,50,57-63] which arises when two resonant vibrations interact with the rotational motion of the molecule, and couples the vibrational angular momentum operator to the principal angular momentum operator. The component of the Coriolis operator directed along the axis has the form

$$\zeta_{ab}^{(\alpha)}[(\omega_a/\omega_b)^{1/2}q_b p_a - (\omega_b/\omega_a)^{1/2}q_a p_b]P_\alpha/I_{\alpha\alpha},$$

[57] S. Reichman, P. C. Johnson, and J. Overend, *Spectrachim. Acta* **25A**, 245 (1969).
[58] Y. Morino, J. Nakamura, and S. Yamamoto, *J. Mol. Spectrosc.* **22**, 34 (1967).
[59] C. DiLauro and I. M. Mills, *J. Mol. Spectrosc.* **21**, 386 (1966).
[60] R. L. Dilling and P. M. Parker, *J. Mol. Spectrosc.* **22**, 178 (1967).
[61] R. L. Dilling and P. M. Parker, *J. Mol. Spectrosc.* **25**, 340 (1968).
[62] J. J. Garing, H. H. Nielsen, and K. N. Rao, *J. Mol. Spectrosc.* **3**, 496 (1959).
[63] J. M. Hoffman, H. H. Nielsen, and K. N. Rao, *Z. Electrochem.* **64**, 607 (1960).

where $\zeta_{ab}^{(\alpha)}$ is the Coriolis coupling factor; ω_a and ω_b and the two perturbing frequencies; q_a and p_a are, respectively, the normal coordinate identified with ω_a and the linear momentum conjugate to q_a; P_α is the component of the total angular momentum directed along the axis α; and $I_{\alpha\alpha}$ is one of the principal moments of inertia. When ω_a and ω_b are fairly well separated, the Coriolis operator contributes a term to B_v proportional to $(\omega_a - \omega_b)^{-1}$. When $\omega_a \simeq \omega_b$, so that resonance occurs, the contribution becomes a first-order effect and produces a strong convergence of the rotational lines in the two bands ω_a and ω_b. For example, Fig. 21 shows the result of a computer simulation of a strong

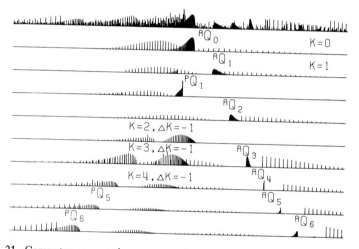

FIG. 21. Computer-generated spectrum showing the effect of a strong Coriolis resonance on the spectrum of Fig. 12. Comparison of the subband structure indicates the rather large effects which such accidental resonances may have on a particular spectrum.

Coriolis resonance between the upper states of the bands shown in Figs. 12 and 13. That is, Fig. 21 shows the band of Fig. 12 in strong Coriolis resonance with the band of Fig. 13 at a lower frequency. Figure 22 shows the same band under various resolution conditions.

In recent years, the use of band contour methods for the study of resonances in spectra of axially symmetric molecules has proven to be a valuable tool. Such an approach has proven extremely productive in the study of l-type doubling and l-resonance.[51,64]

[64] F. N. Masri and W. E. Blass, *J. Mol. Spectrosc.* **39**, 21 (1971); **39**, 98 (1971).

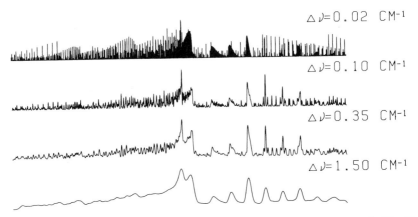

$\triangle\nu=0.02$ CM^{-1}

$\triangle\nu=0.10$ CM^{-1}

$\triangle\nu=0.35$ CM^{-1}

$\triangle\nu=1.50$ CM^{-1}

FIG. 22. The effect of instrumental resolution $\Delta\nu$ on the spectrum shown in Fig. 21 is illustrated. Comparison of the $\Delta\nu = 1.5$ cm^{-1} spectrum of this figure with that of Fig. 17 is particularly interesting.

2.2.3. Experimental Considerations[65-69]

2.2.3.1. Introduction. Infrared spectroscopy is concerned with the observation of transitions of physical systems in the energy range of 30 to 10,000 cm^{-1}, which is equivalent to approximately 0.004 to 1.25 eV. For example, 3000 cm^{-1} in the near infrared is equivalent to $\sim$0.37 eV. Thus any physical system that has transitions in this energy region falls within the range of infrared spectroscopy. In the following, particular emphasis is placed on moderate-to high-resolution spectroscopy characteristically required for the study of gas phase systems. On the other hand, the discussion is applicable to any system for which transitions in the infrared energy range are to be observed. Furthermore, for spectroscopy beyond $\sim$500 cm^{-1}, i.e., below $\sim$0.063 eV, special techniques are mandatory and fall in the realm of far infrared spectroscopy and will not be considered herein.

The physical system to be studied directs the researcher to a particular approach to the study of its infrared spectrum. If the transitions of a

[65] J. F. James and R. S. Sternberg, "The Design of Optical Spectrometers." Chapman and Hall, London, 1969.

[66] S. P. Davis, "Diffraction Grating Spectrographs." Holt, New York, 1970.

[67] G. K. T. Conn and D. G. Avery, "Infrared Methods." Academic Press, New York, 1960.

[68] A. Hadni, "Essentials of Modern Physics Applied to the Study of the Infrared." Pergamon, Oxford, 1967.

[69] James E. Stewart, "Infrared Spectroscopy," Dekker, New York 1970.

system in a limited spectral region are of interest, then instruments that measure infrared spectra over a limited spectral region will often provide the best results in the most economical fashion. In such cases laser spectroscopy or the use of a Fabry–Perot interferometer should be investigated. On the other hand, for observation of transitions of a physical system over a moderate to large spectral region (greater than several reciprocal centimeters) either a dispersive spectrometer employing a prism or grating or a Michelson interferometric spectrometer (currently called a "Fourier transform spectrometer") is generally the most satisfactory instrument. For all but the lowest-resolution requirements, the prism spectrometer is of little interest. However, in the event that a prism spectrometer provides satisfactory resolution, it is often the simplest and most economical instrument to use.[65,70,71] Excellent commercial prism spectrometers are widely available to the extent that it is difficult to justify the fabrication of another "prototype" prism spectrometer in any but the most unusual circumstances.

Such reasoning leaves the grating spectrometer and the Fourier transform spectrometer as the subjects of the following discussion. In order to understand the principles of operation of an infrared spectrometer it is necessary to consider the observable, that is an absorption or emission transition, in some detail. Figure 23a illustrates an idealized pair of

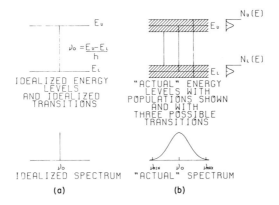

FIG. 23. Molecular energy levels, both (a) idealized and (b) typical are illustrated showing typical allowed transitions and the resulting spectrum.

[70] R. Kingslake, in "Applied Optics and Optical Engineering" (R. Kingslake, ed.), Vol. V, p. 1. Academic Press, New York, 1969.

[71] R. Metzler, in "Applied Optics and Optical Engineering" (R. Kingslake, ed.), Vol. V, p. 47. Academic Press, New York, 1969.

energy levels and an idealized absorption transition and its resulting spectrum. However, since energy levels are in fact not of zero width but have a finite width that depends upon the lifetime of the state, collisions, and other factors, an actual pair of levels shown in Fig. 23b may result in numerous transitions such as the three illustrated. Because, in infrared spectroscopy one simultaneously observes equivalent transitions of a large number of elements of the sample, the various observed transitions between a "single" pair of energy levels at any particular time gives use to the spectral line profile shown in Fig. 23b. In Fig. 23, E refers to energy of the state and N to the number of molecules in a sample as a function of energy. The subscripts U and L refer to upper and lower states, respectively.

2.2.3.2. General Considerations

2.2.3.2.1. GRATING MONOCHROMATOR.[72-81] The attempt to produce an idealized absorption spectrometer, i.e., an idealized source of tunable, monochromatic radiation (or an idealized monochromatic radiation measuring device for emission spectra—the same device actually) has been approached by the plane grating monochromator. If white radiation illuminates slit S_1 in Fig. 24, then M_1 collimates the radiation so that a superposition of plane waves falls upon the grating. The grating selects radiation of wavelength $\lambda = d(\sin \theta_i - \sin \theta_d)$ to be imaged at slit S_2 via M_2. Fundamentally the grating acts as a filter for the optical signal from S_1 to S_2. The filter is not perfect and so radiation of adjacent frequencies near the center frequency is also passed from S_1 to S_2. Therefore, nearby frequencies as well as ν_0 fall in the exit slit with the grating set to pass ν_0. Similarly, this also means that if the grating is

[72] F. Kneubühl, *Appl. Opt.* **8**, 505 (1969).

[73] D. H. Rank, G. D. Saksena, G. Skorinko, D. P. Eastman, T. A. Wiggins, and T. K. McCubbin, Jr., *J. Opt. Soc. Amer.* **49**, 1217 (1959).

[74] T. K. McCubbin, Jr., R. P. Grosso, and J. D. Mangus, *Appl. Opt.* **1**, 431 (1962).

[75] R. H. Hunt, C. W. Robertson, and E. K. Plyler, *Appl. Opt.* **6**, 1295 (1967).

[76] A. R. H. Cole, A. A. Green, G. A. Osborn, G. D. Reese, *Appl. Opt.* **9**, 23 (1970).

[77] R. A. Hill, *Appl. Opt.* **8**, 575 (1969).

[78] D. F. Eggers and M. A. Peterson, *Appl. Opt.* **8**, 589 (1969).

[79] G. W. Stroke, *in* "Progress in Optics" (E. Wolf, ed.), Vol. II, p. 3. North Holland Publ., Amsterdam, 1963.

[80] W. T. Welford, *in* "Progress in Optics" (E. Wolf, ed.), Vol. IV, p. 241. North Holland Publ., Amsterdam, 1965.

[81] D. Richardson, *in* "Applied Optics and Optical Engineering" (R. Kingslake, ed.), Vol. V, p. 17. Academic Press, New York, 1969.

set at ν_0 and illuminated by a zero-width monochromatic signal of frequency ν_0, not all of the ν_0 radiation passes through the exit slit. It is possible to represent the effect of the monochromator on the input signal as a convolution of the input signal with an instrument function $g(\nu' - \nu)$. That is to say, for a white radiation input $I(\nu)\,d\nu$ at S_1 of Fig. 24, the radiation at the exit slit for the monochromator set at ν' is given by

$$\int I(\nu)g(\nu' - \nu)\,d\nu \qquad (2.2.46)$$

when the monochromator is set to pass ν' from S_1 to S_2.

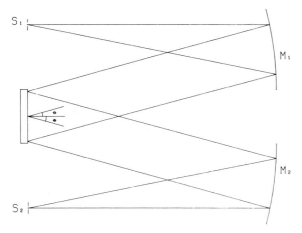

FIG. 24. Typical grating monochromator (θ_1 is the angle of incidence, θ_d is the angle of diffraction).

It is easier to consider at this point the effect of a spectrometer on an emission signal. Note, however, that the results are in fact true for an absorption spectrum although this may not be readily apparent at first. Figure 25a represents the effect of an idealized spectrometer on an idealized monochromatic emission line. The function $\delta(\nu' - \nu)$ is called the *instrument function* or *instrument profile* of the spectrometer. [In electrical signal theory $\delta(\nu' - \nu)$ is called the *impulse response* of a filter.] Notice that the spectrometer output is the convolution of the input spectral intensity $I\delta(\nu)$ and the instrument function $\delta(\nu - \nu')$. Figure 25b represents the effect of a idealized spectrometer on an realizable spectral line $F(\nu)$. Finally, Fig. 25c represents the actual type of result obtained for a spectral line $F(\nu)$ and a realizable spectrometer with instrument function $g(\nu - \nu')$. For example, a high-resolution instrument

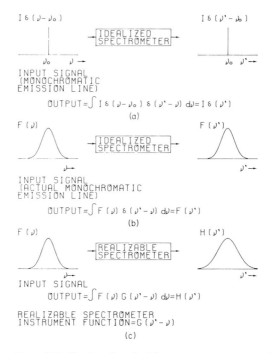

FIG. 25. The effect of idealized and realizable spectrometers on various input spectral lines.

is sometimes *assumed* to have an instrument function of Gaussian form, i.e.,

$$g(\nu - \nu') = k \exp[(\nu - \nu')^2/(2\sigma)^2]. \qquad (2.2.47)$$

Thus a pure monochromatic line $I\delta(\nu' - \nu_0)$ would be output as

$$Ik \exp[(\nu - \nu_0)^2/(2\sigma)^2] = \int_0^\infty I\delta(\nu' - \nu_0) \exp[(\nu - \nu')^2/(2\sigma)^2] \, d\nu'. \qquad (2.2.48)$$

One result of approaching the generalized instrument in this fashion is the following: If the effect of the spectrometer is to convolute the data signal with an instrument function that has an effect comparable to a filter in electrical signal theory, then it is possible, in theory at least, to construct an inverse filter to remove the effect of the instrument from the output data. In fact, the inverse filtering might well be done digitally using a computer.[65,82] Although inverse filtering is possible in principle

[82] E. A. Robinson, "Statistical Communication and Detection." Hafner, New York, 1967.

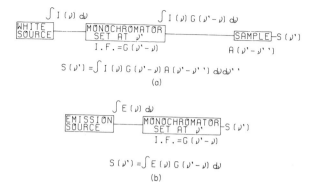

FIG. 26. Generalized (a) absorption and (b) emission spectrometers of the dispersive type. Characteristic signals are represented at several stages.

and directs the researcher's attention to new pathways of instrumental improvement, the actual problem is by no means simple.

2.2.3.2.2. GENERALIZED DISPERSIVE SPECTROMETER.[65] A generalized dispersive absorption spectrometer may be represented as in Fig. 26a, while a dispersive emission spectrometer is represented in Fig. 26b. In order to visualize the effect of a monochromator, consider Fig. 26a. The output of the monochromator looks very much like $g(\nu' - \nu)$ especially if the input white source intensity is essentially constant over the range where $g(\nu' - \nu)$ is significantly larger than zero. Thus to the extent that $g(\nu' - \nu)$ approaches a delta function, the attempt to realize an ideal zero width tunable radiation source is accomplished. It should be emphasized that the differences between an emission and absorption spectrometer are primarily formal.

2.2.3.2.3. GENERALIZED INTERFEROMETRIC SPECTROMETER.[72,83-91] Figure 27a represents an interferometric spectrometer including the Fourier

[83] D. M. Hunten, *Science* **162**, 313 (1968).

[84] G. Horlick, *Appl. Spectrosc.* **22**, 617 (1968).

[85] J. Connes, *Rev. Opt.* **40**, 45, 116, 171, 231 (1961).

[86] J. Connes, H. Delouis, P. Connes, G. Guelachuili, J. Maillard, and G. Michel, *Nouv. Rev. Opt. Appl.* **1**, 1 (1970).

[87] J. Pinard, *Ann. Phys.* **4**, 147 (1969).

[88] P. Jacquinot, *Appl. Opt.* **8**, 497 (1969).

[89] H. A. Gebbie, *Appl. Opt.* **8**, 501 (1969).

[90] P. Jacquinot and B. Roizen-Dossier, *in* "Progress in Optics" (E. Wolf, ed.), Vol. III, p. 29. North Holland Publ., Amsterdam, 1964.

[91] G. A. Vanasse and H. Sakai, *in* "Progress in Optics" (E. Wolf, ed.), Vol. VI, p. 261. North Holland Publ., Amsterdam, 1967.

transform step resulting in a spectrum $E'(\nu')$. The actual signal at the detector is an interferogram (as a function of path difference). Following the Fourier transform a spectrum similar to those shown in Figs. 17–20 is obtained. One possible "effective" instrument function is $\sin \nu/\nu$. It is sufficient at present to state that numerous possible instrument functions exist and are actually a function of controllable factors in the interferometric spectrometer. For comparison purposes the equivalent representation for a dispersive spectrometer is presented in Fig. 27b. Note that the observed spectrum is the actual signal seen by the detector in the dispersive spectrometer.

EMISSION SOURCE — INTERFEROMETRIC SPECTROMETER — INTERFERO-GRAM — FOURIER TRANSFORM — SPECTRUM

EFFECT OF INTERFEROMETRIC SPECTROMETER
INCLUDING FOURIER TRANSFORM

$$E'(\nu') = \int E(\nu) [\sin(\nu'-\nu)] \, d\nu$$
(a)

EMISSION — DISPERSIVE SPECTROMETER — SPECTRUM

$G(\nu'-\nu)$

$$E'(\nu') = \int E(\nu) G(\nu'-\nu) \, d\nu$$
(b)

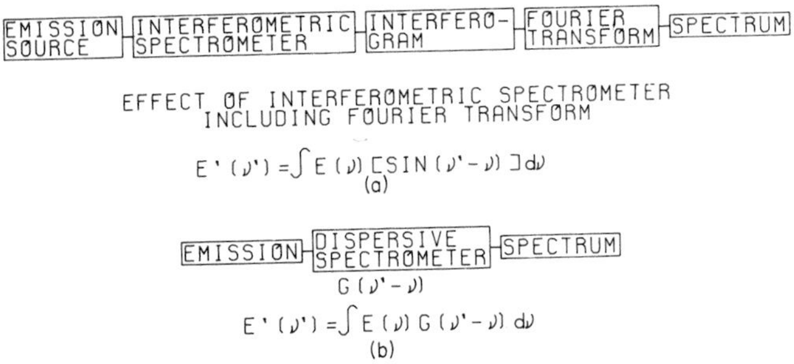

FIG. 27. Generalized (a) interferometric spectrometer and for comparison (b) a dispersive spectrometer.

2.2.3.2.4. RESOLUTION. The theoretical limits of resolution for gratings, prisms, and interferometers are illustrated in Fig. 28. There are other representations of the resolution limit of gratings, for example, but that which is presented is the most straightforward to use. Note that the resolution limit for an interferometer assumes an "instrument function" of $\sin \nu/\nu$. In practice these limits can be approached only with extreme difficulty and attention to numerous instrumental details that can degrade resolution.

2.2.3.2.5. THE SPECTROMETRIC SYSTEM. In order to approach spectrometer design or use on a sound basis it is necessary to consider the complete spectrometric system from source to the final output spectrum. The requirements which a one-shot, single-function experiment places on a spectrometric system are vastly different from those requirements placed on a system by a chemical laboratory running dozens of quality control spectra per day. Both sets of requirements are again different from those placed on a system by a state-of-the-art ultrahigh-resolution

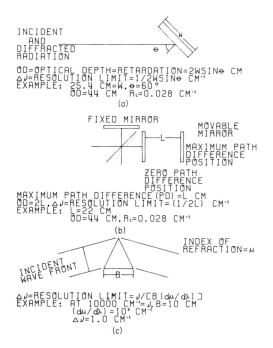

INCIDENT
   AND
DIFFRACTED
RADIATION

OD=OPTICAL DEPTH=RETARDATION=2WSINθ CM
Δν=RESOLUTION LIMIT=1/2WSINθ CM⁻¹
EXAMPLE: 25.4 CM=W,θ=60°
         OD=44 CM Rₗ=0.028 CM⁻¹
(a)

FIXED MIRROR
                            MOVABLE
                            MIRROR
                            MAXIMUM PATH
                            DIFFERENCE
                            POSITION
                   ZERO PATH
                   DIFFERENCE
                   POSITION
MAXIMUM PATH DIFFERENCE (PD) =L CM
OD=2L,Δν=RESOLUTION LIMIT=(1/2L) CM⁻¹
EXAMPLE: L=22 CM
         OD=44 CM,Rₗ=0.028 CM⁻¹
(b)

                         INDEX OF
                         REFRACTION=μ
INCIDENT
WAVE FRONT
              B
Δν=RESOLUTION LIMIT=ν/[B(dμ/dλ)]
EXAMPLE: AT 10000 CM⁻¹=ν,B=10 CM
         (dμ/dλ)=10³ CM⁻¹
         Δν=1.0 CM⁻¹
(c)

FIG. 28. Resolution determination of (a) gratings, (b) interferometers, and (c) prisms.

spectrometer which may require many man years to develop. Upon completion such a system should be able to be used at maximum spectral throughput to recover the development-time investment of the researchers involved.

Figure 29 represents a number of possible system schemes in block form. The system following branches **a** and **c** is the familiar dispersive spectrometer available commercially. For simple, one-time experiments, however, this system is readily assembled and yields usable data in the shortest possible time period. Systems, commerical and otherwise, that take this form often use a separate data reduction system in order to obtain data suitable for computer input to analysis programs. Such systems generally have the lowest hardware investment (relative to resolution, etc.) but require considerable time input on the researcher's part especially in the data reduction step. The comparable interferometric system, paths **b** and **c**, is not very useful due to the difficulty of accurately encoding the interferogram for processing.

It is particularly interesting to note that an interferometric spectrometer following paths **b**, **d**, **g**, and **i** with resolution ∼0.35 cm⁻¹ is commercially

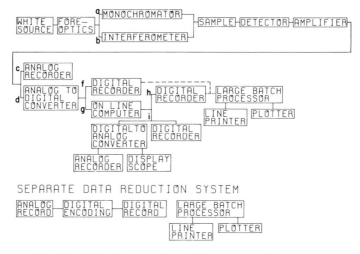

FIG. 29. Block diagram of possible spectrometric systems.

available. In addition, several state-of-the-art ultrahigh-resolution spectrometers following paths **a, d, g, i** exist in various stages of development. It is apparent that this path places a large initial programming burden on the developers of such a system. However, applicable "packaged" data acquisition systems are available from several commercial sources, which can significantly reduce the program development problems.

2.2.3.3. Grating Spectrometers.[72] At present there are two significantly different approaches to high-resolution spectroscopy using plane gratings. On the one hand consider a spectrometer with a fine-ruled grating of 300 to 600 grooves/mm, 25 cm wide by 12.5 cm tall, blazed at, say, 26° used in the first few orders with a collimator focal length of 1 to 10 m. (For illustrative purposes, a focal length of 5 m is assumed.) On the other hand consider a system using a like-dimensioned coarse echelle-ruled grating of 30 to 79 grooves/mm blazed at ∼63° used in high orders[53] (say, 5–15, and with a 5-m focal length for illustrative purposes). At first glance these would seem to be very similar approaches to the task of recording infrared spectra at high resolution. However, there are numerous differences, consideration of which should enable the reader to grasp some of the important but seldom mentioned ideas and problems involved in infrared grating spectroscopy. The fine-ruled grating would normally be used from 20 to 50° and the coarse grating from 50 to 70°. The fine grating presents only a minimal problem of overlapping orders. For example at 30°, 3000, 6000, and 9000 cm^{-1}

radiation is present at the exit slit in orders 1–3. Simple filters or detector response limits may easily be used to eliminate detection of undesirable radiation. The case of the coarse grating is not as simple. When 3000 cm⁻¹ in order 6 is present at the exit slit, radiation at $500n$ cm⁻¹, $n = 1, 2, 3, \ldots$ is also present. Filters are generally not used to eliminate this problem of overlapping orders. Rather, a prism monochromator is often used to limit radiation incident on the detector to that due to a single grating order. This "tandem prism spectrometer" is variously placed before the monochromator or following the monochromator with advantages to both positions. Use of continuously variable filter wheels is an attractive possibility for such order sorting.

Thus one major difference between fine and coarse grating systems is the method and cost of the solution of the overlapping order problem.

The theoretical resolution limit is another difference for two 25-cm-wide gratings—one used near 30°, the other near 63°. The theoretical resolution limits are 0.04 and 0.022 cm⁻¹, respectively. Thus the coarse echelle used at high angles of incidence exhibits superior resolution compared with the fine grating used at low angles. On the other hand, for a 5-m focal length system the fine grating at blaze is an $f/23$ system whereas the coarse echelle at blaze is an $f/45$ system. Therefore, one seems to pay for the increase in resolution of the coarse grating system by a decrease in light gathering power. As a result, given that one desires to use the largest grating available, the astronomer interested in weak sources would likely choose the 30° blaze whereas the absorption spectroscopist concerned with highest resolution would choose the 63° echelle.

This latter consideration of the aperture of the monochromator places some immediate limitations on monochromator designs for the high blaze-angle system. Typical monochromators of the Pfund type using pierced mirrors are completely impractical for high blaze-angle systems. The aperture of the high angle system is extremely small and the pierced mirror design eliminates a large portion of the center of the optical system.

With the above considerations in mind, consider three basic optical designs: the off-axis paraboloidal collimator, the Pfund, and the Ebert.

2.2.3.3.1. THE OFF-AXIS PARABOLOID COLLIMATOR.[65,77,78] Figure 30 shows a typical optical layout for a grating spectrometer in which an off-axis paraboloid is used as a collimator. This is a convenient design requiring few reflections and gives excellent resolving power depending upon the excellence of figure of the principal mirror. Generally, the focal length of the collimator is between 1 to 10 m. The grating section is preceded by a prism monochromator which serves as a filter to select a narrow spectral range which passes into the grating section. This is important because of the confusion which results from the overlapping of orders produced by the grating. Provided that suitably coarse gratings and detectors with suitable windows are available, it is possible to work to long wavelengths with such an instrument. The prism monochromator drive may be locked to the grating drive so that the energy is always maximum for a given grating position. For the near infrared and a fine-ruled grating it is convenient to replace the fore-prism system by narrow-passband filters. This reduces the number of reflection surfaces and the losses incurred by their use, and considerably simplifies the optical and mechanical details.

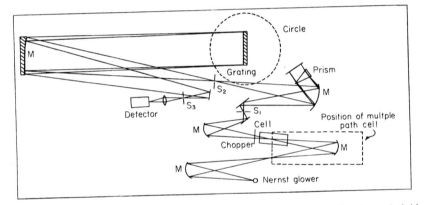

FIG. 30. Optical layout for a prism–grating spectrometer using an off-axis paraboloid as a collimator.

The grating drive must be very accurately made and smoothly operating so that the full resolving power of the large gratings may be exploited. Many designs[53] for bearing and shaft assemblies have been made, but the one most often used is a cone with about an 11° taper and a reasonably large bearing surface. The cone may be driven either by a large spur and worm gear, or by a long lever arm pulled by a metal tape attached to a split nut and screw mechanism, or if a linear wavelength drive is preferred by a sine bar arrangement. In any case, it is usual and advantageous to have many possible speeds so that the grating may first of all be rotated rapidly from central image position to the angle corresponding to the wavelength of the spectrum, and secondly may be rotated slowly for scanning the spectrum. It is desirable that one of the speeds be very slow to take full advantage of available resolving power. When an amplifier with high gain, tuned with a narrow passband for low noise is used, the drive must be slow enough to permit the detector–amplifier–recorder system to respond to the changes in energy on the detector.

A method has been described[53] for extremely smooth scanning of a very small spectral region in which a narrow wedge prism inserted in the beam just in front of the exit slit is translated along the direction of the beam. This scheme together with a Fabry–Perot interferometer makes it possible to take full advantage of the resolving power, and to obtain very precise frequency determinations of observed lines.

The problems attendant upon atmospheric absorption are more disturbing in grating spectrometry than in prism spectrometry. Whereas, in the latter case, many of the bands are not well resolved, or resolved at all, the interference is essentially related to lack of energy, or uncertain backgrounds. In a grating spectrometer, the water vapor or carbon dioxide bands are resolved into sharp rotation lines, many of which are totally absorbing for the path lengths normally encountered in a spectrometer. There is, therefore, a rapidly varying background superposed on the spectrum being investigated which must be subtracted out. The only completely satisfactory method of eliminating atmospheric interference is to place the spectrometer in an evacuable chamber. However, a simple and sometimes useful scheme is to enclose the spectrometer in an airtight box equipped with activated alumina drying towers. The air within the spectrometer is then continuously circulated and dried. A large fraction of the water vapor and carbon dioxide may thus be removed making it possible to work over the whole useful range of the instrument. A disadvantage in the air-drying approach is that the frequency of spectral lines must then be corrected for the index of refraction of air in the infrared.[53]

2.2.3.3.2. THE PFUND ARRANGEMENT.[65] Figure 31 shows a typical optical layout of the mirror system in an infrared spectrometer—originally proposed by Pfund and Hardy—in which paraboloidal mirrors may be used on axis in order to obtain excellent images throughout. It is simpler to produce large excellent on-axis paraboloids than off-axis ones and is consequently somewhat less expensive. As may be seen from Fig. 31, the entrance and exit slits lie on the axes of the two paraboloids. The first provides a parallel beam reflected to the grating by the first flat mirror, and the diffracted beam is reflected to the focusing paraboloid

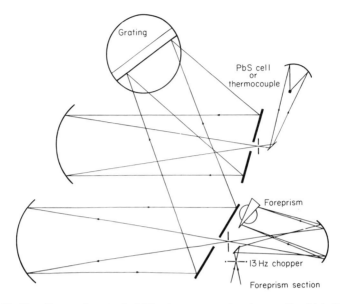

Grating

PbS cell
or
thermocouple

Foreprism

13 Hz chopper

Foreprism section

FIG. 31. Ray diagram for a typical Pfund arrangement; prism-grating high dispersion infrared spectrometer (courtesy of the Department of Physics, The Ohio State University).

by the second flat mirror. Thus, as the grating turns, the spectrum produced is swept across the exit slit and ultimately received by the detector.

2.2.3.3.3. THE EBERT ARRANGEMENT. In 1952 Fastie[92] independently rediscovered a principle first stated by Ebert[93] in 1889 and apparently not used since. This optical arrangement is shown in Fig. 32. In this system one large spherical mirror is used as both collimator and telescope with the slits situated on opposite sides of the grating. The principal advantage of this system is that it corrects for image aberrations in spherical systems except for astigmatism, and this defect, too, can be corrected by curving the slits, thus producing a superior optical system at lower cost. Another advantage is that by using longer curved slits more light may be passed through the instrument than for the previous two designs, thus permitting narrower slits. This arrangement can also easily be enclosed in a vacuum chamber for the elimination of atmospheric absorption.

The Ebert arrangement has gained much favor among spectroscopists and several instruments have been built with excellent results.

[92] W. G. Fastie, *J. Opt. Soc. Amer.* **42**, 641, 647 (1952).
[93] H. Ebert, *Ann.* [4] **38**, 489 (1889).

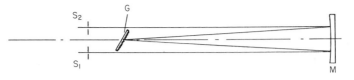

FIG. 32. Ray diagram for the Ebert arrangement.

2.2.3.3.4. MISCELLANEOUS ITEMS.

2.2.3.3.4.1. *Double Passing in Grating Spectrometers.* The scheme of double passing may be applied to advantage in grating spectrometers. The mirror arrangement necessary to double pass varies with the design of the instrument, and illustrations will not be given here. Gain in resolving power may be achieved so long as sufficiently large detector signal-to-noise ratio can be maintained, and so long as the optical components are sufficiently good.

2.2.3.3.4.2. *Diffraction Gratings.* High-quality diffraction gratings were first ruled by Rowland at the Johns Hopkins University between 1882 and 1901. In 1910 the echelette principle was discovered by R. W. Wood in which, by properly shaping the grooves, a large fraction of the energy could be concentrated at a given angle thus making the grating much more efficient at specifically chosen wavelengths. Since then a number of grating ruling engines have been built, and many gratings of varying degrees of success were made, and used by spectroscopists. In the decade following World War II, the technique of producing excellent replica gratings of large dimensions has been developed to a high degree. The master gratings are ruled on optical flats heavily coated with aluminum by vacuum evaporation, and for a grating with 600 grooves/mm about 2.5 cm of ruling can be accomplished in a 24-hour day. To produce the right groove shape the diamond used in ruling the groove must be carefully ground, and painstakingly adjusted. The replicas are made by transferring in a thin plastic mounted on another optical flat the rulings of the master grating and covering this impression with a vacuum evaporated coating of aluminum. Such replicas are reported to be at least as good as the original, or master, because the ridges of the master ruling, which are apt to have a wiry edge, become the valleys of the replica. Sizes up to about 20×40 cm are available.

2.2.3.3.4.3. *Resolving Power of Grating Spectrometers.* As has been previously stated, if the requirement of the investigation demands large resolving power, it is necessary to resort to gratings. A considerable number of individually constructed grating spectrometers have sufficiently good optical components and detecting systems to provide resolution of spectral lines separated by from 0.1 to 0.2 cm^{-1} throughout the regions to about 20 μm. A few are capable of resolving an order of magnitude better than this, but it is a fairly difficult task, as it requires the ultimate in excellence of the optics, the scanning system, and the detector–amplifier–recorder system. Improvements over this have been made by Jaffe and associates[94] by combining a grating spectrometer with a Fabry–Perot interferometer. In so doing the free spectral range is extremely limited, and the system is only used to scan narrow regions of a few tenths of a reciprocal centimeter. The spectrometers previously described may be used over wide regions.

2.2.3.4. Fourier Transform Spectrometer

2.2.3.4.1. INTRODUCTION.[65]   In recent years, the development of high-speed digital computers has facilitated the practical development of Fourier transform spectroscopy. The Fourier transform spectrometer is a multiplex spectrometer, i.e., each element of the spectrum is observed simultaneously and the resulting spectrometer output—an interfero-gram—must be decoded to obtain the frequency spectrum that is sought. In a spectral region such as the infrared where the observing device is detector-noise limited (as opposed to source-noise limited, for example) the signal-to-noise advantage gained by frequency multiplexing as in the Fourier transform spectrometer is equal to $N^{1/2}$, where there are N elements in the spectrum. (If the resolution of a spectrometer is Δv and the spectrum to be recorded covers a region $v_2 - v_1$, then $N = (v_2 - v_1)/\Delta v$.) This means that a Fourier transform spectrometer may enjoy a S/N advantage of $N^{1/2}$ over a grating spectrometer if both are using identical recorders and electronics, identical input radiant flux, and identical instrument throughput efficiency. In practical terms, this advantage can mean faster acquisition of data for the interferometer or recovery of a less intense spectrum in the same elapsed time. In addition, the Fourier transform spectrometer enjoys throughput advantage since its entrance "slit" is actually an aperture of modest size compared to the entrance slit of a grating spectrometer.

2.2.3.4.2. THE BASIC SPECTROMETER.   The basic Fourier transform spectrometer is built around a Michelson interferometer with one movable

[94] J. H. Jaffe, D. H. Rank, and T. A. Wiggins, *J. Opt. Soc. Amer.* **45**, 636 (1955).

mirror and one fixed mirror. Let l_1 be the path length to the fixed mirror from the beam splitter and l_2 the path length to the movable mirror (l_2 is thus variable). Then, after passing through the interferometer, the detector sees a resultant amplitude of

$$(A/2^{1/2}) \exp i(\theta_1 + \theta_2)\{\exp[2\pi i \nu(2l_1)] + \exp[2\pi i \nu(2l_2)]\}$$

for a resultant intensity of

$$I(d) = (A^2/2)[1 + \cos(2\pi\nu d)],$$

where the path difference $d = l_2 - l_1$. Thus the transmitted intensity is a cosine function of the path difference d. If the input radiation is given by an intensity function $S(\nu)$, then the detector output for a given path difference d is

$$I(d) = \tfrac{1}{2}\int_0^\infty S(\nu)\,d\nu + \tfrac{1}{2}\int_0^\infty S(\nu)\cos(2\pi\nu d)\,d\nu$$

and the quantity $J(d) = 2I(d) - I(0)$ is the interferogram containing the information defining the spectrum $S(\nu)$.

Recovery of the spectrum $S(\nu)$ from the interferogram $J(d)$ involves Fourier transformation of the interferogram. Although this is not a trivial matter in practice, much work has been done especially by J. Connes.[85]

In addition, the hardware demands placed on the researcher are stringent. However, much has been accomplished in this area: the work of P. Connes[86] is of particular note. P. Connes has built an interferometric spectrometer with a maximum path difference of 1 m. Thus the theoretical maximum resolution is 0.005 cm^{-1} over its near infrared operating range. The largest gratings available for use in dispersive spectrometers (40 cm wide blazed at 63°) have theoretical resolution limits of 0.0075 cm^{-1} in double pass configuration. However, as Gebbie has said, each time an interferometric spectrometer is used the mirror must be moved with extreme precision whereas the equivalent high precision is required only once for a grating, i.e., when it is ruled.

There is little doubt, however, that insofar as infrared spectroscopy remains detector-noise limited and the multiplex advantage therefore holds, interferometric spectrometers will continue to gain favor as a standard spectroscopic tool. The exception to this statement is found in the study of line shapes of relatively isolated spectral lines where the multiplex advantage is of lesser consequence.

2.2.3.5. Sources of Infrared Radiation[67,69]

2.2.3.5.1. NERNST FILAMENT. A convenient source of continuous radiation for use throughout the entire infrared spectrum is the Nernst filament, which approaches the radiation characteristics of a blackbody. The Nernst filament is a mixture of several rare earth oxides. When used in vacuum some filaments give off a fine white powder which collects on the mirrors producing bad scattering effects. For normal operation the Nernst filament has its radiation peak between 1.5 and 2.0 μm.

2.2.3.5.2. GLOBAR. Another convenient source for general use is the Globar. The Globar is a cylindrical rod of carborundum about 6 mm in diameter and 5 cm long capped with silver electrodes. It operates at about 1200°C for a power of 250 W, and must be cooled by means of a water jacket. Its long life makes it a very useful source, although it is not as good an emitter as the Nernst filament.

2.2.3.5.3. GAS MANTLE. Although seldom used now, the gas mantle composed of thorium oxide, and heated to about 1800°C by means of a gas flame, is an extremely efficient source, especially beyond 10 μm.

2.2.3.5.4. QUARTZ MERCURY ARC. For extremely long wavelengths (40–800 μm), the quartz mercury arc is the best available source.

2.2.3.5.5. TUNGSTEN FILAMENT. In the very near infrared region a substantial gain can be achieved over the aforementioned sources by the use of a tungsten ribbon–filament lamp operated at 20 to 30 A in a glass envelope with a quartz window. In addition, quartz–iodine lamps with a tungsten filament are commercially available and exhibit excellent long life characteristics.

2.2.3.5.6. ZIRCONIUM ARC. Another source of considerable brilliance and usefulness is the concentrated zirconium point arc. While the other sources mentioned are elongated and can be used to illuminate a slit, this source approximates a point source. It has been used as a general type source for the near infrared, and for the source in formation of Fabry–Perot interferometer fringes for calibration purposes.

2.2.3.5.7. CARBON ARC.[94] Strong[95] and his collaborators have reinvestigated the use of the positive crater of the carbon arc operated at about 3900°C. They have designed a special source with which the

[95] C. S. Rupert and J. Strong, *J. Opt. Soc. Amer.* **39**, 1061 (1949).

energy output is appreciably improved in the 3–15 μm region. In fact, a comparison of this source with a Globar operated at 1175 K showed a gain of about sixfold in this region which increased the resolving power by a factor of three through narrowing of the spectrometer slits. The principal disadvantages of this source are its bulky shape, necessity for automatically feeding the carbon rods, relatively short life, and the fact that it cannot be used in a vacuum.

2.2.3.5.8. CARBON ROD FURNACE.[96] An excellent source of infrared radiation is the carbon rod furnace in which a carbon rod is resistively heated to approximately 3000 K. For general high-resolution absorption spectroscopy this source is in almost universal use.

2.2.3.6. Optical Materials for the Infrared.[67-69,97-102] Among the materials that transmit infrared radiation, and that can be used as dispersive media, the most frequently used is sodium chloride. It is the commonest of the halogen salts of the alkali metals, and has satisfactory dispersive properties for most problems in infrared spectroscopy met with in the ordinary course of events. In general it may be said that all the alkali halides transmit infrared radiation and may be fashioned into prisms suitable for various spectral ranges. These materials, sometimes found in nature as medium-sized crystals of optical quality, are now made synthetically into crystals of almost any desired size. Many of these materials are fairly hygroscopic and should be used in a dry atmosphere to prevent their surfaces from fogging, and thereby scattering short-wavelength light excessively. Sodium chloride transmits well from the visible to about 650 cm⁻¹ and has fairly good dispersive properties throughout its useful range (which coincides with the region in which most vibrational absorptions occur), and has the added advantage of being fairly inexpensive. Other materials of this class may be used in specific—and somewhat limited—spectral regions for the purpose of gaining dispersion and resolving power. For example, it may be seen from Fig. 33 that a prism of lithium fluoride will have an observed resolution of about 4 cm⁻¹ at 3000 cm⁻¹, while a prism of sodium chloride

[96] R. Spanbauer, P. E. Fraley, and K. N. Rao, *Appl. Opt.* **2**, 340 (63).
[97] D. E. McCarthy, *Appl. Opt.* **2**, 591 (1963).
[98] D. E. McCarthy, *Appl. Opt.* **2**, 596 (1963).
[99] D. E. McCarthy, *Appl. Opt.* **4**, 317 (1965).
[100] J. M. Bennett and E. J. Ashley, *Appl. Opt.* **4**, 221 (1965).
[101] H. E. Bennett, M. Silver, and E. J. Ashley, *J. Opt. Soc. Amer.* **53**, 1089 (1963).
[102] S. S. Ballard and J. S. Browder, *Appl. Opt.* **5**, 1873 (1966).

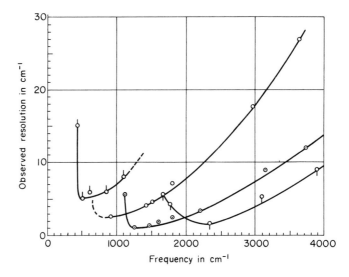

FIG. 33. Graph of spectral resolution of prisms made of NaCl (○), CaF₂ (●), KBr (⬨), and LiF (◯) versus frequency [R. C. Gore, R. S. MacDonald, V. Z. Williams, and J. U. White, *J. Opt. Soc. Amer.* **37**, 73 (1947)].

will have an observed resolution of only 17.5 cm⁻¹ at the same wave number in the spectrum. However, a LiF prism becomes opaque at about 1660 cm⁻¹ and is, therefore, more limited in its general usefulness. Table XI lists the materials of this class and their useful ranges.

Besides the alkali halides there are a number of other very satisfactory materials that can be used both as window materials and as dispersive media. These too are listed in Table XI, but some brief comments will be made about them individually.

Quartz is very satisfactory both as prism and window material to about 3300 cm⁻¹ where the transmission begins to fall off sharply. It remains opaque from 4.0 to 40.0 μm where the transmission again increases markedly, and it is again used in the far infrared principally as window and filter material.

Calcium fluoride with its cutoff at about 1100 cm⁻¹ has its region of maximum dispersion at about 1250 cm⁻¹ (8.0 μm). Thus a combination of NaCl, LiF, and CaF₂ prisms would provide excellent coverage of the near infrared vibration region (to 650 cm⁻¹, 15.0 μm).

It is possible to manufacture sapphire windows of considerable size. Sapphire is quite transparent to 1600 cm⁻¹ (6.5 μm) in thicknesses suitable for windows, and has the advantage of being resistant to mechanical and thermal shocks, and not subject to corrosion.

TABLE XI. Principal Optical Materials for Prisms and Windows in the Infrared[a]

Material	Low-frequency cutoff (cm^{-1})[b]		Comments
Glass	5000	(2.0)	Excellent for prisms and windows
Quartz	3300	(3.0)	Excellent for prisms and windows
LiF	1660	(6.0)	Excellent for prisms and windows
Sapphire	1600	(6.5)	Excellent for windows—resists mechanical and thermal shock
CaF$_2$	1100	(9.0)	Excellent for prisms and windows
NaCl	650	(15.0)	Best over-all coverage in infrared; hygroscopic; is easily handled
KCl	500	(20.0)	Hygroscopic; not used much presently
AgCl	455	(22.0)	Ductile—light sensitive; reflection losses; avoid contact with metals other than Ag; in general good window material
KBr	400	(25.0)	Hygroscopic; fractures easily; excellent prism material to 25 μm
KI	330	(30.0)	Similar to KBr
KRS-5 (Tl-Br-I)	250	(40.0)	Ductile; toxic; reflection losses; prism shape may alter with time; resists corrosion; good window material
CsBr	250	(40.0)	Hygroscopic; fractures easily
CsI	200	(50.0)	Similar to CsBr

[a] The materials are listed in order of their approximate low-frequency cutoff values.
[b] Values in parenthesis are in micrometers.

During World War II silver chloride, and a material called KRS-5 composed of a mixture of thallium bromide and thallium iodide, became popular as infrared materials because they are not too susceptible to attack by water vapor. Indeed, both materials are reasonably resistant to corrosive substances. The principal disadvantages of silver chloride are its softness—which causes it to be easily scratched and causes the windows to deform when used on cells under partial vacuum: and its sensitivity to light—which causes it to turn black and opaque after long exposure to light. It is also necessary to use silver or inert gaskets with silver chloride because of the reaction that occurs to free silver from the silver chloride. This material can, however, be used over relatively long periods in direct contact with Monel or nickel without ill effects occurring provided that no moisture is present—or the humidity is low.

KRS-5 is a heavy, deep red material with an extremely high refractive index, which, as for AgCl, causes it to have large reflective losses. Its principal virtues are that it transmits nearly to 40.0 μm in thicknesses suitable for windows, and that it is not readily attacked by water vapor or subject to corrosion. It is, however, quite soft and care must be exercised not to distort prisms made from it. It is also toxic so care must be used in grinding or polishing not to inhale any of the powder that results. With the availability of CsBr and CsI, which transmit to about 45 and 50 μm, respectively, KRS-5 has lost much of its usefulness in spectroscopy.

Diamond has come into fairly wide use in small pieces as window material for use in detectors for the far infrared (e.g., the Golay detector). Various other substances such as germanium, selenium, silicon, arsenic trisulfide glass, etc., may be used as window materials in special cases.

2.2.3.7. Infrared Detectors[67,72,103–106]

2.2.3.7.1. INTRODUCTION. The resolving power of spectroscopic instruments is dependent upon the energy available from the source, the efficiency of the optical system, and the sensitivity of the detector. Until World War II the detector most used in infrared spectroscopy was the thermopile, or, more simply, the thermocouple. It was generally constructed of thin bismuth–antimony and bismuth–tin alloy wires welded together as a junction with a slit-shaped gold–black or platinum–black coated receiver attached to the junction. This device was mounted in a slender, cylindrical, evacuable case closed with an NaCl or KBr window to pass the infrared. Such a thermocouple would have a time constant of the order of 1 sec, and it was necessary to use dc amplifiers and galvanometers to measure the radiation falling on the detector.

Because various military uses of infrared detection could only be satisfied with detectors having considerably shortened time constants without sacrifice of sensitivity, a considerable government-sponsored effort was put into research and development of new types of infrared detectors as well as for improved thermocouple materials and design. This program resulted in development of a number of new detectors

[103] R. A. Smith, *Appl. Opt.* **4**, 631 (1965).

[104] *Appl. Opt.* **4**, No. 6 (1965).

[105] H. Levinstein, *in* "Applied Optics and Optical Engineering" (R. Kingslake, ed.), Vol. II, p. 310. Academic Press, New York, 1965.

[106] H. L. Hackforth, "Infrared Radiation." McGraw-Hill, New York, 1960.

(photoconductive) whose principle of operation depends upon changes in their solid state properties with incident infrared radiation.

Detector performance depends not only upon the detector but also on the experimental circumstances. Jones[107] has developed criteria for comparing performance of detectors, some of which are listed below.

2.2.3.7.1.1. *Responsivity* (R_v). Responsivity, dependent upon input signal form and modulation frequency, is defined as

$$R_v = \text{rms output voltage/rms power incident upon detector.} \quad (2.2.49)$$

2.2.3.7.1.2. *Detectivity.* Detectivity D is defined as

$$D = R_v/\text{rms noise-voltage from detector} = 1/\text{NEP}, \quad (2.2.50)$$

where NEP is the noise equivalent power. Generally, detectors are compared by the use of D^*, which is the detectivity D reduced to unit bandwidth of the noise and unit area of the detector

$$D^* = (A \, \Delta f)^{1/2} D, \quad (2.2.51)$$

where D is detectivity measured with a noise equivalent bandwidth Δf (a characteristic of the amplification *system*) and A is the detector area. Thus D^* is measured in centimeters (hertz)$^{1/2}$ per watt and is often specified as D^* (chopping frequency, bandwidth of the amplifier) for broadband detectors and D^* (chopping frequency, blackbody temperature, bandwidth of the amplifier) for narrow-band detectors.

2.2.3.7.1.3. *Detector Summary.* Table XII summarizes some salient characteristics of a number of available types of detectors discussed by Kneubühl.[72]

2.2.3.7.2. PHOTOCONDUCTIVE DETECTORS.

2.2.3.7.2.1. *Lead Sulfide.* Of the various available photoconductive detectors, lead sulfide is in general use. For spectroscopic purposes these detectors are made slit shaped. Lead sulfide detectors for the infrared have a useful range of from 0.7 to somewhat more than 3.5 μm, and are of the order of 100 times more sensitive than a good thermocouple. Their peak sensitivity is usually near 2.0 μm. Cooling often increases the sensitivity and extends the useful wavelength limit. For cooling, the

---

[107] R. C. Jones, *in* "Advances in Electronics," Vol. 5, p. 1. Academic Press, New York, 1953.

TABLE XII. Generalized Characteristics of Infrared Detectors

Detector	Operating temperature (K)	λ Cutoff (μm)	D^*	Time constant (sec)
Golay cell	300	Window limited	10^9	10^{-2}
Bolometer	300	Window limited	10^7–10^8	10^{-3}
Thermopiles thermocouples	300	Window limited	10^7–10^9	3×10^{-2}
Pyroelectric	123/300	No limit	10^7–10^9	10^{-5}
InSb	77/300	5.5	10^9–10^{10}	10^{-5}–10^{-7}
InAs	77/196/300	3.5	10^8–10^9	10^{-6}
PbS	193/300	2.5–4	10^8–10^{10}	4×10^{-3}–5×10^{-5}
PbSe	77/193/300	4.7–7	10^8–10^9	5×10^{-5}
PbTe	77	6		10^{-5}
Ge:Zn	4.2	40	4×10^9	10^{-8}
Ge:Au	77	9–10	2×10^9	10^{-6}
Ge:Hg	35/40	14–15	4×10^9	10^{-6}
Ge:Cu	5/4.2	30	10^9–10^{10}	10^{-6}
HgCdTe	77	13	5×10^9	10^{-6}

sensitive surface is deposited so it will be in contact with the cooling device—as, for example, on the inner surface of a Dewar flask. The flask may be equipped with a sapphire window, which is transparent well beyond the long wavelength limit of the detector. The response time of lead sulfide detectors is of the order of 100 μsec. Chopping speeds of at least 500 Hz may, therefore, be used making it possible to design excellent amplifying systems to be used in conjunction with them.

2.2.3.7.2.2. Lead Telluride. In general, the properties of lead telluride detectors are similar to those of lead sulfide. However, cooling is much more important, as it appears to be difficult to make PbTe cells having much sensitivity at room temperature.

2.2.3.7.2.3. Lead Selenide. Detectors of lead selenide are most sensitive and have their longest wavelength response at low temperatures. The technique of making good PbSe cells is said not to be as difficult as for PbTe. Spectral response curves indicate a sensitivity peak of about 7.0 μm for a temperature of 20 K.

2.2.3.7.2.4. Germanium. Doped germanium detectors have seen extremely rapid development in recent years. As easily seen from Table XII the doped Ge detectors cover a wide frequency range at high values of D^*.

2.2.3.7.3. THERMAL DETECTORS.[108,109]

2.2.3.7.3.1. *Thermocouples.* In the decade or more prior to World War II the thermocouple described in Section 2.2.3.7.1, in conjunction with various types of dc amplifier–galvanometer combinations, was the standard detection system, and galvanometer readings were observed and manually recorded for specific wavelength settings of the prism or grating of the spectrometer. As has previously been indicated, it was, thus, a very slow process to observe, record, and finally plot the data of infrared spectra. Except for a resonance amplifier designed by Firestone, automatic recording was virtually unknown. Research into new thermoelectric materials, and methods of producing the elements, and better receiver blackening have, however, made it possible to develop thermocouples of extremely high sensitivity and sufficiently short time constant so that chopping speeds from 10 to 15 Hz are practicable. Although varying with the type of thermocouple, minimum detectable powers for 1-Hz bandwidth of the order of a few times 10^{-11} W have been claimed by the makers. Various designs have been executed by Hornig and O'Keefe, Harris, Roess and Dacus, Cary and George, Schwartz, and others, all of which may be used with chopping speeds between 5 and 15 Hz.

2.2.3.7.3.2. *Bolometers.* The bolometer, first used in 1880 by Langley, is a device the responsivity of which depends upon the first derivative of the resistance with respect to temperature. Bolometers are essentially of the types: metal strip (platinum, nickel, gold, etc.); thermistor (semiconductor); and the superconducting (columbium nitride).

Metal strip bolometers have been produced by a variety of methods of obtaining thin films or wires, e.g., by rolling, sputtering, electroplating, and vacuum evaporation. In general, they have time constants of 5 to 15 msec. Nickel, which has one of the largest temperature coefficients of resistance of the metals, is frequently used, and minimum detectable powers of about 3×10^{-9} W at 30 Hz and bandwidth of 1 Hz have been reported.

The thermistor (semiconductor), described by Wormser,[110] Brattain, and Becker,[111] is a bolometer made of a mixture of the oxides of Ni, Mn, and Co—a combination having a very large temperature coefficient of re-

---

[108] R. A. Smith, F. E. Jones, R. P. Chasmar, "The Detection and Measurement of Infrared Radiation." Oxford Univ. Press, London and New York, 1957.

[109] V. Z. Williams, *Rev. Sci. Instrum.* **19**, 135 (1948).

[110] E. M. Wormser, *J. Opt. Soc. Amer.* **43**, 15 (1953).

[111] J. A. Becker and W. H. Brattain, *J. Opt. Soc. Amer.* **36**, 354 (1946).

sistance. The material is fashioned into flakes of about 10 μm thickness and is backed by glass or quartz. Such bolometers have time constants of 5 to 8 msec and 2 to 5 msec for glass and quartz, respectively. Becker and Moore[112] have quoted a minimum detectable power of 7×10^{-11} W for chopping at 15 Hz at 1-Hz bandwidth. It appears that a good thermistor is comparable with a good thermocouple.

During World War II, D. H. Andrews, R. M. Milton, and W. de Sorbo[113] developed a bolometer operating on the principle that at low temperatures certain substances become superconducting, and therefore have a very steep dR/dT in a narrow temperature range. The material used was columbium nitride, which becomes superconducting at the triple point of H_2, between 14 and 15 K. The bolometer is mounted on a Cu block that projects into a cylindrical chamber filled with liquid hydrogen. The liquid hydrogen chamber is shielded by a cylindrical shell-like chamber filled with liquid nitrogen, and the whole apparatus is enclosed in a vacuum chamber with an infrared window. By pumping at an appropriate rate on the hydrogen, the temperature of the triple point may be maintained. This detector has an extremely short time constant—about 0.5 msec, and chopping speeds of 1000 Hz may be used with small loss over dc sensitivity. For a 360-Hz chopping speed at 1-Hz bandwidth, Andrews and Milton have reported a minimum detectable power of 2.1×10^{-11} W. The principal disadvantages are the bulky character of the cryostat in which it must be mounted, and the difficulty of pumping at the proper rate to maintain the hydrogen at the triple point over the long periods necessary for recording spectra.

2.2.3.7.3.3. *The Golay Cell.* A thermal detector invented by M. Golay, which operates on an entirely different principle from either the thermocouple or the bolometer, has proved to be a real competitor to them as to convenience, sensitivity, and possible chopping speeds. This detector, shown in Fig. 34, operates on the pneumatic principle. It is composed of a small xenon-filled cylindrical chamber closed with two thin windows. The front window, which has an infrared absorbing film deposited on it, transfers the energy to the gas causing it to expand. The rear window, normally a plane, but distensible, mirror, bulges a little because of the gas expansion and thus alters the optical constants of a point source, line grid, photocell receiver amplifying system so that more or less light from the source reaches the photocell. The signal from the

[112] J. A. Becker and H. R. Moore, *J. Opt. Soc. Amer.* **36**, 354 (1946).

[113] D. H. Andrews, R. M. Milton and W. de Sorbo, *J. Opt. Soc. Amer.* **36**, 518 (1946).

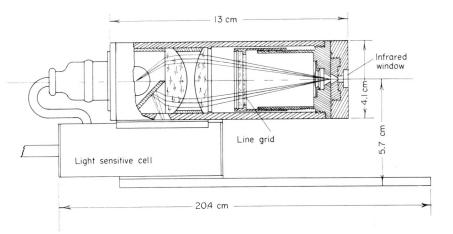

FIG. 34. Schematic diagram of the Golay detector (courtesy of Eppley Laboratories).

photocell is further amplified electronically in the same manner as with other detectors. By design these detectors can be made to give peak response at various chopping speeds from 10 to 100 Hz. The detector is black and can be used throughout the near and far infrared and as a microwave detector.

2.2.3.8. Miscellaneous Items[65,67,68]

2.2.3.8.1. SLITS. One of the most important parts of any spectrograph is the entrance and exit slit system. All modern slits are bilateral so that the center of the slit remains fixed when the slit jaws open and close. Slits are generally designed with ground prismatic edges and so mounted that they present the flat surface to the incident beam. The jaws must be accurately parallel so that a slit width can be defined. For the infrared —depending on the optics—the slits may be of various lengths, and capable of opening at least to 1 mm. The mechanism must be very precise so that it is possible to repeat slit settings with considerable accuracy. This is particularly important when intensity measurements are to be made. Most spectrographs for the infrared—whether prism or grating— have one of the slits (usually the entrance) made with curved jaws, because the prism angle, or the grating space, is not a constant to rays from all points along the slit. A curved image, therefore, results from a straight slit. If the entrance slit is curved to counteract this image curvature, the final image is straight. It should be remembered that the curvature is opposite in gratings from that in prisms.

2.2.3.8.2. FILTERS. A serious problem in the optical systems of spectrographs is that of scattered radiation–principally of wavelengths shorter than 2 μm. The problem is particularly annoying where quantitative work or intensity measurements are attempted. In grating spectrographs there is an additional problem that occurs because of overlapping orders.

Scattering light can be removed by one or a combination of the following techniques: clean mirrors, lenses, and slits; careful masking of components and properly placed baffles; double monochromatizing; powder filters with appropriate particle sizes; germanium filters with short-wave cutoff; *reststrahlen* plates and coarse gratings used as mirrors at longer wavelengths. There has recently been considerable research into the theory and production of interference-type filters of infrared materials with evaporated films laid down in such a manner as to make relatively narrow passband filters. These can be made to transmit 85–95% of the radiation over quite a narrow range of frequencies, which can be specified. For special purposes they are very useful.

2.2.3.8.3. SAMPLING TECHNIQUES. Infrared spectroscopy is useful for obtaining information about molecules in all the states of matter—gaseous, liquid, and solid. Each state, however, requires the use of a different technique for preparing the sample to be examined.

The gaseous phase is perhaps the simplest in that all that is required is a vacuum-tight cell of appropriate length made from glass or metal. The cell is closed with windows. Windows may be attached to the cells with a number of available waxes, or Glyptal—or by gaskets and pressure flanges making them more easily removed when regrinding or repolishing the highly hygroscopic salts becomes necessary. Gaseous cells used at both low and high temperatures have been designed. Where long absorbing paths are necessary it is frequently useful to employ a multiple traversal cell similar to a design given by White[114,115] in which a short cell can be made equivalent to a very long cell by folding the optical beam many times. It must be realized, however, that possibly serious energy losses may result from loss of beam aperture and reflection in the many traversals.

Some substances are corrosive, and it has been found beneficial to use stainless steel, Monel, or nickel in the construction of cells and gas handling apparatus. Such systems are actually simpler to construct and

[114] J. U. White, *J. Opt. Soc. Amer.* **32**, 285 (1942).
[115] T. H. Edwards, *J. Opt. Soc. Amer.* **51**, 98 (1961).

clean up than glass. An added problem is that cells must be closed with infrared transparent windows, many of which are subject to attack by the samples to be studied. The attacking of the windows manifests itself in a variety of ways; by fogging, etching, production of spurious window bands, or even vapor impurities. Of the materials listed in Table XI, sapphire LiF, CaF$_2$, AgCl, and KRS-5 have been found to be satisfactory in most cases.

Because windows are so susceptible to attack some comment should be made about processes useful in cleaning them. The subject of grinding and polishing has been treated in detail—for example, by Strong[116]—and in books on mirror making. However it is frequently not necessary to resort to extensive grinding and polishing unless the window is seriously damaged. If a window is fogged, it is often sufficient to polish on a lap stretched over a plate glass surface using filtered rouge or gamma-alumina as an abrasive and absolute alcohol as a lubricant. If a window is etched badly, fine grinding is necessary before polishing. This can readily be accomplished with carborundum powder 2/F and absolute alcohol on a good piece of plate glass. Silver chloride and KRS-5 require special treatment. It has been noted that cleaved surfaces of such materials as rock salt, potassium bromide, etc., are less easily attacked than are polished ones.

The infrared spectrum of a liquid sample is also fairly easy to obtain. As the density of liquids is much greater than that of gases, the cell thickness must be much smaller—of the order of hundredths of millimeters. The window material must not be soluble in the sample. Thus water solutions cannot be used with NaCl, KBr, CsBr, etc. Solvents such as CS$_2$ and CCl$_4$ are useful because they have few absorptions and do not attack cell windows. Liquid cells are generally made by clamping together two windows separated by a spacer of proper thickness. Because so many liquids have high volatility it is important that the cells be tightly sealed. Spacers are frequently made of very thin lead sheets amalgamated with mercury. When pressed between the plates this produces a tight seal. Cells are also frequently designed to accept the ground glass end of an hypodermic syringe, so that samples may be recoverable. Liquid cells are available commercially for various purposes, such as microcells for extremely small samples and variable length cells that can be set accurately to a desired length by a micrometer head.

[116] J. Strong, "Procedures in Experimental Physics." Prentice Hall, Englewood Cliffs, New Jersey, 1947.

The solid state is somewhat more difficult to study from the point of view of sample preparation. If the solid is crystalline or occurs in pieces of some size, it must either be cleaved into extremely thin slices, or ground extremely thin to prevent obscuration of detail in regions of absorption. Thin slices may be placed in the beam as any other sample. Very thin layers can be deposited by vacuum evaporation, or by evaporation from solvents onto transparent windows which can then be placed in the beam. Deposits of this sort, however, have a tendency to scatter the radiation, and another procedure in common use which reduces the scattering greatly is to mull the powder in some mineral oil, such as Nujol, or in a fluorocarbon oil, and spread the mixture evenly on a rocksalt or potassium bromide window. Certain regions are, however, obscured because of absorption by the oils—for example with Nujol the regions of 1375, 1450, and 2900 cm^{-1} have strong C—H absorption: and with the fluorocarbon oil the region 900–1250 cm^{-1} contains intense C—F bands. As these regions do not overlap the two mulls complement each other. A technique of solid sample preparation has been developed in which the finely powdered sample is thoroughly mixed with finely powdered KBr, placed in an evacuated die, and subjected to great pressure. This process results in a small clear tablet or pellet in which scattering by the powder is greatly minimized. Such a pellet may readily be mounted and inserted into the spectrometer beam. As much or as little sample may be put into the mix as is found necessary to bring out the bands in the spectrum, although best results are obtained with concentrations of about 1% or less. The important field of solids at low temperatures has also been developing recently, and special absorption cells have been designed for this purpose. Generally, such cells are variations of the following sort: a transparent window, seated in good contact with a copper block which projects into a liquid-nitrogen bath, is situated in a vacuum chamber closed with transparent windows. A small amount of vapor admitted to the cell then freezes on the cold window providing an absorbing layer of solid.

In general, qualitative and quantitative results can be obtained in all three states, although quantitative measurements are much more difficult and less accurately done with solid samples.

2.2.3.8.4. ATMOSPHERIC ABSORPTION. Under normal circumstances the atmosphere contains a great deal of water vapor and carbon dioxide, both of which absorb intensely over extensive regions of the infrared.

The water vapor fundamental bands, which are the most intense, are centered at 3700 and 1595 cm^{-1} and, depending upon moisture content of the air, extend several hundred reciprocal centimeters in each direction from band center. There are several less intense overtones and combinations at about 5550 and 3200 cm^{-1} as well as pure rotation lines beginning in the vicinity of 650 cm^{-1} and running toward lower wave number units.

Carbon dioxide has its most intense fundamental band at about 2350 cm^{-1} and another fundamental band at 650 cm^{-1} with several lesser bands between. Thus it may be seen that interference in several spectral regions important for analytical purposes may be somewhat troublesome unless efforts are made to eliminate these atmospheric absorptions. In general, the interference is greater in high dispersion measurements than in prism work because of accidental overlapping of sample vibration–rotation lines with individual 100% absorbed water vapor lines.

The most complete removal of atmospheric absorption may be accomplished by installing the spectrometer in a vacuum chamber, and indeed this has been done for several commercial instruments, as well as for several custom-built instruments. There are, however, certain inconveniences which result from evacuation, and in some cases instruments are designed so that dry and CO_2-free air may be circulated throughout the instrument housing. Such air may be provided by incorporating activated alumina towers with blowers in the instrument housing. Another effective, though less convenient, method of drying the air is to place trays of P_2O_5 and Ascarite inside the spectrometer housing while causing the air to circulate over the trays. In all cases it is of course important that the covers be tightly sealed to prevent leakage.

Another device that serves to obviate the atmospheric absorption is the double-beam arrangement in which only differences between the two beams are observed by the detector. Although the atmospheric bands interfere somewhat with observations, they do, however, provide many well-known wavelength standards which are useful for internal calibration.

2.2.3.8.5. CALIBRATION. Each instrument with its dispersing element must be calibrated in either wavelength units (micrometers) or frequency units (reciprocal centimeters). There exists a considerable number of molecules whose spectra have been accurately measured under high resolution which provide wave number standards useful for the calibration of both prism and grating spectrometers. As the dispersion of prisms is quite nonlinear, it is desirable to have points as close together as can conveniently be arranged. The diatomic molecules HF, HCl, HBr, and HI whose band centers are located at about 3970, 2860, 2560, and 2220 cm^{-1}, respectively, are all resolvable into rotational lines with LiF and CaF_2 prisms. Calibration curves with Littrow mirror positions plotted versus reciprocal centimeters of the lines of these well-known bands may be made for each prism, and positions of unknown bands accurately determined. Carbon monoxide, with its overtone band at 4260 cm^{-1}, also provides many excellent calibration points. As has already been mentioned, H_2O vapor and CO_2 vibration–rotation lines are well known and widely distributed. There is also a large number of easily obtainable polyatomic molecules which may be used, as for example CH_4 (3000 and 1300 cm^{-1}), C_2H_2 (3300, 1328, 700 cm^{-1}), NH_3 (1000 cm^{-1}), and H_2O [pure rotation lines (650 to lower reciprocal centimeters)]. For the very near infrared there exist many atomic emission lines from mercury, neon, etc., which are very accurately known and may be used.

Calibration of grating spectrometers is a considerably more difficult problem because precision of an entirely different order is obtainable. Rao et al.[53] have treated the precision calibration of high-resolution grating instruments in detail. Up to several years ago the usual procedure was to use a few well-known frequencies, the measured angles between them, and central image and compute an instrument grating constant. This constant was then used to calculate the frequencies of the measured lines. Even when grating circles were carefully calibrated, really high precision was not possible, and band centers measured on different instruments did not agree exactly with each other. Two better methods are now widely used to provide extraordinary precision. In one a noble gas emission spectrum or a molecular absorption spectrum is used to calibrate a carefully temperature-controlled Fabry–Perot interferometer fringe system. This fringe system is recorded simultaneously with the unknown spectrum on a two-pen recorder. The fringe system is then used as a comparison spectrum. In the second method a similar calibration spectrum is simultaneously recorded with the unknown spectrum. As the frequencies of these lines are extremely well known, and can be quite easily identified, and as they are well distributed when using coarse ruled gratings this method too provides excellent accuracy. Careful consideration of the index of refraction[117] of air must be made in precise measurements of infrared lines when the grating is in air and not vacuum.

[117] B. Edlén, J. Opt. Soc. Amer. 43, 339 (1953).

2.3. Electronic Spectroscopy*

2.3.1. Introduction

The scope of electronic spectroscopy has become so great that any attempt to discuss it requires a number of arbitrary decisions—some imposed by the space and time available and some imposed by the limitations of the author. The first edition of this chapter by G. Wilse Robinson appeared in 1962. At that time it provided an unequaled source of information on experimental techniques and on the interpretation of polyatomic molecular spectra. Since then the field has continued to expand, and a large number of reviews and texts have been written which consider in detail many of the topics of interest. Consequently, this chapter will not develop some topics thoroughly, but rather it will introduce them and then provide comments and references to appropriate reviews which were of particular use to the author. Approximately one third of the chapter is devoted to experimental techniques, and the remaining portion deals primarily with the basic descriptions and analyses of the electronic spectra of small molecules in the gas phase. Limitations of time and space led to the author's decision not to include discussions of the following topics, all of which are closely related and of considerable importance to electronic spectroscopy: (1) the effects of electric and magnetic fields on molecular energy levels and spectra, (2) theoretical and experimental descriptions of transition probabilities (and their relation to radiationless transitions), and (3) the study of electronic spectra of condensed phase systems.

2.3.2. Experimental Apparatus and Techniques

2.3.2.1. Dispersing Instruments

2.3.2.1.1. GRATINGS. Reflection grating instruments have been the preferred method of obtaining molecular spectra for many years. They offer the advantages of moderately high resolution, a minimum of serious

---

* Chapter 2.3 is by C. Weldon Mathews.

dependence on transmission properties (as is encountered with prisms), moderate speed, and convenient dispersion characteristics. These advantages have been enhanced markedly by improvements in the production of high-quality gratings which maximize the light diffracted at a particular angle (the *blaze angle*), minimize random and periodic ruling errors that produce spectra at multiple undesired angles (*grating ghosts*), and maximize reflectivity of the grating by the use of better reflecting coatings. An excellent introduction to the technology, theory, and use of diffraction gratings is available in the "Diffraction Grating Handbook,"

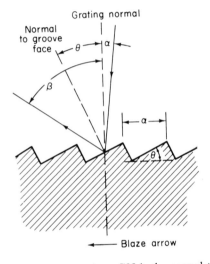

FIG. 1. Basic parameters of grating, where GN is the normal to the grating surface and BN is the normal to the groove facet in the blaze direction [reproduced from the "Diffraction Grating Handbook." Bausch and Lomb, 1970, with the kind permission of the publishers].

published by Bausch and Lomb,[1] as well as the book by Davis.[2] A review by Stroke[3] provides a much more detailed and rigorous treatment of the technical problems associated with the production of high-quality gratings.

It will be convenient to review a few of the basic properties of gratings before discussing the more common mounts for their use. Figure 1 defines the parameters a (grating spacing), α (angle of incidence), β

[1] "Diffraction Grating Handbook." Bausch and Lomb, Rochester, New York, 1970.
[2] S. P. Davis, "Diffraction Grating Spectrographs." Holt, New York, 1970.
[3] G. W. Stroke, Diffraction Gratings, *in* "Handbuch der Physik," Vol. 29. Springer, New York, 1967.

(angle of diffraction), and θ (the blaze angle) with GN and BN the grating and blaze normals, respectively. For arbitrary values of a, α, and β, constructive interference will occur when the grating equation is obeyed

$$m\lambda = a \mid \sin \alpha \pm \sin \beta \mid, \qquad (2.3.1)$$

where λ is the diffracted wavelength and m is the diffracted order (an integer that is usually small). The plus sign is used when α and β are on the same side of the grating normal, while the minus sign is used when they are on opposite sides of the normal (which is the case shown in the figure). Maximum intensity is attained when the diffracted beam is also reflected from a groove surface, i.e., $\theta = \frac{1}{2}(\alpha \pm \beta)$. Available blaze angles range from about 5 to about 64°, and the art of producing flat surfaces at reproducible angles results in a very sharply defined blaze angle. The advantage of a well-blazed grating rests in its ability to direct most of the light incident on the grating into a single direction, i.e., it minimizes the light "lost" by diffraction into inaccessible orders.

Many spectrograph designs are such that the angle of incidence is approximately equal to the angle of diffraction ($\alpha = \beta$, referred to generally as the *Littrow configuration* by Bausch and Lomb), in which case the grating equation simplifies to

$$m\lambda = 2a \sin \beta. \qquad (2.3.2)$$

In this case one can obtain an expression for the angular dispersion*

$$d\beta/d\lambda = m/a \cos \beta \qquad (2.3.3)$$

which may be rewritten, using Eq. (2.3.2), as

$$d\beta/d\lambda = (1/\cos \beta)(2 \sin \beta/\lambda) = (2/\lambda) \tan \beta. \qquad (2.3.4)$$

In spectrographs which use photographic detection, the *plate factor* or reciprocal dispersion $d\lambda/dl$ (e.g., angstroms per millimeter) is a more useful quantity. It may be obtained from the angular dispersion after determining the relation between dl and $d\beta$. In most cases $dl = f \, d\beta$, where f is the focal length of the instrument (the radius of curvature of

---

* The more general expression, Eq. (2.3.1), may be used to derive this relation (α is constant), or one must introduce an appropriate factor of two when using Eq. (2.3.2) since the dispersion applies for a stationary entrance slit.

the grating for an Eagle mount). Use of this conversion factor yields the expression

$$d\lambda/dl = (a/fm) \cos \beta, \qquad (2.3.5)$$

or, using Eq. (2.3.2),

$$d\lambda/dl = (\lambda/2f) \cot \beta. \qquad (2.3.6)$$

Equations (2.3.4) and (2.3.6) emphasize the increased dispersion gained by using the grating at high angles. They also demonstrate that the dispersion will be the same on a single instrument at a fixed grating angle and wavelength, regardless of the values of a and m.

Plane gratings designed to take full advantage of the increased dispersion at high angles are called *echelles* and typically have a blaze angle of about $63°$. Rao[4] has summarized their use recently. They are now available in sizes up to 210×410 mm (8×16 in.) with the promise of even larger ones in the future.[5] These gratings typically are ruled with 316 or 79.1 grooves/mm—numbers that are convenient multiples of the wavelength of a helium–neon laser used to control the groove spacing. When these echelles are used to record wavelengths in the visible and ultraviolet regions of the spectrum, the order of the diffracted light [m of Eq. (2.3.1)] will be between 10 and 60. Use of such high orders presents two problems: separation of adjacent orders and the elimination of grating ghosts, whose relative intensities increase with the square of the order. The former problem is solved by the addition of a predispersion system, i.e., a small monochromator, in front of the spectrograph. A number of such systems have been described (see pp. 212, 214) which use prisms or gratings and where the dispersion is parallel or perpendicular to the entrance slit of the main spectrograph. The second problem has been reduced considerably by improved ruling techniques, which is out of the hands of the spectroscopist, of course. It is important, however, with any grating to determine the positions of the most intense ghosts and to be aware of their possible presence, especially when emission spectra are being studied. The manufacturer usually can give some information regarding the positions and relative intensities of Lyman ghosts. [These arise because of different effective values of a in Eq. (2.3.1) and therefore appear to produce spectra in "nonintegral" orders.]

---

[4] K. N. Rao, Large Plane Gratings for High-Resolution Infrared Spectrographs, *in* "Molecular Spectroscopy: Modern Research" (K. N. Rao and C. W. Mathews, eds.). Academic Press, New York, 1972.

[5] G. R. Harrison and S. W. Thompson, *J. Opt. Soc. Amer.* **60**, 591 (1970).

The ability of a grating (and spectrograph) to separate adjacent spectral lines is measured in terms of resolution or resolving power. *Resolution* normally refers to the separation $\Delta\lambda$ or $\Delta\nu$ measured in wavelength (angstroms) or energy (reciprocal centimeters), of two spectral lines of equal intensity that are just resolved. *Resolving power* is then defined as

$$R = \lambda/\Delta\lambda = \nu/\Delta\nu. \tag{2.3.7}$$

The theoretical resolving power of a grating is $R = mN$,

$$R = m \times N = mW/a, \tag{2.3.8}$$

where N is the total number of lines (illuminated) on the grating, m is the spectral order, and W is the width of the grating. Equation (2.3.2) can be used to eliminate m in Eq. (2.3.8) to give the following expression of resolving power (appropriate when $\alpha = \beta$)

$$R = (1/\lambda)2Na \sin \beta. \tag{2.3.9}$$

The ruled width of the grating W is simply Na; therefore

$$R = (1/\lambda)2W \sin \beta. \tag{2.3.10}$$

In practice, the echelle gratings come much closer to their theoretical resolving power in all orders than do the finer-ruled gratings (1200 grooves/mm) blazed for angles between 5 and 25°. For concave gratings, Eq. (2.3.10) must be modified to include considerations of W_{opt}, the optimum width of the grating (see p. 211).

2.3.2.1.2. GRATING MOUNTS. Extensive comparisons of the various grating and prism mountings are available in various texts[2,6-8] and will not be repeated here. The book by Samson offers a particularly good summary of current use and theory of both concave-grating and plane-grating spectrographs. Nevertheless, there are two grating mounts which deserve special consideration in view of their widespread current use. One of these is the Czerny–Turner mount, which uses a plane grating, and the other is the Eagle mount, which uses a concave grating.

[6] J. A. R. Samson, "Techniques of Vacuum Ultraviolet Spectroscopy." Wiley, New York, 1967.

[7] R. A. Sawyer, "Experimental Spectroscopy," 3rd ed. Dover, New York, 1963.

[8] G. R. Harrison, R. C. Lord, and J. R. Loofbourow, "Practical Spectroscopy." Prentice-Hall, Englewood Cliffs, New Jersey, 1948.

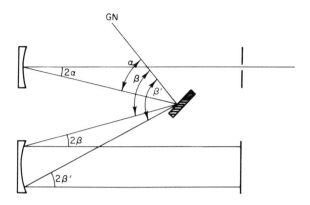

FIG. 2. Czerny–Turner spectrograph with S as the entrance slit, C as the collimating mirror, G as the grating, F as the focusing (or camera) mirror, and P as the focal plane.

The Czerny–Turner mount[4,9] uses two concave mirrors and a plane grating, as indicated in Fig. 2. Normally the slit, two mirrors, grating, and detector system (photographic plate or exit slit) are in the same plane (the in-plane mount). The very similar Ebert–Fastie mount[10,11] uses only one concave mirror as both collimator and camera mirror, and the light may pass over and under the grating in some instruments (the out-of-plane mount). To a first approximation these instruments are stigmatic and require only a rotation of the grating in order to change wavelengths. Thus it is ideal for use with large plane gratings efficiently blazed for angles near 63°, i.e., the echelles. Fastie[12] found the serious problem of coma could be minimized if the following relation of angles were maintained

$$a/b = \cos^3 \beta / \cos^3 \alpha. \qquad (2.3.11)$$

A number of spectrographs of this type have been built with focal lengths between 0.3 and 10 m, and a 5-m instrument is available commercially. A minimum focal length of about 4 m appears essential[13] in order to obtain maximum resolving power without being limited by photographic grain with a 125×254 mm echelle grating. With the newer, larger echelles (already up to 375 mm) the longer focal length instruments

[9] M. Czerny and R. F. Turner, *Z. Phys.* **61**, 792 (1930).
[10] H. Ebert, *Wiedemann Ann.* **38**, 489 (1889).
[11] W. G. Fastie, *J. Opt. Soc. Amer.* **42**, 641, 647 (1952).
[12] W. G. Fastie, U. S. Patent 3011391 (1961).
[13] G. G. Chandler, *J. Opt. Soc. Amer.* **58**, 895 (1968).

will be essential. In addition to providing the ability to use faster photographic plates, they also may prove more suitable for use with image intensifiers or other two-dimensional electrooptic devices of somewhat limited resolving capabilities. The relation expressed in Eq. (2.3.11) was explored in more detail by Rosendahl[14] who used ray-tracing techniques to calculate the aberrations. Similar calculations have been conducted for the Czerny–Turner mount by Leo,[15] Shafer et al.,[16] Chandler,[13] Rouse et al.,[17] and Reader.[18] The last three papers contain considerations of a 4-m instrument built at University College, London. The paper by Reader also includes a discussion of a 3.34-m instrument built at the NBS, Washington. The last two papers compare geometric theory with ray-tracing techniques and correct an apparent inconsistency raised by Chandler.

Concave-grating spectrographs have long been favorite instruments for photographic recording of spectra since they permit large regions of the spectrum to be photographed simultaneously. The additional advantage of a single reflecting surface is of particular importance for wavelengths less than 2500 Å, where reflectivities decrease markedly.* The properties of a concave-grating spectrograph may be viewed most conveniently by recalling that the spectrum will be in focus only when the slit, grating, and plateholder remain on the Rowland circle. A number of spectrographic mounts incorporate the concave grating and fulfill the above requirements, but the most common ones are the Eagle mount and the grazing-incidence mount (where the angle of incidence approaches 90°) for use at wavelengths less than 500 Å. A series of papers by Namioka[19–21] provides an extensive theory of the optical properties

[14] G. R. Rosendahl, J. Opt. Soc. Amer. **52**, 412 (1962).

[15] W. Leo, Z. Angew. Phys. **8**, 196 (1956); see also the following for considerations of prism mounts, W. Leo, Z. Instrumentenk. **66**, 240 (1958); **70**, 9 (1962).

[16] A. B. Shafer, L. R. Megill, and L. Droppleman, J. Opt. Soc. Amer. **54**, 879 (1964).

[17] P. E. Rouse, Jr., B. Brixner, and J. V. Kline, J. Opt. Soc. Amer. **59**, 955 (1969).

[18] J. Reader, J. Opt. Soc. Amer. **59**, 1189 (1969).

[18a] R. E. Miller, J. Opt. Soc. Amer. **60**, 171 (1970).

[19] T. Namioka, J. Opt. Soc. Amer. **49**, 446 (1959).

[20] T. Namioka, J. Opt. Soc. Amer. **49**, 460 (1959).

[21] T. Namioka, J. Opt. Soc. Amer. **49**, 951 (1959).

* The use of MgF_2 overcoatings has made it practical to use plane-grating spectrographs as low as 1200 Å. Miller (Ref. 18a) reports an investigation of the N_2 emission spectrum in the region 1383–1810 Å (32nd–43rd orders) with a 5-m Fastie–Ebert spectrograph. Coincidence errors were found to be less than ± 0.005 cm^{-1}.

of concave-grating instruments. Much of this theory has been summarized by Samson.[6]

The basic design for an Eagle mount is shown in Fig. 3, which also demonstrates the rather complicated adjustments that must be made when the angle of incidence α is changed. Since it is convenient to maintain the entrance slit S and optical axis OA fixed, it is necessary to rotate the grating, rotate the plateholder P, and adjust the grating-to-slit distance for each value of α. Note that the in-plane mount presents some mechanical problems of interference between the slit and plateholder when efforts are made to keep $\alpha \approx \beta$. Since most Eagle mounts

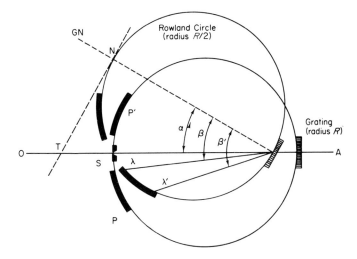

FIG. 3. Eagle spectrograph, where S is the entrance slit. Two positions are shown for the grating and plateholders, P and P' (see text for details).

have the plateholder only on one side of the slit, there is also the necessity of choosing whether the plateholder is pivoted toward the grating (location P with $\alpha < \beta$) or away from the grating (location P' with $\alpha > \beta$). The choice depends on a balance between resolution, intensity, and mechanical convenience. Alternatively the slit and plateholder may be placed on opposite sides of the Rowland plane (the off-plane mount), in which case $\alpha \approx \beta$. Namioka[20] gives a detailed theoretical evaluation of such an instrument, as well as some comparisons with the 6.6 m off-plane instrument described by Wilkinson.[22] A number of the commercial version of this instrument have been built, although the present trend

[22] P. G. Wilkinson, *J. Mol. Spectrosc.* **1**, 288 (1957).

favors the in-plane mount. The primary operational modification of the off-plane mount is that the slit must be rotated slightly with respect to the rulings on the grating. The angle of rotation may be as large as a few degrees, depending on the angle of incidence and the ratio of the distance of the slit from the Rowland plane to the radius of curvature of the grating.

One of the major problems with concave-grating instruments is the presence of astigmatism. This aberration is associated with the production of a line on the photographic plate for every point on the slit. This results in a loss of intensity and an inability to juxtapose two spectra by using masks in front of the slit. The theory is included in the treatment by Namioka[20] who also gives references to earlier work. The astigmatism may be eliminated under certain circumstances by using both the horizontal and vertical focus (see discussion by Samson,[6] pp. 19–23). If a point light source were located at N in Fig. 3, the horizontal focus would be at the entrance slits and the vertical focus would be along the line perpendicular to the grating normal, i.e., at point T in Fig. 3. Therefore, vertical masking at point T could be used to produce sharply defined masking at the photographic plate. Likewise, it is possible to use other optical elements to cancel the astigmatism of the spectrograph under certain conditions. These developments are of particular importance for concave-grating instruments since the achievement of maximum resolving power requires a much more complex optimization of parameters. For example, Eq. (2.3.10) must be modified to include an expression for the optimum width of the grating that depends on the angles of incidence and diffraction. The net effect is that operating at higher angles may lead to a considerable *reduction* of resolving power.

The largest Eagle spectrograph is the 10.7-m vacuum instrument designed and built at the National Research Council of Canada, and described by Douglas[23] and Douglas and Potter.[24] Slightly modified versions are available from the Jarrell–Ash Company, one of which is in use at National Bureau of Standards, Washington. A similar 6.6-m instrument also is available from the McPherson Instrument Company. These and other manufacturers market a number of smaller normal-incidence spectrographs and monochromators for use in the visible and vacuum ultraviolet, as well as grazing-incidence instruments for use at wavelengths below 500 Å.

[23] A. E. Douglas, *J. Opt. Soc. Amer.* **49**, 1132 (1959).
[24] A. E. Douglas and J. G. Potter, *Appl. Opt.* **1**, 727 (1962).

The separation of overlapping orders is a problem common to all grating instruments, since a large number of sets of m and λ can satisfy Eq. (2.3.1) for a fixed value of a, α, and β. For example, the first order of 6000 Å will coincide exactly with second order of 3000 Å, third order of 2000 Å, fourth order of 1500 Å, etc. In many cases it is possible to eliminate all but the desired wavelengths by using combinations of suitable light sources, detectors of limited sensitivity, and filters. For example, the second order of 3000 Å may be recorded with no difficulty by using a photographic film insensitive to 6000 Å and a filter that absorbs all light with wavelengths less than about 2500 Å. (In fact, air would eliminate the 1500-Å region and greatly reduce the intensity of the 2000-Å region.) Observation of the fourth order of 1500 Å, on the other hand, presents a much more difficult problem. In such cases and in ones requiring a very narrow band pass (for example, when $m > 5$) the most convenient and efficient method involves the use of a predispersion system. Several successful systems have been reported,[25-27] all of which incorporate a simple prism or grating monochromator placed between the light source and the entrance slits of the spectrograph. When the dispersion is perpendicular to the spectrograph slit, these systems permit only the desired wavelengths into the main spectrograph. Proper matching of the dispersion and resolution (as well as polarization of the light) can result in an efficient "filter" of the desired wavelength and band pass. With stigmatic instruments operating in high orders it is sometimes convenient to rotate the predisperser so that the dispersion is parallel to the spectrograph slit. In other words, different wavelengths enter the main spectrograph at different heights along the slit. The resulting *echellogram* contains a two-dimensional dispersion—along the length of the plate from the main spectrograph and different orders along the height of the slit image. Consequently the technique offers the advantage of covering very broad spectral regions at very high resolution and dispersion. The measurement of such plates is rather tedious,[28] but it has been used both for atomic and molecular spectra. A recent description by Liller[29] of an echelle spectrograph for stellar spectroscopes serves as an example of the current use of such instruments.

[25] A. E. Douglas and G. Herzberg, *J. Opt. Soc. Amer.* **47**, 625 (1957).

[26] J. Reader, L. C. Marquet, and S. P. Davis, *Appl. Opt.* **2**, 963 (1963).

[27] H. E. Blackwell, G. S. Shipp, M. Ogawa, and G. L. Weissler, *J. Opt. Soc. Amer.* **56**, 665 (1966).

[28] F. L. Moore, Jr., and F. Benjamin, *J. Opt. Soc. Amer.* **62**, 762 (1972).

[29] W. Liller, *Appl. Opt.* **9**, 2332 (1970).

The use of interferometers to disperse light has increased markedly in the past few years, owing partly to their use in an active or passive role in many laser systems. The development of precisely controlled reflective layers[30] and of piezoelectric controls of the spacings[31] have greatly enhanced their applications. The description and references given by Kessler and Crosswhite, Volume 4B, Section 4.1.1, of this series is a particularly appropriate introduction to the field. The recent observations by Innes and Kroll[32] of the nuclear quadrupole hyperfine structure of I_2 illustrate the application of a Fabry–Perot interferometer to line-shape analyses of molecular spectra. Innes *et al.*[33] recorded the rotational fine structure of the s-tetrazine-d_0 band at 18,913 cm^{-1}. They then fitted peak positions, line widths, and relative intensities by band contour calculations to obtain five of the possible six inertial constants.

2.3.2.1.3. LASERS. This technique was virtually nonexistent 10 years ago, whereas it now offers opportunities and challenges to almost all areas of optical investigation. The field is far too extensive to be reviewed in this section, but fortunately a number of textbooks and reviews are available (see, for example, those by Lengyel,[34] Bloom,[35] Demtröder,[36] Snavely,[37] Röss,[38] and Moore[38a]).

The impact of laser techniques on electronic spectroscopy already has been immense, and it promises to become even greater with the advent of tunable lasers. A few applications of lasers will be mentioned throughout this chapter, but in this section it is important to recognize that tunable lasers can now be considered as an alternative to monochromators for some applications within the wavelength range 2500–10,000 Å. Com-

---

[30] H. A. Macleod, "Thin-Film Optical Filters." American Elsevier, New York, 1969.

[31] W. P. Mason, Properties of Transducer Materials, *in* "American Institute of Physics Handbook," 3rd ed. McGraw-Hill, New York, 1972.

[32] M. Kroll and K. K. Innes, *J. Mol. Spectrosc.* **36**, 295 (1970).

[33] K. K. Innes, A. Y. Khan, and D. T. Livak, *J. Mol. Spectrosc.* **40**, 177 (1971).

[34] B. N. Lengyel, "Lasers," 2nd ed. Wiley (Interscience), New York, 1971.

[35] A. L. Bloom, "Gas Lasers." Wiley (Interscience), New York, 1968.

[36] W. Demtröder, "Laser Spectroscopy." Springer-Verlag, New York, 1971.

[37] B. B. Snavely, *Proc. IEEE* **57**, 1374 (1969).

[38] D. Röss, "Lasers: Light Amplifiers and Oscillators." Academic Press, New York, 1969.

[38a] C. B. Moore, "Lasers in Chemistry," *Ann. Rev. Phys. Chem.* **22**, 387 (1971).

mercially available dye lasers* and parametric oscillator lasers† offer continuously tunable monochromatic light sources for wavelength ranges as large as 3000 Å, with line widths as small as 0.001 Å. Such sources appear especially promising for use in studies of resonance fluorescence, photoelectron effects, absorption spectra within well-defined regions, and in optical pumping systems.

Discrete-line lasers (primarily the helium–neon and argon–ion lasers) already have been used for studies of fluorescence spectra, e.g., for Na_2,[39–41] Li_2,[42,43] K_2,[44,45] I_2,[46–48] and glyoxal.[49] In some cases the technique also has produced data on radiative lifetimes and energy transfer processes. The papers by Zare[50] and Drullinger and Zare[51] provide stimulating discussions of the information available through experiments involving the optical pumping of molecules, especially when selective excitation of particular levels or orientations is possible.

The above studies were possible only because the available laser lines happened to overlap a particular molecular absorption line. Even with the variety of laser transitions available (see, for example, Lengyel,[34] chapter 9) and with the successful use of frequency-doubling and Raman-shifting techniques (see, for example, the papers by Calvert and co-workers[52]), the discrete-line lasers do not yet provide a continuous, wide-range selection of wavelengths. The parametric oscillator does meet many of these needs, but it has not been widely used thus far.

[39] W. Demtröder, M. McClintock, and R. N. Zare, *J. Chem. Phys.* **51**, 5495 (1969).
[40] M. McClintock, W. Demtröder, and R. N. Zare, *J. Chem. Phys.* **51**, 5509 (1969).
[41] S. E. Johnson, K. Sakurai, and H. P. Broida, *J. Chem. Phys.* **52**, 6441 (1970).
[42] Ch. Ottinger, R. Velasco, and R. N. Zare, *J. Chem. Phys.* **52**, 1636 (1970).
[43] R. Velasco, Ch. Ottinger, and R. N. Zare, *J. Chem. Phys.* **51**, 5522 (1969).
[44] W. J. Tango, J. K. Link, and R. N. Zare, *J. Chem. Phys.* **49**, 4264 (1968).
[45] W. J. Tango and R. N. Zare, *J. Chem. Phys.* **53**, 3094 (1970).
[46] G. R. Hanes and C. E. Dahlstrom, *Appl. Phys. Lett.* **14**, 362 (1969).
[47] J. I. Steinfeld, J. D. Campbell, and N. A. Weiss, *J. Mol. Spectrosc.* **29**, 204 (1969).
[48] K. Sakurai and H. P. Broida, *J. Chem. Phys.* **53**, 1615 (1970).
[49] W. Holzer and D. A. Ramsay, *Can. J. Phys.* **48**, 1759 (1970).
[50] R. N. Zare, *J. Chem. Phys.* **45**, 4510 (1966).
[51] R. E. Drullinger and R. N. Zare, *J. Chem. Phys.* **51**, 5532 (1969).
[52] H. W. Sidebottom, *et al.*, *J. Amer. Chem. Soc.* **93**, 2587 (1971); **94**, 13 (1972).

* Avco Everett Research Laboratory, Everett, Massachusetts 02149; Coherent Radiation, Inc., Palo Alto, California, 94304; Spectra Physics, Inc., Mountain View, California 94040; and Synergetics Research, Inc., Princeton, New Jersey 08540.

† Chromatix, Mountain View, California 94040 and Garching Instrumente GMBH, Freisinger Landstrasse 25, 8046 Garching, West Germany.

The development of dye lasers provides a particularly promising solution to the above problems of cost and lack of flexibility. The light from a dye laser typically is a continuum that may extend over a region of about 500 Å. When a dispersive element (prism, grating, filter, etc.) is included *within* the resonant cavity, the output will be narrowed to an extent dictated by the dispersive element. The most important property, however, is that only a small fraction of the output power is lost by narrowing the spectral distribution of the laser beam. For example, Strome and Webb[53] reduced the band pass (full-width-at-half-maximum intensity) from 38 to 1.7 Å and maintained 56% of the total output power. Many of the techniques and early papers of dye lasers are summarized by Lengyel,[34] pp. 253–260, and in the review by Snavely.[37] A few interesting examples of papers published after these reviews will serve as an indication of the rapid developments in the field. In 1970 Walther and Hall[54] reported their use of birefringent filters to produce line widths of 0.01 Å with a flash-pumped dye laser. In the same year Peterson *et al.*[55] reported a continuously operating dye laser that was powered by a 1-W argon-ion laser. In addition, Capelle and Phillips[56] reported the use of a pulsed nitrogen laser to produce lasing with several dyes over the range 4140–6420 Å with a band pass of about 3 Å. In 1971 Hänsch *et al.*[57] also used a pulsed nitrogen laser to pump a dye laser in order to study the saturation spectrum[38a] (thereby eliminating the Doppler widths) of the sodium-D lines. Their laser system produced a band pass of 7 MHz and a pulse length of 30 nsec with an external Fabry–Perot interferometer. A recent paper by Hänsch[58] described their system in more detail and provides a more complete set of references to recent papers dealing with experimental techniques of dye lasers. Moore and Yeung[59] have developed a system that uses a frequency-quadrupled ruby laser to pump a dye laser, which is then mixed with a portion of the frequency-doubled output of the ruby laser. The continuously tunable output of this laser (in the 3500-Å region) was used to investigate the fluorescence spectrum of formaldehyde.

[53] F. C. Strome, Jr., and J. P. Webb, *Appl. Opt.* **10**, 1348 (1971).

[54] H. Walther and J. L. Hall, *Appl. Phys. Lett.* **17**, 239 (1970).

[55] O. G. Peterson, S. A. Tuccio, and B. B. Snavely, *Appl. Phys. Lett.* **17**, 245 (1970).

[56] G. Capelle and D. Phillips, *Appl. Opt.* **9**, 2742 (1970).

[57] T. W. Hänsch, I. S. Shahin, and A. L. Schawlow, *Phys. Rev. Lett.* **27**, 707 (1971).

[58] T. W. Hänsch, *Appl. Opt.* **11**, 895 (1972).

[59] C. B. Moore and E. S. Yeung, *J. Amer. Chem. Soc.* **93**, 2059 (1971).

Despite the immense amount of progress, each of the above systems has problems with repetition rate, stability, reproducibility, or wavelength range. Nevertheless, it is obvious that dye lasers will be an important tool for electronic spectroscopy.

2.3.2.2. Standards

2.3.2.2.1. DISPERSION OF AIR.   The single most useful quantity derived from electronic spectra is the precise measure of the energy difference between two states, i.e.,

$$\Delta E = h\tilde{\nu}, \tag{2.3.12}$$

where $\tilde{\nu}$ is the frequency of the associated radiation. In practice it is more convenient to measure the associated wavelength λ, which emphasizes the convenience of expressing the energy in terms of reciprocal wavelength, indicated by the wave number ν

$$\Delta E = h\tilde{\nu} = hc/\lambda = hc\nu. \tag{2.3.13}$$

Since c is defined as the speed of light in vacuum, these relations are correct only when the quantities are measured in a vacuum, i.e., λ_{vac} and ν_{vac}. Wavelength measurements made in air may be converted to the equivalent vacuum wavelengths and vacuum wave numbers by the relation

$$\nu_{vac} = 1/\lambda_{vac} = 1/n\lambda_{air}, \tag{2.3.14}$$

where n is the refractive index of air for that wavelength. Therefore, it is important to know the refractive index of air to a precision comparable to that obtainable in wavelength measurements (i.e., about one part in 10^8).

In 1952 the Joint Commission on Spectroscopy recommended the adoption of the formula given by Edlén[60] for the dispersion of *standard air*,

$$(n - 1)10^8 = 6432.8 + \frac{2{,}949{,}810}{146 - \sigma^2} + \frac{25{,}540}{41 - \sigma^2}, \tag{2.3.15}$$

where σ is the wave number expressed in *reciprocal micrometers* ($= 10^4$ cm^{-1}). Standard air is defined as dry air that contains 0.03% by volume CO_2 at a temperature of 15°C and at standard atmospheric pressure (760 Torr and 0°C and $g = 980.665$ cm/sec^2). Edlén provided a critical evalua-

---

[60] B. Edlén, *J. Opt. Soc. Amer.* **43**, 339 (1953).

tion of existing measurements of the refractive index of air and deduced his formula from the measurements he felt were most consistent, primarily the results of Barrell and Sears,[61] Koch,[62] and Traub.[63] In 1960 Svensson[64] repeated measurements of the dispersion of air for wavelengths between 2302 and 6907 Å and suggested slight modifications to the formula given by Edlén,

$$(n - 1)10^8 = 6686.68 + \frac{2{,}875{,}204}{144 - \sigma^2} + \frac{24{,}816}{40.9 - \sigma^2}. \qquad (2.3.16)$$

The difference between the two formulas amounts to less than 0.1×10^{-8} between 3020 and 10,140 Å. For wavelengths below 2860 Å the difference reaches a maximum deviation of about 1×10^{-8} in the range 2000–2300 Å. Even in this range the difference amounts to only about 2×10^{-5} Å. Both authors discuss the correction (easily as large as 1×10^{-4} Å) that should be made for measurements in nonstandard air; primarily for changes in density, CO_2 content, and water content. Svensson examined in detail the effect of drying air and found that some of the previous experimental differences could be ascribed to the variation in composition, primarily of H_2O and O_2, which resulted from the drying techniques used by different experimenters. Although Edlén's formula was based on measurements between 2300 and 6440 Å, it has been found to be in excellent agreement with measurements of the refractive index of air for wavelengths to 2.0 μm. Rank et al.[65] measured the refractivity of air in the range 3651–15,300 Å and found agreements to be about one part in 10^5. Schlueter and Peck[66] found no significant deviations between 7000 Å and 2.0 μm.

In practice, it is most convenient to incorporate the Edlén formula in a computer program which also fits the wavelength standards and interpolates all other measurements. When fewer conversions are required, the *Tables of Wavenumbers*[67] provide convenient tables for converting

[61] H. Barrell and J. E. Sears, Jr., *Trans. Roy. Soc. (London)* **A238**, 1 (1939).
[62] J. Koch, *Arkiv. Mat. Astron. Fys.* **8**, No. 20 (1912).
[63] W. Traub, *Ann. Phys.* **61**, 533 (1920).
[64] K.-F. Svensson, *Arkiv. för Fysik* **16**, 361 (1960).
[65] D. H. Rank, G. D. Saksena, and T. K. McCubbin, Jr., *J. Opt. Soc. Amer.* **48**, 455 (1958).
[66] D. J. Schlueter and E. R. Peck, *J. Opt. Soc. Amer.* **48**, 313 (1958).
[67] C. D. Coleman, W. R. Bozman, and W. F. Meggers, Tables of Wavenumbers, Vol. I, 2000 Å to 7000 Å. NBS Monograph 3, U.S. Govt. Printing Office, Washington, D.C., 1960; Tables of Wavenumbers, Vol. II, 7000 Å to 1000 μ. NBS Monograph 3, U.S. Government Printing Office, Washington, D.C., 1960.

from λ_{air} to ν_{vac}. These books also include values for the refractive index and for the difference in λ_{air} and λ_{vac}, as well as a brief discussion of the most useful corrections required for measurements made in non-standard air.

2.3.2.2.2. WAVELENGTH STANDARDS. Ideally all wavelengths should be measured by direct comparison with the primary standard of length, which is defined in terms of the $2p_{10}-5d_5$ transition of ^{86}Kr. Although this is not practical for grating spectrographs, it is normal practice in interferometric techniques to measure all wavelengths with respect to a single known standard, which may be the primary standard. In practice it is most convenient to adopt secondary standards which have been measured precisely by several different laboratories with interferometric techniques if possible. Edlén[68] has suggested the designation of Class A and Class B secondary standards where the wavelengths are known to ± 0.0001 and 0.001 Å, respectively. Littlefield[69] has suggested the distinction be based on wave number accuracies of 0.001 cm^{-1} for Class A standards and 0.01 cm^{-1} for Class B standards rather than Edlén's wavelength accuracies. The characteristics of good secondary standards have been discussed and summarized by Meggers.[70] These include that they be sharp lines which may be reproduced conveniently in many laboratories, that their appearance and energies not depend critically on excitation conditions, and that they be uniformly spaced over the entire wavelength region with uniform intensity. Needless to say, there is no single set of ideal secondary standards, but a large number of atomic spectra have been measured carefully during the past few years, which result in an extensive set of atomic systems that may serve as secondary standards in most regions of the spectrum.

Summaries of the wavelengths that have been recommended as secondary standards may be found in the reports of Commission 14 of the International Astronomical Union. These have been summarized and tabulated in terms of λ_{air}, λ_{vac}, and ν_{vac} by Mathews and Rao.[71] The secondary standards fall naturally into a number of convenient categories; (1) iron arc spectra, (2) inert gas spectra, (3) miscellaneous

[68] B. Edlén, Trans. Int. Astron. Un. 10, 211 (1960).
[69] T. A. Littlefield, Trans. Int. Astron. Un. 11B, 208 (1962).
[70] W. F. Meggers, J. Anal. Chem. 28, 616 (1956).
[71] C. W. Mathews and K. N. Rao, Tables of Standard Data, in "Molecular Spectroscopy: Modern Research" (K. N. Rao and C. W. Mathews, ed.). Academic Press, New York, 1972.

spectra of heavy atoms such as mercury and cadmium, (4) vacuum ultraviolet standards, and (5) thorium spectra. Iron arc spectra have been the most widely used because of the convenience of producing them and because of the abundance of lines between 2500 and 4500 Å. The problems of pressure broadening, arc instability, and asymmetric line shapes have been reduced by the current availability of iron hollow-cathode discharge tubes. Although the spectrum from hollow-cathode discharges differs slightly from that of arcs, the line list and micro-densitometer traces published in Crosswhite,[72] and hopefully to be up-dated as an NBS Monograph, render their use particularly convenient. These sources have the additional advantage that precision measure-ments[73-75] have been made of their spectra.

The spectrum of thorium provides an interesting and useful example of the careful evaluation needed prior to the adoption of an additional set of secondary standards. (Thorium has not yet been recommended as a source for secondary standards.) In 1955 Meggers[76] pointed out the deficiencies of the existing wavelength standards based on the iron spec-trum produced by an arc in air. He suggested that the thorium emission spectrum would yield narrower lines (because of its higher mass and the absence of hyperfine structure) more uniformly spaced over a broader wavelength region. In 1958 Meggers and Stanley[77] first reported inter-ferometric measurements of wavelengths between 3289 and 6992 Å using an electrodeless discharge through thorium iodide and helium. Since that time a number of additional sets of interferometrically mea-sured thorium wavelengths have been published[78-84] as a result of ex-

[72] H. M. Crosswhite, The Spectrum of Iron I. Johns Hopkins Spectroscopic Rep. No. 13, Baltimore, Maryland, 1958. See also, "American Institute of Physics Handbook," 3rd ed. McGraw-Hill, New York, 1972.

[73] R. W. Stanley and G. H. Dieke, J. Opt. Soc. Amer. 45, 280 (1955).

[74] J. Blackie and T. A. Littlefield, Proc. Roy. Soc. (London) A234, 398 (1956).

[75] R. W. Stanley and W. F. Meggers, J. Res. Nat. Bur. Std. (U.S.) 58, 41 (1957).

[76] W. F. Meggers, Trans. Int. Astron. Un. 9, 225 (1955).

[77] W. F. Meggers and R. W. Stanley, J. Res. Nat. Bur. Std. (U.S.) 61, 95 (1958).

[78] A. Davison, A. Giacchetti, and R. W. Stanley, J. Opt. Soc. Amer. 52, 447 (1962).

[79] A. Giacchetti, M. Gallardo, M. J. Garavaglia, Z. Gonzalez, F. P. J. Valero, and E. Zakowicz, J. Opt. Soc. Amer. 54, 957 (1964).

[80] T. A. Littlefield and A. Wood, J. Opt. Soc. Amer. 55, 1509 (1965).

[81] W. F. Meggers and R. W. Stanley, J. Res. Nat. Bur. Std. (U.S.) 69A, 109 (1965).

[82] F. P. J. Valero, J. Opt. Soc. Amer. 58, 484 (1968).

[83] F. P. J. Valero, J. Opt. Soc. Amer. 58, 1048 (1968).

[84] D. Goorvitch, F. P. J. Valero, and A. L. Clúa, J. Opt. Soc. Amer. 59, 971 (1969).

perimental work in four different laboratories. The range of wavelengths measured is between 2565 and 12,381 Å. The wavelengths were measured by reference to standard sources of ^{198}Hg, ^{86}Kr, or internal thorium wavelengths, and the spectra were produced by several different versions of the electrodeless discharge, as well as by a cooled hollow-cathode discharge. Extensive evaluations of the random and systematic errors between all measurements have been made by Giacchetti et al.[85] and Valero.[86] The former also presents a list of all wavelengths judged suitable for Class B standards. This list includes 1375 ThI lines which have been fitted with an average standard deviation of 0.0014 cm^{-1}, and an additional 181 accurately measured wavelengths, which result in a total of 1556 lines that may be adopted as secondary standards. Despite the absence of any significant shift among the different experiments, it has been recommended that additional measurements be made, particularly with the cooled hollow cathode source described by Littlefield and Wood,[80] and that they be made by direct reference to the primary standard. The listing of grating measurements by Zalubas[87] and the photographs by Junkes and Salpeter[88] will be of particular value for initial identifications.

Wavelength standards in the vacuum ultraviolet present an especially difficult experimental problem. The use of interferometers below 2000 Å is not practical at the present because of the low reflectivity and transmission of most materials in that region. As an alternative, the best standards below 2000 Å are calculated line positions that have been derived from energy levels determined precisely by measurements of other transitions at longer wavelengths. In addition, measurements with large grating spectrographs have yielded a large number of very useful reference spectra (but not with the accuracy necessary for secondary standards). In most cases these lines have been measured in higher orders by comparison with longer wavelength standards of lower orders. This technique is quite satisfactory as long as it can be demonstrated that the grating equation is obeyed, i.e., that coincidence errors do not exist or

[85] A. Giacchetti, R. W. Stanley, and R. Zalubas, J. Opt. Soc. Amer. 60, 474 (1970).

[86] F. P. J. Valero, J. Opt. Soc. Amer. 60, 1675 (1970).

[87] R. Zalubas, New Description of Thorium Spectra. Nat. Bur. Std. U.S., Monograph 17, U.S. Govt. Printing Office, Washington, D.C., 1970.

[88] J. Junkes and E. W. Salpeter, "Spectrum of Thorium from 9400 Å to 2000 Å." Specola Vaticana, Città del Vaticano, Rome, 1964.

that they can be corrected.* The presence of a few well-measured lines in the same order as the lines to be measured may be used to check for and correct such coincidence errors (which appear to be much less a problem with modern gratings). The procedures and difficulties have been summarized effectively by Edlén,[89] who also gives a condensed finding list of atomic lines between 2000 and 17 Å, as well as a summary of the existing standards. Samson,[6] (Chapter 10) presents a more recent summary of references of this topic, as well as a reproduction of Edlén's finding list of light elements.

The identification of atomic lines in all wavelength regions often is of vital importance in evaluating an emission or absorption system. For wavelengths between 2000 and 10,000 Å, the MIT wavelength tables[90] are particularly useful. When possible, precision measurements may be obtained for some lines in more recent sources (see, for example, the work of Zaidel' et al.[91]), since the MIT tables are correct only to 0.1 or possibly 0.01 Å for most lines. In the vacuum ultraviolet, identification of atomic lines may be facilitated by the use of the photographs by Junkes et al.[92] and by the more extensive tables of atomic wavelengths by Kelly.[93]

2.3.2.2.3. INTENSITY STANDARDS. Measurements of relative intensities are necessary in the standard detection system of any spectrograph or spectrometer. The problem of obtaining *precise* measurements of intensities (relative or absolute) at any wavelength is a more difficult problem. In this case, the detector must have a known response to the

[88a] G. Herzberg, *J. Mol. Spectrosc.* **33**, 147 (1970).

[89] B. Edlén, *Rep. Progr. Phys.* **26**, 181 (1963).

[90] G. R. Harrison, "M.I.T. Wavelength Tables," revised ed. M.I.T. Press, Cambridge, Massachusetts, 1969.

[91] A. N. Zaidel', V. K. Prokof'ev, S. M. Raiskii, V. A. Slavnyi, and E. Ya. Schreider, "Tables of Spectral Lines." IFI/Plenum, New York, 1970 (transl. from the 1969 Russian edition published by Nauka Press, Moscow).

[92] J. Junkes, E. W. Salpeter, and G. Milazzo, "Atomic Spectra in the Vacuum Ultraviolet from 2250 Å to 1100 Å," Part One: Al, C, Cu, Fe, Ge, Hg, Si, (H_2). Specola Vaticana, Città del Vaticano, Rome, 1965.

[93] R. L. Kelly, Vacuum Ultraviolet Emission Lines, UCRL 5612. Univ. of California, Lawrence Radiat. Lab., 1960.

---

* Herzberg (Ref. 88a) reports that, for a good grating, coincidence errors should be less than ± 0.002 Å. He also discusses the standards used in the study of H_2 in the region 880–810 Å.

radiation, or it must be calibrated against a source of known intensity. Since there are very few absolute detectors, and even those that exist are not satisfactory at all wavelengths or at low intensities, most intensity measurements are made by comparison to an intensity standard. Most standards refer ultimately to a "blackbody" radiator (or emitter). An ideal blackbody radiator would be a large hollow sphere that is perfectly black inside (i.e., it would completely absorb or emit all wavelengths). Radiation could enter or leave the sphere through a small hole in the wall. The absolute intensity and spectral distribution of such a system would be defined in terms of its equilibrium temperature by Planck's radiation law. There are descriptions for the construction of such systems,[94,95] but they are not well suited for routine use. In practice, standard tungsten filament lamps[96-98], which have been carefully calibrated against such sources, may be obtained from the National Bureau of Standards. The "color temperature" of these lamps (at a specified operating voltage) is the temperature at which a blackbody source would produce the same spectral distribution as the lamp.

Although photographic plates serve very well as detectors, especially at low light levels, great care must be exercised if they are to be used for precise measurements of intensities. Sawyer[7] (Chapter 10) describes several methods of calibrating photographic plates, and outlines the precautions and procedures that must be followed in order to maintain the calibration of such plates. Photoelectric detectors are much more suitable for intensity measurements because their characteristics may be determined and maintained more easily. Even with photoelectric detectors it is important that the operating conditions (temperature, applied voltage, and signal current) be maintained within the calibrated range. The use of thermopile–galvanometer detectors and chemical actinometers has been discussed in considerable detail by Calvert and Pitts.[99] The thermopile and photoionization detectors also are discussed in detail by Samson[6] (Chapter 8), especially as they apply to intensity measurements in the vacuum ultraviolet. Samson also discusses the important development of synchrotron radiation as a source of continuum of predictable intensity and spectral distribution from the far infrared to the X-ray

[94] W. W. Coblentz, *Nat. Bur. Std. (U.S.) Bull.* **10**, 1 (1914).
[95] F. Anacker and R. Mannkopff, *Z. Phys.* **155**, 1 (1959).
[96] D. B. Judd, *J. Res. Nat. Bur. Std. (U.S.)* **44**, 1 (1950).
[97] L. E. Barbrow, *J. Opt. Soc. Amer.* **49**, 1122 (1959).
[98] M. W. P. Cann, *Appl. Opt.* **8**, 1645 (1969).
[99] J. G. Calvert and J. N. Pitts, Jr., "Photochemistry." Wiley, New York, 1967.

region. Recent papers by Pitz[100,101] demonstrate the use of synchrotron radiation for intensity calibrations in the region near 2000 Å. Buckley[102] has used a tungsten filament with a sapphire window as a calibration source in the region 1500–2700 Å.

The reviews by Nicodemus[103] and Rutgers[104] provide leading references to recent papers, and they are especially helpful in clarifying the nomenclature used in this field.

2.3.2.3. Light Sources

2.3.2.3.1. EMISSION SOURCES. Flames provide a convenient source of many band systems.[105] Molecular emitters may be formed directly from the primary combustion materials or by the addition of a substance into an established flame. The species produced tend to be neutral diatomic molecules (such as CH, NH, OH, C_2, and CN), although several polyatomic molecules have been observed as well (such as BO_2, C_3, NH_2, and CHO). The observed transitions characteristically are between the ground and low-lying excited states of the molecule. The primary disadvantages of such flames are their low intensity, normally high operating pressure, and moderately low excitation energies.

The use of flames has received considerable attention recently as a method of producing gas-phase chemical lasers, such as the HCl laser reported by Kasper and Pimentel.[106] Most such systems involve supersonic jets that permit two highly reactive species to mix.

Arc and spark discharges (see Sawyer,[7] pp. 21–26) provide much more energetic sources. Typical conditions for an arc in air would be for an electrode gap of about 10–15 mm, operated at 100 to 250 V dc with a current of about 5 A. A typical spark condition would require higher voltages (as high as 15 kV) often with a capacitor and intermittent contact to "condense" the discharge in order to obtain higher excitation energies. These sources are not particularly useful for discrete molecular emission spectra. The effective temperatures may be as high as 3000°C in an arc, thus allowing a wider range of substances to be vaporized into

[100] E. Pitz, *Appl. Opt.* **8**, 255 (1969).

[101] E. Pitz, *Appl. Opt.* **10**, 813 (1971).

[102] J. L. Buckley, *Appl. Opt.* **10**, 1114 (1971).

[103] F. E. Nicodemus, *J. Opt. Soc. Amer.* **59**, 243 (1969).

[104] G. A. W. Rutgers, *Appl. Opt.* **10**, 2595 (1971).

[105] R. Mavrodineanu and H. Boiteux, "Flame Spectroscopy." Wiley, New York, 1965.

[106] J. V. V. Kasper and G. C. Pimentel, *Phys. Rev. Lett.* **14**, 352 (1965).

the arc. Arcs have been a common source of diatomic oxide and hydride spectra when they are operated in oxygen- or hydrogen-rich atmospheres.

Electrical discharge tubes offer a wider range of operating conditions than the simpler flames or arcs, so that these devices are used much more frequently today. Initial disadvantages of low intensity have been improved for many of them and they can provide considerably more stable sources suitable for photoelectric detection and statistical reduction of background noise. Although these glow discharges may be sustained by dc, ac, or rf power sources, the dc discharge has been studied extensively and serves as a model for other types. In a "normal discharge" several different characteristic regions may be distinguished between the cathode and the anode: (1) Aston dark space, (2) cathode glow, (3) cathode dark space, (4) negative glow, (5) Faraday dark space, (6) positive column (often with striations), (7) anode glow, and (8) the anode dark space. The position and extent of these regions depend critically on the nature and pressure of the gas. Some depend on the geometry of the discharge tube or of the electrodes. The most energetic excitations will be found in the negative glow, since most of the discharge voltage drop is found between it and the cathode (voltage gradients of several hundred volts per centimeter are typical in the cathode dark space). It is fairly common to observe highly excited states of both neutral and ionized atoms and molecules in the negative glow. The positive column, on the other hand, produces much less energetic excitations (typical voltage gradients of only a few volts per centimeter), normally of neutral species. Even excitations in the positive column tend to be more energetic excitations than may be obtained in a flame or arc. One of the main disadvantages of the positive column is that it tends to be an extended discharge, which may spread out over an extensive volume. This may be overcome to a certain extent by forcing the discharge through a restriction (e.g., a a typical tube might be 10–20 cm long with a bore of 3 to 5 mm) and viewing the discharge end on.

The *hollow cathode* discharge is a special case of the glow discharge which takes advantage of the tendency of the negative glow to remain closely associated with the cathode surface. Paschen[107] was the first to use such a discharge, although the version used by Schüler[108] is more convenient for higher currents. Under a given set of conditions, for

[107] F. Paschen, *Ann. Phys.* **50**, 901 (1916).
[108] H. Schüler, *Phys. Z.* **22**, 264 (1921).

example, the negative glow will be located a constant distance from the cathode surface—regardless of the shape of the electrode. If the cathode is formed into two parallel plates, there will be a negative glow associated with each; and if they are brought close enough together, their negative glows will merge. Such a discharge will behave as a hollow cathode discharge if the cathodes are reasonably large. In practice it is more convenient for the cathode to be a hollow cylinder and to adjust the gas pressure so that the negative flow fills the inside of the cylinder. Such discharges are characterized by high excitation energies and by a tendency of the material on the walls of the cathode to be "sputtered" into the discharge. Location of the anode is not important, although there usually is a small visible discharge near the anode (which does not dissipate much energy). The starting voltage for hollow cathode discharge lamps normally is about 500–700 V and the operating voltage is about 150–400 V depending primarily on the cathode material. Since these are constant voltage lamps, the current must be limited by an external device, such as a resistor, to a value that is determined by the ability to dissipate the heat generated at the cathode. Air-cooled lamps normally can operate with currents between 5 and 100 mA, whereas water-cooled lamps may handle currents of several amperes. When moderate currents ($\sim$30 mA) are used and the cathode is cooled to liquid nitrogen temperatures these lamps can yield extremely sharp and reproducible spectra. They have the further advantage of "sputtering" metal atoms from the surface of the cathode into the discharge, thus providing a very convenient source of atoms in the gas phase. These properties, in addition to their high stability, make them well suited for use as sources in atomic absorption spectroscopy. Commercial lamps serve very well as routine sources of iron reference spectra, thus effectively replacing the iron arcs. Cooled hollow cathode lamps also have been recommended as appropriate sources for interferometric studies of secondary standards.[109]

Electrodeless discharges have been used with increasing regularity as sources of atomic and molecular spectra and as sources of metastable atoms or molecules for chemical studies.[110] These discharges usually are maintained by high-frequency fields *outside* the actual discharge tubes, which explains in part their utility and convenience. Commercial micro-

[109] H. M. Crosswhite, G. H. Dieke, and C. S. Legagneur, *J. Opt. Soc. Amer.* **45**, 270 (1955).

[110] "Chemical Reactions in Electrical Discharges," Adv. in Chem. Series 80. American Chem. Soc., Washington, 1969.

wave generators* (often medical diathermy units) operating at 2.45 GHz (2450 MHz) with about 125-W output can be coupled effectively to quartz or Pyrex tubes (typically 13-mm o.d.) with the resonance cavities described and evaluated in detail by Feshenfeld et al.[111] Commercial versions[†] of these cavities are available. These cavities can be tuned to optimize light output and to minimize the power reflected to the magnetron over fairly broad ranges of gas pressure (e.g., between 0.1 and 100 Torr for helium). A reflected power meter located between the magnetron and the resonance cavity greatly simplifies the tuning adjustments and ensures a minimum of reflected power. This tuning is essential since excess reflected power can damage the magnetron.

Discharges powered by the medical diathermy units may not be sufficiently stable for some photoelectric detection applications. This instability reflects the poor regulation of the high voltage power supplies in these units. The problem can be eliminated by modification of the power supplies or by purchasing microwave generators (such as those marketed by the Jarrell–Ash Company and Raytheon Corporation) that already have adequate regulation. The microwave discharge can be modulated or pulsed (see, for example, Callear et al.[112]) but these techniques have not yet gained widespread use.

An electrodeless discharge operating at 2.45 GHz tends to be localized to about 5–10 cm of the 13-mm o.d. tube, except at lower pressures, where it may spread out over a 30–50-cm length of the tube. The spectra produced in these discharges tend to be intermediate in energy between those produced in a hollow cathode and those obtained in a positive column. The excitation process is quite different. Ions are seldom seen with these discharges, yet they are so disruptive that few polyatomic molecules can be studied with them. In addition, they do provide convenient sources of continua in the vacuum ultraviolet as well as spectra

[111] F. C. Fehsenfeld, K. M. Evenson, and H. P. Broida, Rev. Sci. Instrum. **36**, 294 (1965).

[112] A. B. Callear, J. Guttridge, and R. E. M. Hedges, Trans. Faraday Soc. **66**, 1289 (1970).

---

* The Burdick Corporation, Milton, Wisconsin; Raytheon Co., Production Equipment Department, Norwalk, Connecticut 06856; Jarrell–Ash Division, Fisher Scientific Co., Waltham, Massachusetts 02154; and Aztec Instruments, Inc., South Norwalk, Connecticut 06854.

† Opthos Instrument Co., Rockville, Maryland 20850 and Aztec Instruments, Inc., South Norwalk, Connecticut 06854.

involving highly excited states. The excitation processes seem to be rather complex since a rather wide variety of spectra may be obtained by control of the gas pressure, composition, flow rate, and discharge power input. The construction details given by McNesby and Okabe[113] and by Okabe[114] for atomic sources and by Wilkinson and Byram[115] for rare-gas continua emphasize the pitfalls to successful operation of such lamps.

Electrodeless discharges operating at lower frequencies, e.g., 10–20 MHz, produce spectra more analogous to those of the positive column. These discharges tend to fill the available volume, often extending over a length of 50 to 100 cm. The output powers may be between 100 and 1000 W, but the power density in the discharge and the intensity is much lower than with the 2.45-GHz discharges. These discharges (see, for example, Robinson's papers[116–118]) have been used to produce the emission spectra of benzene, formaldehyde, and related molecules. Plato[119] and Garton et al.[120] have reported their use for the production of vacuum ultraviolet radiation.

2.3.2.3.2. SOURCES FOR ABSORPTION SPECTRA. All of the techniques described in the previous section may be used to produce a continuum over some wavelength range. The real problem then is to find the most convenient source for a particular application. The simplest source is a blackbody radiator, for example, a tungsten filament lamp, where the intensity as a function of wavelength is given by the Planck radiation law. For temperatures below 5000 K, the intensity is very low for wavelengths less than 3000 Å, although such a source is very useful for the visible and near-infrared regions (4000–12,000 Å). A typical tungsten ribbon lamp may operate at 6 to 100 V and 25 to 200 W in an inert atmosphere. The tungsten–halogen lamps provide much greater intensity, although their filaments are coiled and less suitable for imaging on a spectrograph slit. Blackbody radiation also is present (whether wanted or not) in

[113] J. R. McNesby and H. Okabe, in "Advances in Photochemistry" (W. A. Noyes, Jr., G. S. Hammond, and J. N. Pitts, Jr., eds.), Vol. 3. Wiley (Interscience), New York, 1964.

[114] H. Okabe, J. Opt. Soc. Amer. 54, 478 (1964).

[115] P. G. Wilkinson and E. T. Byram, Appl. Opt. 4, 581 (1965).

[116] G. W. Robinson, Can. J. Phys. 34, 699 (1956).

[117] G. W. Robinson, J. Chem. Phys. 22, 1147 (1954).

[118] G. W. Robinson, J. Chem. Phys. 22, 1384 (1954).

[119] M. Plato, Z. Naturforsch 19a, 1324 (1964).

[120] W. R. S. Garton, M. S. W. Webb, and P. C. Wildy, J. Sci. Instrum. 34, 496 (1957).

TABLE I. Continuum Light Sources for Absorption Spectroscopy[n]

Source	Region (Å)	Remarks[a]
Zirconium	5500–12,000	About 20 atomic lines between 7000 and 9000 Å.[b]
Tungsten incandescent, Pyrex envelope	3000–12,000	May be operated considerably above normal rating for high intensity, short lifetime operation.[c]
A–H6 mercury lamp	2800–4000	Pressure about 110 atm; broadened atomic lines and continuum; useful continuum only at very high dispersion.[d]
High-pressure xenon	2000–9000	Some atomic lines, especially for $\lambda > 4000$ Å.[e]
Hydrogen	1300–3500	Good continuum; weak in vacuum ultraviolet compared with rare gas discharges.[f]
Xenon	1470–2100	Simple condensed discharge or electrodeless
Krypton	1240–1800	sources using 2450-MHz excitation; con-
Argon	1070–1600	densed source much more intense.[g,h]
Neon	750–950	2450-MHz excitation or condensed discharge. Not useful since range more easily covered with He continuum.[i,j]
Helium	600–900	2450-MHz excitation or condensed discharge.[k,l,m]
Lyman continuum	270–9000	

[a] See W. C. Price, *Advan. Spectrosc.* **1**, 56 (1959), for more details concerning sources for the vacuum ultraviolet.

[b] W. D. Buckingham and C. R. Deibert, *J. Opt. Soc. Amer.* **36**, 245 (1946). Sylvania Electric Products, Inc., New York, New York.

[c] Commercially available from a number of manufacturers.

[d] General Electric Company, Cleveland, Ohio.

[e] T. Heller, *Z. Astrophys.* **38**, 55 (1955); Hanovia Chemical and Manufacturing Co., Newark, New Jersey.

[f] Many references: see, for example, G. B. Kistiakowsky, *Rev. Sci. Instrum.* **2**, 549 (1931); and P. L. Hartman and J. R. Nelson, *J. Opt. Soc. Amer.* **47**, 646 (1957). Commercially available from a number of manufacturers.

[g] Y. Tanaka, *J. Opt. Soc. Amer.* **45**, 663, 710 (1955); P. G. Wilkinson and Y. Tanaka, *ibid.* **45**, 344 (1955); P. G. Wilkinson, *ibid.* **45**, 1044 (1955).

[h] Y. Tanaka, A. S. Jursa, and F. J. Le Blanc, *J. Opt. Soc. Amer.* **48**, 304 (1958).

[i] T. Takamine, S. Suga, Y. Tanaka, and P. Imotani, *Sci. Papers Inst. Phys. Chem. Res.* (*Tokyo*) **35**, 447 (1939).

[j] J. J. Hopfield, *Astrophys. J.* **72**, 133 (1930).

[k] G. Collins and W. C. Price, *Rev. Sci. Instrum.* **5**, 423 (1934); see also T. A. Brix and G. Herzberg, *Can. J. Phys.* **32**, 110 (1954).

[l] W. R. S. Garton, *J. Sci. Instrum.* **36**, 11 (1959).

[m] P. A. Warsop, *Spectrochim. Acta* **16**, 575 (1960).

[n] Reproduced from the chapter by G. W. Robinson, Electronic Spectra, *in* "Methods of Experimental Physics," Academic Press, New York, 1962.

flames, arcs, and discharge tubes. Table I provides a brief summary of a few of the more common light sources used for absorption spectroscopy. A more efficient lamp may use an electrical discharge to produce a transition between molecular electronic states, one of which is repulsive. Such molecular continua usually cover a more limited spectral region and they often will have discrete atomic or molecular transitions superimposed. For example, a discharge (positive column, hollow cathode, or electrodeless) through H_2 or D_2 produces a very intense continuum between 1300 and 3500 Å. Similarly the Hopfield continuum[121-123] of helium extends from 600 to 1000 Å.* Tanaka and Huffman have studied this continuum extensively and developed much-improved experimental techniques.[124-130] Their lamp assembly and discharge circuit are suitable for excitation of other rare-gas continua (at higher pressures) that extend to longer wavelengths (see, for example, papers by Hopfield, Huffman et al., Takamine et al., and Tanaka et al.[121,131-135]).† Wilkinson[120,136,137]

[121] J. J. Hopfield, *Phys. Rev.* **35**, 1133 (1930).

[122] J. J. Hopfield, *Phys. Rev.* **36**, 784 (1930).

[123] J. J. Hopfield, *Astrophys. J.* **72**, 133 (1930).

[124] Y. Tanaka, A. S. Jursa, and F. J. LeBlanc, *J. Opt. Soc. Amer.* **48**, 304 (1958).

[125] R. E. Huffman, Y. Tanaka, and J. C. Larrabee, *Appl. Opt.* **2**, 617 (1963); *J. Appl. Phys. (Japan) Suppl. I* **4**, 494 (1965).

[126] R. E. Huffman, J. C. Larrabee, Y. Tanaka, and D. Chambers, *J. Opt. Soc. Amer.* **55**, 101 (1965).

[127] R. E. Huffman, J. C. Larrabee, and D. Chambers, *Appl. Opt.* **4**, 1145 (1965).

[128] R. E. Huffman, W. W. Hunt, Y. Tanaka, and R. L. Novack, *J. Opt. Soc. Amer.* **51**, 693 (1961).

[129] R. E. Huffman, Y. Tanaka, and J. C. Larrabee, *J. Opt. Soc. Amer.* **52**, 851 (1962).

[130] R. E. Huffman, J. C. Larrabee, and Y. Tanaka, *Appl. Opt.* **4**, 1581 (1965).

[131] M. E. Levy and R. E. Huffman, *J. Quant. Spectrosc. Radiat. Transfer* **9**, 1349 (1969).

[132] M. E. Levy and R. E. Huffman, *Appl. Opt.* **9**, 41 (1970).

[133] T. Takamine, T. Suga, Y. Tanaka, and G. Imotani, *Sci. Papers Inst. Phys. Chem. Res. (Tokyo)* **35**, 447 (1939).

[134] Y. Tanaka, A. S. Jursa, and F. J. LeBlanc, *J. Opt. Soc. Amer.* **47**, 105 (1957).

[135] Y. Tanaka, *J. Opt. Soc. Amer.* **45**, 710 (1955).

[136] P. G. Wilkinson, *J. Opt. Soc. Amer.* **45**, 1044 (1955).

[137] P. G. Wilkinson and Y. Tanaka, *J. Opt. Soc. Amer.* **45**, 344 (1955).

---

* Herzberg reports the use of a pulse unit supplied by Velonex, Santa Clara, California, which resulted in a considerable improvement in the intensity of their Hopfield continuum lamp.

† The "Hinteregger Type Lamp" (McPherson Instrument Corp., Acton, Massachusetts) and the "VUV Arc Radiation Source" (Photochem Industries, Fairfield, New Jersey) are commercial versions of the lamps described in these references.

has described a series of sealed microwave discharge lamps which provide very convenient sources of continua throughout the region 1900 to about 600 Å. These inert gas continua offer the additional advantage of simplifying the separation of orders, since the continuum from any single gas is quite restricted, as shown in Fig. 4 and listed in Table I. They may be made by the techniques described by Wilkinson[115] or purchased commercially.* Huffman et al.[130] found that the relative intensities of the microwave powered lamps were much lower than the condensed spark discharges, even if an 800-W microwave generator was used in place of the more conventional 125-W generator.

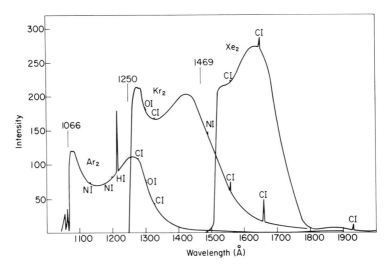

FIG. 4. Rare gas continua of argon, krypton, and xenon. Gas pressures were about 200 Torr and they were excited by a 125-W microwave generator [reproduced from the paper by P. G. Wilkinson and E. T. Byram, *Appl. Opt.* **4**, 581 (1965), by permission of the publishers].

When high-energy electrons are accelerated by changing their trajectories with external fields, radiation will be produced. The intensity distribution as a function of wavelength will resemble blackbody radiations, but in this case it is determined by the energy and acceleration of the electrons. This technique provides radiation from the X-ray region into the infrared region, but it is unique in providing a smooth continuum in the vacuum ultraviolet. The first spectroscopic use of synchrotron

---

* Ophthos Instrument Co., Rockville, Maryland.

radiation was reported by Elder et al.[138,139] in 1947, based on experiments with the 70-MeV General Electric Synchrotron. Subsequent reports by Tomboulian and co-workers,[140–142] Codling and Madden,[143–148] Cauchois et al.,[149] and Stebbings and Taylor[150] indicate the utility of the technique and summarize the necessary theoretical and experimental considerations. Unfortunately this is a case where the spectrograph must be taken to the light source rather than the reverse.

High-pressure xenon lamps are arc sources (20 V and up to 200 A) that operate at pressures of about 15 atm and provide continua between about 1900–9000 Å. They are most efficient in the visible region, but they do have some broad linelike peaks superimposed on the continuum. They offer an additional advantage of being nearly point sources, i.e., most of the light is produced in an arc that is only a few millimeter high and about 1 mm wide. Johnson[151] has described the construction of such lamps from standard high-vacuum components, and he concludes that gases other than xenon are useful when it is desirable to shift the intensity distributions to shorter wavelengths (useful to 1100 Å with argon).

A modification of the lamp is available from Varian,* in which a parabolic reflector is built axially around the light source. This design permits virtually all of the light to be directed into the optical system. The technique is also incorporated with external reflectors in solar simulators.

All of the lamps mentioned above (except the condensed discharge) normally are operated continuously. The use of pulsed light sources,

[138] F. R. Elder, A. M. Gurewitsch, R. V. Langmuir, and H. C. Pollock, *Phys. Rev.* **71**, 829 (1947).

[139] F. R. Elder, R. V. Langmuir, and H. C. Pollock, *Phys. Rev.* **74**, 52 (1948).

[140] P. L. Hartman and D. H. Tomboulian, *Phys. Rev.* **91**, 1577 (1953).

[141] D. H. Tomboulian and P. L. Hartman, *Phys. Rev.* **102**, 1423 (1956).

[142] D. H. Tomboulian and D. E. Bedo, *J. Appl. Phys.* **29**, 804 (1958).

[143] R. P. Madden and K. Codling, *Phys. Rev. Lett.* **10**, 516 (1963).

[144] K. Codling and R. P. Madden, *Phys. Rev. Lett.* **12**, 106 (1964).

[145] R. P. Madden and K. Codling, *J. Opt. Soc. Amer.* **54**, 268 (1964).

[146] K. Codling and R. P. Madden, *Appl. Opt.* **4**, 1431 (1965).

[147] R. P. Madden and K. Codling, *Astrophys. J.* **141**, 364 (1965).

[148] K. Codling and R. P. Madden, *J. Appl. Phys.* **36**, 380 (1965).

[149] Y. Cauchois, C. Bonnelle, and G. Missoni, *C. R. Acad. Sci. Paris* **257**, 409, 1242 (1963).

[150] W. L. Stebbings and J. W. Taylor, *Int. J. Mass Spectrom. Ion Phys.* **6**, 152 (1971).

[151] P. M. Johnson, *J. Opt. Soc. Amer.* **60**, 1669 (1970).

* Varian, Palo Alto, California 94303.

on the other hand, offers two possible advantages: higher peak powers and time resolution. The Lyman discharge is characteristic of such lamps and has been used with considerable success for absorption studies between about 700–9000 Å. It has been the subject of studies[152–158] and has undergone considerable modifications. However, in its simplest form it is a capacitor discharge through a very small restriction in the shortest possible time. For example, a 2-μF capacitor charged to about 15 kV and discharged within 5 μsec through a 3-cm-long capillary of about 3-mm bore produces the desired continuum when viewed down the capillary. If the capillary is made of quartz, the discharge may be viewed from the side (which eliminates many of the emission lines originating outside the capillary) for wavelengths greater than about 3500 Å. Capillary materials and gas composition are not critical; however, argon at a pressure of about 20 Torr is a convenient gas for precisely timed discharges. Ceramic, quartz, or boron nitride provide reasonably strong capillaries. In order to obtain the maximum current density, and therefore maximum intensity, it is essential to minimize the resistance and inductance of the circuit. The resistance of the discharge itself is quite small, so that the flash duration may be estimated as about $1/2f$, where f is the ringing frequency,

$$f = \frac{1}{2\pi} \left(\frac{1}{2C} - \frac{R^2}{4C^2} \right)^{1/2} \approx \frac{1}{2\pi} \left(\frac{1}{LC} \right)^{1/2}. \tag{2.3.17}$$

It may not be practical to reduce the duration of a high-current flash below about 3 μsec because of the time required to quench the discharge. The Garton source[159–161] uses a tube with a larger bore, but apparently depends on high current densities to "pinch" the discharge into a narrow channel. The major advantage of this source is the extended life of the discharge tubes. The system does require very careful design in order to minimize the total inductance in the discharge circuit. This includes the use of specially designed capacitors of very low inductance. The fast

[152] T. Lyman, *Astrophys. J.* **60**, 1 (1924).

[153] T. Lyman, *Science* **64**, 89 (1926).

[154] G. Collins and W. C. Price, *Rev. Sci. Instrum.* **5**, 423 (1934).

[155] R. E. Worley, *Rev. Sci. Instrum.* **13**, 67 (1942).

[156] P. Brix and G. Herzberg, *Can. J. Phys.* **32**, 110 (1954).

[157] M. Morlais and S. Robin, *C. R. Acad. Sci. Paris* **258**, 862 (1964).

[158] R. Goldstein and F. N. Mastrup, *J. Opt. Soc. Amer.* **56**, 765 (1966).

[159] W. R. S. Garton, *J. Sci. Instrum.* **30**, 119 (1953).

[160] W. R. S. Garton, *J. Sci. Instrum.* **36**, 11 (1959).

[161] J. E. G. Wheaton, *Appl. Opt.* **3**, 1247 (1964).

magnetic compression of plasmas may result from axial currents (the "z-pinch") or from circular currents (the "θ-pinch"). Recent reports by Niemann and Klenert,[162] Daby et al.,[163] and by Kuswa and Stallings[164] explore the possible applications of these pinch devices for the production of high-intensity light sources.

2.3.2.3.3. FLASH PHOTOLYSIS. The light sources to be considered in this section are those that produce photodissociation of the parent species into molecular fragments. They normally are used along with one of the light sources discussed in previous paragraphs (usually the Lyman lamp for spectroscopic studies). A number of excellent reviews are available (see, for example, Porter[165,166] and Calvert and Pitts[99] (pp. 590–592, 710–723)) on the history and development of the flash photolysis technique. Briefly the early investigations were carried out about 1950 at Cambridge by Norrish and Porter,[167–169] at California Institute of Technology by Davidson et al.,[170] and at the National Research Council of Canada by Ramsay and Herzberg.[171,172] The system developed by Claesson and Lindqvist,[173,174,174a] and shown schematically in Fig. 5, marked a significant increase in the amount of energy delivered to the flash lamps (up to 33 kJ). These authors, in particular, provide considerable details on the construction of the system. Recent systems[175] incorporate one or two flash tubes about 2 m long which produce a flash lasting

[162] E. G. Niemann and M. Klenert, *J. Appl. Opt.* **7**, 295 (1968).

[163] E. E. Daby, J. S. Hitt, and G. J. Mains, *J. Phys. Chem.* **74**, 4204 (1970).

[164] G. Kuswa and C. Stallings, *Rev. Sci. Instrum.* **41**, 1429 (1970).

[165] G. Porter, Flash Photolysis, *in* "Techniques of Organic Chemistry" (S. L. Friess, E. S. Lewis, and A. Weissberger, eds.), 2nd ed., Vol. VIII, Part II. Investigations of Rates and Mechanisms of Reactions. Wiley (Interscience), New York, 1963.

[166] G. Porter, Flash Photolysis, *in* "Photochemistry and Reaction Kinetics" (P. G. Ashmore, N. S. Dainton, and T. M. Sugden, eds.). Cambridge Univ. Press, London and New York, 1967.

[167] R. G. W. Norrisch and G. Porter, *Nature (London)* **164**, 658 (1949).

[168] G. Porter, *Proc. Roy. Soc. (London)* **A200**, 284 (1950).

[169] G. Porter, *Discuss. Faraday Soc.* **9**, 60 (1950).

[170] N. Davidson, R. Marshall, A. E. Larsh, Jr., and T. Carrington, *J. Chem. Phys.* **19**, 1311 (1951).

[171] G. Herzberg and D. A. Ramsay, *Discuss. Faraday Soc.* **9**, 80 (1950).

[172] D. A. Ramsay, *J. Chem. Phys.* **20**, 1920 (1952).

[173] S. Claesson and L. Lindqvist, *Arkiv. Kem.* **11**, 535 (1957).

[174] S. Claesson and L. Lindqvist, *Arkiv. Kem.* **12**, 1 (1958).

[174a] S. Claesson, L. Lindqvist, and R. L. Strong, *Arkiv Kem.* **22**, 245 (1963).

[175] J. W. C. Johns, S. H. Priddle, and D. A. Ramsay, *Discuss. Faraday Soc.* **22**, 90 (1963).

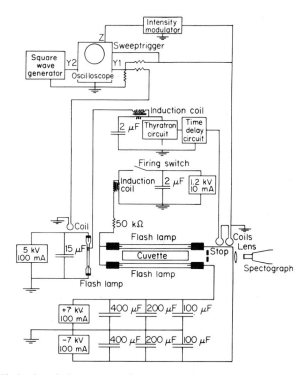

FIG. 5. Flash photolysis apparatus [reproduced from the paper by S. Claesson and
L. Lindqvist, *Arkiv. Kem.* **11**, 535 (1957), by kind permission of the authors and of the
journal].

about 50 μsec. As with the Lyman lamps, the flash duration is roughly
$\pi(LC)^{1/2}$, so that it is imperative to minimize the inductance in the discharge circuit.* Since the total energy is given by $E = \frac{1}{2}CV^2$, decreasing
the capacitance (at a fixed voltage) decreases both the flash duration and
the flash intensity.† With all flashlamps, especially those dissipating a
large amount of energy, ion recombination rates in the flash tube may
establish a lower limit of 2 to 50-μsec duration.

---

* Low-inductance capacitors are available from suppliers such as: Aerovox Corp.,
740 Belleville Avenue, New Bedford, Massachusetts 02741; McGraw–Edison, Power
Systems Division, Canonsburg, Pennsylvania 15317; Maxwell Laboratories, Inc., 9244
Balboa Avenue, San Diego, California 92123; and Tobe Deutschmann Laboratories,
Canton, Massachusetts 02021.

† Most systems currently in use are designed for a maximum voltage of about 20 kV,
although Claesson *et al.*[174a] have described a 50 kV system.

Data are available[176] for the maximum energy per discharge for straight flashlamps made of quartz. Actual lamp construction or operating conditions may differ from those used in establishing the reported limits, therefore each lamp must be tested with a considerable degree of caution. A 35-mm o.d. tube might operate typically at about 50–70 Torr when filled with argon (or other rare gases). Typically the flash lamp would be placed parallel to a quartz absorption cell containing the parent molecules and both tubes would be surrounded by a reflector housing. The reflectors might be cylindrical or elliptical in shape, and their inner surface should be highly reflective throughout the entire visible and ultraviolet. Good results may be obtained by "smoking" the surface with MgO produced by burning a magnesium ribbon in air. When the MgO is sufficiently thick (i.e., *almost* ready to flake off) these diffuse coatings reflect about 98% of the light between 2400 and 8000 Å.[177] Eastman Kodak Company* now markets high-purity $BaSO_4$, which may be applied as a paint or in packed form. They claim that either of these surfaces is more reflective than the best MgO surfaces, especially at wavelengths below 2500 Å after repeated exposure to ultraviolet light.

Flash photolysis of compounds that do not absorb at wavelengths greater than 2000 Å is extremely difficult because of the limited transmission of the quartz flash and absorption tubes and because of the limited reflectivity of the reflectors. Several techniques have been used to minimize some of these problems. These techniques include the use of sapphire tubes (1450–2000 Å),[178,179] or LiF windows with vacuum arcs (1050–2000 Å).[180,180a] Other special techniques of flash photolysis (but not in the vacuum ultraviolet yet) include the use of pulsed microwaves by Callear and co-workers,[181,182] flash pyrolysis by Nelson and co-workers,[183,184]

[176] D. E. Permlan, *Rev. Sci. Instrum.* **38**, 68 (1967).
[177] W. E. K. Middleton and C. L. Sanders, *J. Opt. Soc. Amer.* **41**, 419 (1951).
[178] L. S. Nelson and D. A. Ramsay, *J. Chem. Phys.* **25**, 372 (1956).
[179] L. S. Nelson, *J. Opt. Soc. Amer.* **46**, 768 (1956).
[180] W. Braun, A. M. Bass, and A. E. Ledford, Jr., *Appl. Opt.* **6**, 47 (1967).
[180a] W. Braun, A. M. Bass, D. D. Davis, and J. D. Simmons, *Proc. Roy. Soc. A* **312**, 417 (1969).
[181] A. B. Callear and R. E. M. Hedges, *Trans. Faraday Soc.* **66**, 605, 615 (1970).
[182] A. B. Callear, J. Guttridge, and R. E. M. Hedges, *Trans. Faraday Soc.* **66**, 1289 (1970).
[183] L. S. Nelson and J. L. Lundberg, *J. Phys. Chem.* **63**, 433 (1959).
[184] L. S. Nelson and N. A. Kuebler, *J. Chem. Phys.* **33**, 610 (1960).

* "White Reflectance Paint," Eastman Kodak Co., Rochester, New York.

and pulsed lasers by Novak and Windsor[185] and more recently by Porter and Topp.[186] Flash photolysis with lasers offers the additional advantage of a time duration of only tens of nanoseconds.

In the more conventional flash photolysis system, the absorption cell normally includes a multiple pass mirror system (see Section 2.3.2.4.5) to increase the optical path length for greater sensitivity. Thirty traversals is not difficult with such systems when used for wavelengths greater than about 3500 Å.

2.3.2.3.4. PULSE-DISCHARGE SYSTEMS. This system is quite similar to the flash photolysis system, except that the initial discharge through the flash lamp is replaced by a discharge directly through the parent compound in the absorption cell. In this case the molecular fragments

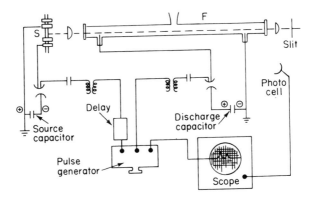

FIG. 6. Pulse-discharge system [reproduced from the paper by G. Herzberg and A. Lagerqvist, *Can. J. Phys.* **46**, 2363 (1968), with the kind permission of the authors and the National Research Council of Canada].

are produced by a much more complicated process that may involve photolysis, vacuum ultraviolet photolysis, electron bombardment, and positive-ion bombardment. Therefore, this technique is especially useful for the production of highly excited states (or of ions), and it is not limited to molecules that happen to absorb at a particular wavelength region. Figure 6 presents a schematic diagram of a system used by Herzberg[187] for the production of the absorption spectra of C_2^- (as well as N_2^+, CO^+, and CO_2^+). The amount of energy discharged typically will be

[185] J. R. Novak and M. W. Windsor, *J. Chem. Phys.* **47**, 3075 (1967).
[186] G. Porter and M. R. Topp, *Proc. Roy. Soc. (London)* **A315**, 163 (1970).
[187] G. Herzberg and A. Lagerqvist, *Can. J. Phys.* **46**, 2363 (1968).

much smaller (less than 20 J) than in a flash photolysis system of comparable length (up to 30 kJ). As a result the duration of the discharge may be reduced considerably (less than 1 µsec in some cases). A technique somewhat related to the above involves pulse radiolysis in which a pulsed beam of electrons at moderately high energy is directed into the absorption cell. The results reported by Firestone and co-workers[188] of their use of a Febetron* to produce spectra interpreted as transitions between excited states of inert gas dimers appears quite promising as a more general tool.

2.3.2.4. Miscellaneous Instrumentation

2.3.2.4.1. WAVELENGTH MEASUREMENTS. In principle, wavelengths could be determined from the grating equation [Eq. (2.3.1)] from precise measurements of the necessary angles and a single known wavelength (to determine the grating spacing). This procedure is used for preliminary estimates of wavelengths and it was used extensively at one time as the primary method of measuring wavelengths. However, precise wavelength determinations normally are made by direct comparison with the standard reference spectra described in Section 2.3.2.2.2. Superposition of such a reference spectrum on the spectrum to be measured must be done with considerable care in order to avoid systematic shifts between the two. On photographic instruments this is normally accomplished by the use of masks in front of the slit (for stigmatic instruments) or in front of the plateholder. It is convenient to have the reference spectra both above and below the sample spectrum, with slight overlaps where they meet.

The positions of both spectra may be measured with an optical comparator, which provides two functions: an ability to position an indicator on the desired spectral feature and an ability to measure that position to about ± 0.0005 mm (± 0.5 µm). The simpler comparators use a microscope (with a crosshair as indicator) mounted over a traveling stage that is driven by a precision screw. The position of the spectral feature on the photographic plate is then measured in terms of the number (and fractions) of revolutions of the screw. The semiautomatic comparators described by Tomkins and Fred[189] render this demanding job a great deal less tedious. The advantage of the Tomkins–Fred system is that it

---

[188] S. Arai and R. F. Firestone, *J. Chem. Phys.* **50**, 4575 (1969).
[189] F. S. Tomkins and M. Fred, *J. Opt. Soc. Amer.* **41**, 641 (1951).

---

* Field Emission Corp., McMinnville, Oregon 97128.

displays a line profile on an oscilloscope in a manner that makes it very simple to measure line spectra very quickly and reproducibly. Commercial versions of this instrument are available* with digital output of the position of the plate for automatic recording on printed tape, punched tape, or punched cards. These comparators also may be used conveniently to measure plate density at each position, or as microdensitometers. This facility has been used to greatest advantage in the fully automatic measuring machines described by Steinhaus et al.[190] and by Ditzel and Giddings.[191]

The Abbe comparator (see Sawyer,[7] p. 230) avoids the limitations of a long precision screw by using a precision scale, usually etched on glass, from which plate positions may be read by direct comparison. A more recent alternative to the precision screw is the *laser interferometer*. Such a modification could result in more precise measurements over a greater length of plate, and it would eliminate possible errors resulting from backlash of the screw. The principal limitation of this technique so far appears to be the high cost of a sufficiently stable laser that also includes a pressure or temperature compensation system. In addition, the fringe counting system must be safeguarded against "losing count," especially at fast counting rates. Commercial versions of such measuring devices are available[†] which resolve ± 0.00001 mm over a range of 60 m at a maximum rate of 300 cm/sec.

The measured positions of the reference lines (l_n) and their known wavelengths (λ_n) are fitted to an equation of the form $\lambda_n = \sum_{i=0}^{N} a_i l_n{}^i$, where N may be as large as 5, but usually is no larger than 3. When the interpolations are performed by hand, it is suitable to fit the data to a linear interpolation (i.e., limit N to 2) and then plot a graph of deviations, as described by Sawyer[7] (p. 249). In cases where few reference lines are available over wide lengths of plate, it may be better to fit the data to an expression that correctly describes the dispersion of the particular instrument.

The need for a large number of well-spaced reference lines leads to the necessity of using standards which originate from orders different from that of the sample spectrum. This procedure leads to two additional

[190] D. W. Steinhaus, R. Engleman, Jr., and W. L. Briscoe, *Appl. Opt.* **4**, 799 (1965).
[191] E. F. Ditzel and L. E. Giddings, Jr., *Appl. Opt.* **6**, 2085 (1967).

* Grant Instruments, Inc., Berkeley, California.
† Hewlett-Packard, Palo Alto, California; Jorway Corp., Westbury, New York; and Perkin–Elmer, Norwalk, Connecticut.

difficulties. The first problem is associated with "coincidence errors"—which have been discussed earlier (Section 2.3.2.2.2). Michelson[192] showed that relative displacements could occur as a result of imperfect gratings that resulted in slightly different line shapes for different wavelengths.* Therefore, it is important to establish whether or not this effect introduces a significant error with each grating that is used. The dependence of the magnitude of the shift on imperfect rulings is consistent with observations that coincidence errors seem to be less a problem with the modern, higher-quality gratings—especially the plane gratings.

The second problem deals with when and where to apply the vacuum correction to wavelengths, i.e., to correct for the dispersion of air, before calculating the energy ν_{vac} (cm^{-1}) for the transitions. Obviously this is not a problem when using an evacuated spectrograph, since all wavelengths must be expressed as λ_{vac}. The problem does arise with a spectrograph operating in standard air, for example, if the desired spectrum occurs near 4500 Å and is photographed in second order, whereas the superimposed reference spectrum occurs near 3000 Å and is photographed in third order. The procedure may be stated simply: First convert all reference wavelengths to "effective" wavelengths in the same order as the spectrum to be measured (near 4500 Å in the example)—and then apply the necessary vacuum correction.

2.3.2.4.2. DETECTORS. Photographic plates and films constitute the oldest and most widely used detectors available to the spectroscopist. Introductory discussions in their use and characteristics are available in a number of texts (see, for example, Sawyer,[7] Chapter 8). More detailed technical data regarding the choices of speed, resolution, and spectral sensitivity may be obtained from the manufacturers. The booklet "Kodak Plates and Films for Science and Industry"[193] provides an outstanding source of such information. Some special problems are associated with

---

[192] A. A. Michelson, *Astrophys. J.* **18**, 278 (1903).

[192a] R. C. Preston, *Opt. Acta* **17**, 857 (1970).

[193] Kodak Plates and Films for Science and Industry, Kodak Publ. No. P-9, Eastman Kodak Co., Rochester, New York, 1967.

---

* A more recent paper by Preston (Ref. 192a) presents calculated line shapes which display this kind of asymmetry. They were calculated from an extension of Stoke's treatment of Rowland ghosts (Ref. 3), based on periodic ruling errors. In this context, asymmetric line shapes are caused by Rowland ghosts (generated by ruling errors of small periodicity) which approach the central diffraction peak. The author is indebted to Dr. R. Dakin, Bausch and Lomb Co., for his help in locating these reports.

the use of photographic materials in the infrared and the vacuum ultra-violet. The infrared-sensitive emulsions are slow at best, and they usually must be hypersensitized just prior to use in order to reduce their greater tendency to fog during storage. They should always be stored at the lowest practical temperature. The procedures developed by Pope and Kirby[194] and by Barker[195] involve hypersensitizing with ammonium hydroxide solution, followed by neutralization with an acetic acid solution, and rapid drying at a low temperature. Methyl alcohol may be used to speed the drying process, although Barker suggests not using it if possible since it leads to a lower gain in sensitivity. Gains of as much as 460 compared to untreated plates (Kodak type I-Z) are reported by Pope and Kirby. Treated plates should be used within a few hours if they are stored at 0°C or within 48 hr if they are stored at −25°C.

Vacuum ultraviolet emulsions present another kind of problem. In this case the gelatin normally used to support the photosensitive materials absorbs strongly below about 2500 Å. Two methods may be used to avoid the problem. The first is to coat the emulsion with a fluorescent material* which converts the vacuum ultraviolet radiation into longer wavelength radiation.[196-198] The second, and more widely applicable method, is to remove most of the gelatin, as was first done by Schumann.[199] Commercially available plates of this type include the Ilford Q plates (available in three speed-resolution classes), Eastman Kodak SWR plates, and Kodak Pathé SC-5 and SC-7 plates. The SC-7 emulsion is a fairly recent development that features a tenfold increase in speed over the SWR emulsion. Each of these emulsions must be treated with extreme care to avoid abrasions because of the absence of the protective gelatin. The low gelatin content also makes them more susceptable to contamination, especially in the developing solutions. The availability of SWR film with raised edges (a deposit of small granules) provides a simple but effective method of reducing abrasions of the film when stored in rolls.

[194] T. P. Pope and T. B. Kirby, *J. Opt. Soc. Amer.* **57**, 951 (1967).
[195] E. S. Barker, *J. Opt. Soc. Amer.* **58**, 1378 (1968).
[196] F. S. Johnson, K. Watanabe, and R. Tousey, *J. Opt. Soc. Amer.* **41**, 702 (1951).
[197] P. Lee and G. L. Weissler, *J. Opt. Soc. Amer.* **43**, 512 (1953).
[198] R. Allison and J. Burns, *J. Opt. Soc. Amer.* **55**, 574 (1965).
[199] V. Schumann, *Wien. Akad. Anzeiger* **23**, 230 (1892).

---

* The most commonly used materials are Eastman Kodak Ultraviolet Sensitizer No. 2 and deposits of sodium salicylate (Ref. 198).

Photoelectric detectors provide considerable advantages over photographic methods for measuring relative intensities or for monitoring a single wavelength. These detectors may incorporate any of a number of photon-induced changes in properties, but the most common device is the photomultiplier tube, which depends on photoemission followed by electron multiplication within the tube. The particular choice of envelope material (for transmission characteristics), cathode material (for spectral sensitivity and efficiency), and detailed tube characteristics will be dictated by the light level, duration, repetition rate if pulsed, and wavelength range. Technical literature from the manufacturers[*] may be consulted for such information. Channel multipliers, such as the Bendix channeltron,[†] seem to offer some attractive possibilities, as is demonstrated in its application in image intensifiers. Solid-state detectors may provide faster response, smaller size, and lower power consumption as some of their advantages over photodiodes. For example, the "Fotofet"[‡] (with a response time of 14 nsec) may be used as a convenient replacement of a rather bulky photodiode with considerable reduction in size of detector and power pack.

2.3.2.4.3. WINDOWS AND FILTERS. Table II contains a summary of the optical properties of a number of materials commonly used as windows in the study of electronic spectra. Quartz provides a chemically inert and mechanically strong window suitable for wavelengths above 2800 Å. Some care must be exercised in choosing quartz for use below 2800 Å, since the transmission in that region depends critically on the chemical purity. The problem is made more difficult by the absence of any uniform nomenclature to distinguish one grade from another, except by trade name. Unless specified otherwise, the transmission of most quartz (10 mm thick) will drop to about 65% at an absorption peak near 2400 Å, increase to approximately 80% near 2250 Å, and then decrease rapidly to less than 50% at 2000 Å. Ultraviolet grades of quartz will eliminate the absorption at 2400 Å and extend the lower wavelength as low as 1700 Å (for 50% transmission). Even with the best quartz, this lower wavelength limit will shift toward longer wavelengths as the temperature increases,

---

[*] EMR Photoelectric, P. O. Box 44, Princeton, New Jersey 08540; Gencom Division, Emitronics, Inc., 80 Express Street, Plainview, New York 11803; and RCA, Electronic Components, Harrison, New York 07029.

[†] Bendix, Electro-Optics Division, 1975 Green Road, Ann Arbor, Michigan 48107. Similar devices are available from EMR Photoelectric (see above list).

[‡] Crystalonics, Inc., 147 Sherman Street, Cambridge, Massachusetts 02140.

TABLE II. Properties of Optical Materials Useful in the Study of Molecular Electronic Spectra

Usable transmission range (approx.)	Remarks[a]
Fused or crystal quartz, good quality[b,c] 1650 A, 3.5 μm	Short wavelength transmission limit highly dependent upon purity.
Pyrex brand glass[d] 2900 Å, 2.5 μm	
Optical glass[b] 3400 A, 1.0 μm	
Calcium fluoride[b,e,f,i] 1250 Å, 10.0 μm	Excellent for work where combined vacuum
Lithium fluoride[e,g,h] 1100 Å, 6.0 μm	ultraviolet and infrared transmission is
Barium fluoride[e,i] 1450 Å, 13.5 μm	necessary; may be used with indium gaskets for low temperature work down to 0 K, if temperature changes are not abrupt.
Synthetic sapphire[j,k] (Al$_2$O$_3$) 1450 Å (1750 Å), 6.0 μm	Broad transmission region; resistant to mechanical stress or low-temperature fracture; useful for high-pressure, low-temperature applications.

[a] Darkening may occur in ultraviolet transmitting materials under prolonged exposure to $\lambda < 2000$ Å light.

[b] G. R. Harrison, R. C. Lord, and J. R. Loofbourow, "Practical Spectroscopy," p. 51. Prentice-Hall, Englewood Cliffs, New Jersey, 1948.

[c] Amersil Quartz Division, Engelhard Industries, Inc., Hillside, New Jersey; Corning Glass Works, Corning, New York.

[d] Glass Color Filters. Corning Glass Works, Corning, New York.

[e] Available from Harshaw Chemical Co., Cleveland, Ohio.

[f] E. G. Schneider, *Phys. Rev.* **45**, 152 (1934).

[g] E. G. Schneider, *Phys. Rev.* **49**, 341 (1936).

[h] Perkin–Elmer Corp., Norwalk, Connecticut.

[i] S. S. Ballard, L. S. Combes, and K. A. McCarthy, *J. Opt. Soc. Amer.* **42**, 684 (1952).

[j] Available from Linde Air Products, New York, New York.

[k] S. Freed, H. L. McMurry, and E. J. Rosenbaum, *J. Chem. Phys.* **7**, 853 (1939); K. Dressler and O. Schnepp, *ibid.* **32**, 1682 (1960), report the higher uv transmission limit which is probably more realistic for currently available sapphire.

reaching a value of 1950 Å at 1000°C, as shown by Edwards.[200] Likewise, the refractive index will depend on the grade of quartz being used, although comparable grades appear to have the same properties, as shown by Brixner[201] in his fit of observed data to a single equation. Wray and Neu[202] have determined the refractive index as a function of temperature (26–826°C) for Corning Code 7940 (fused silica, ultraviolet grade), Corning Code 7913 (Vycor, optical grade), and Corning Code 1723 (Aluminosilicate glass).

The physical and optical properties of crystalline materials (including LiF, BaF_2, and CaF_2) are summarized conveniently in the data available from companies that supply these optical materials (especially Harshaw[203] and Optovac).[204] These materials are used when transmission is required into the vacuum ultraviolet and/or infrared regions. Samson[6] (pp. 180–184) gives a very convenient summary of the optical properties of these materials in the vacuum ultraviolet, as well as some information on other crystalline optical materials. The lower wavelength transmittance limit of all of these crystals (with the possible exception of MgF_2) shifts toward longer wavelengths as the temperature is increased[205–208] within the range 196°–380°C. Transmission of vacuum ultraviolet radiation can deteriorate for a number of other reasons as well: (1) Exposure to high-energy radiation leads to the formation of opaque color centers in all crystals other than BaF_2.[209,210] These usually can be removed by annealing the crystal at about 500°C. (2) Exposure to such radiation also can lead to the deposition of thin layers of materials, apparently from the oil used in the pumping systems.[115,211] These deposits may be removed by repolishing lightly with a fine abrasive. (Note that when this deposit forms on the window of a sealed vacuum ultraviolet lamp, the deposit usually will form on the window away from the discharge, i.e., on the side exposed to the vacuum spectrograph or the absorption cell.) (3)

[200] O. J. Edwards, *J. Opt. Soc. Amer.* **56**, 1314 (1966).

[201] B. Brixner, *J. Opt. Soc. Amer.* **57**, 674 (1967).

[202] J. H. Wray and J. T. Neu, *J. Opt. Soc. Amer.* **59**, 774 (1969).

[203] Harshaw Optical Crystals. Harshaw Chem. Co., Cleveland, Ohio (1967).

[204] Optical Crystals. Optovac, Inc., North Brookfield, Massachusetts.

[205] A. R. Knudson and J. E. Kupperian, *J. Opt. Soc. Amer.* **47**, 440 (1957).

[206] S. A. Yakovlev (English translation) **2**, 396 (1962).

[207] A. H. Laufer, J. A. Pirog, and J. R. McNesby, *J. Opt. Soc. Amer.* **55**, 64 (1965).

[208] R. J. Davis, *J. Opt. Soc. Amer.* **56**, 837 (1966).

[209] P. Warneck, *J. Opt. Soc. Amer.* **55**, 921 (1965).

[210] D. F. Heath and P. A. Sacher, *Appl. Opt.* **5**, 937 (1966).

[211] R. G. Taylor, T. A. Chubb, and R. W. Kreplin, *J. Opt. Soc. Amer.* **55**, 1078 (1965).

Surface reactions of the crystals may lead to a loss of transmission. In the case of LiF, such a reaction occurs with moisture.[208,212] The transmission may be restored by heating to 500°C or by cleaning with ethyl alcohol in an ultrasonic bath.

A wide variety of filters are available for wavelengths greater than 3000 Å. These include the standard tinted glass filters[213] and gelatin filters[214] available, for example, from Corning and Kodak, as well as interference filters, which can be custom designed for the desired band pass properties.* Some care must be exercised in the use of the glass and gelatin filters since they are subject to aging and thermal degradation. In general it is advisable to avoid large thermal gradients and to use the filters in collimated light where possible. Additional filters may be made of appropriate gases or solutions that absorb in the desired wavelength regions. Figure 7 displays the absorption spectra of gas cells containing Cl_2, Br_2, and I_2, which are among the more useful gas filters. Special filter solutions of particular use to the spectroscopist or photochemist have been described in detail by Calvert and Pitts[99] (pp. 728–742), and by Bennett and McBride.[215]

The choice of transmission filters for wavelengths less than 3000 Å is much more restricted since most compounds do absorb strongly in this region. Thin metallic films† may transmit as much as 30% over narrow wavelength regions below 1000 Å (see Samson,[6] pp. 184–200). Although narrow-band interference filters are not available in the vacuum ultraviolet, broad-band filters have been made by Bates and Bradley.[216] They evaporated alternate layers of aluminum and MgF_2 on a fused silica substrate and achieved peak transmissions of 29%. The wavelength

---

[212] D. A. Patterson and W. H. Vaughan, *J. Opt. Soc. Amer.* **53**, 851 (1963).

[213] Corning Color Glass Filters. Corning Glass Works, Inc., Corning, New York (1970).

[214] Kodak Filters for Scientific and Technical Uses. Eastman Kodak Co., Rochester, New York (1970).

[215] H. E. Bennett and W. R. McBride, *Appl. Opt.* **3**, 919 (1964).

[216] B. Bates and D. J. Bradley, *Appl. Opt.* **5**, 971 (1966).

---

* Such filters are available from a variety of companies, such as the following: Baird Atomic, Incorporated, Cambridge, Massachusetts; Bausch and Lomb Optical Company, Rochester, New York; Corion Instrument Corporation, Waltham, Massachusetts; Laser Energy, Incorporated, Rochester, New York; Optics Technology, Incorporated; Palo Alto, California; and Oriel Optics Corporation, Stamford, Connecticut.

† Aluminum foils (transmission between about 150–700 Å, with peak of 50% transmission at 200 Å) are available from Sigmatron, Inc., Goleta, California.

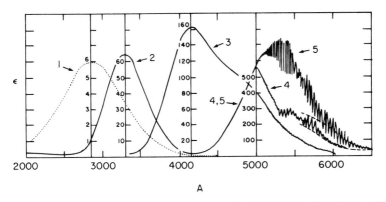

A

FIG. 7. Absorption spectra of the halogens: (1) $F_2(g)$, 25°; (2) $Cl_2(g)$, 18°; (3) $Br_2(g)$, 25°; (4) $I_2(g)$, 70–80°; (5) $I_2(g)$ plus 1 atm air, 70–80° [reproduced from the book by J. G. Calvert and J. N. Pitts, Jr., "Photochemistry." Wiley, New York, 1967, with the kind permission of the authors and the publishers].

of maximum transmission for their filters ranged between 1700 and 2200 Å, depending on the relative thicknesses of the deposited layers. Wood[217,218] describes the preparation of alkali metal films, deposited at liquid air temperatures on the inner surface of an evacuated quartz vessel, which transmit in the vacuum ultraviolet and absorb strongly throughout the visible region. The transmissions extend from at least 1860 Å to the following upper wavelength limits: Li (2050 Å), Na (2100 Å), K (3600 Å), and Cs (4400 Å). Ginter and Tilford[218a] describe the use of an argon gas filter with the Hopfield continuum of helium. This combination permits them to photograph spectra in the fifth, sixth, and seventh orders of 790–950 Å without appreciable overlapping of orders.

2.3.2.4.4. REFLECTING SURFACES. Aluminum is the most commonly used material for coating mirrors and gratings (see, for example, the paper by Hass et al.[219]). Good coatings have reflectivities greater than 90% for wavelengths greater than 2000 Å, and for wavelengths greater than 4000 Å, they appear relatively insensitive to evaporation technique, aging, or radiation. In the visible region, silver has a greater reflectivity,

[217] R. W. Wood, "Physical Optics," 3rd ed., pp. 558–566. MacMillan, New York, 1934.

[218] R. W. Wood, Phil. Mag. 38, 98 (1918).

[218a] M. L. Ginter and S. G. Tilford, Appl. Opt. 11, 958 (1972).

[219] G. Hass, J. Opt. Soc. Amer. 45, 945 (1955).

but it is much less reflective in the ultraviolet. Silver has the additional disadvantage of tarnishing, whereas the aluminum oxide layer which does form on aluminum surfaces protects the surface without seriously affecting the reflectivity. Gold and copper have reflectance greater than 90% above 6000 Å, but are much poorer at shorter wavelengths. For wavelengths greater than about 3500 Å, protective overcoatings such as silicon oxide or beryllium oxide minimize abrasion damage to the first surface coatings and facilitate cleaning. Unfortunately these coatings cannot be used at shorter wavelengths. In a series of papers, Hass and co-workers[220-224] have demonstrated that the reflectivity of aluminum may be enhanced considerably at wavelengths less than 2500 Å with MgF_2. Without the overcoating, the aluminum showed a marked decrease in reflectivity (below 2000 Å) after exposure to air for periods as short as one hour, and the aging continued for at least one month (at which time the reflectivity at 1200 Å was less than 20%). Overcoating with MgF_2 yielded reflectivities of about 80%, with no decrease as a result of exposure to air. Hunter[225] measured the reflectances in the vacuum ultraviolet as a function of the thickness of the MgF_2. The review by Madden[226] summarizes the techniques for preparing these reflective coatings and of measuring the thicknesses of the overcoating. For wavelengths less than about 1000 Å, platinum seems to be the best reflective coating,[227,228] although gold may also be used.[229]

Multilayer reflective coatings may be designed for peak reflectance, as was discussed for transmission filters above. In this case also, the technique is most successful in producing reflectances of over 99% for wavelengths greater than 3500 Å.

2.3.2.4.5. MULTIPLE-PASS CELLS. When very long absorbing path lengths are required, the most convenient method of obtaining them is

[220] G. Hass, W. R. Hunter, and R. Tousey, *J. Opt. Soc. Amer.* **46**, 1009 (1956).

[221] G. Hass, W. R. Hunter, and R. Tousey, *J. Opt. Soc. Amer.* **47**, 1070 (1957).

[222] G. Hass and R. Tousey, *J. Opt. Soc. Amer.* **49**, 593 (1959).

[223] P. H. Berning, G. Hass, and R. P. Madden, *J. Opt. Soc. Amer.* **50**, 586 (1960).

[224] L. R. Canfield, G. Hass, and J. E. Waylonis, *Appl. Opt.* **5**, 45 (1966).

[225] W. R. Hunter, *Opt. Acta* **9**, 255 (1962).

[226] R. P. Madden, *in* "Physics of Thin Films," (G. Hass, ed.), p. 123. Academic Press, New York, 1963.

[227] R. P. Madden and L. R. Canfield, *J. Opt. Soc. Amer.* **51**, 838 (1961).

[228] G. F. Jacobus, R. P. Madden, and L. R. Canfield, *J. Opt. Soc. Amer.* **53**, 1084 (1963).

[229] L. R. Canfield, G. Hass, and W. R. Hunter, *J. Phys.* **25**, 124 (1964).

with the multiple-pass cell first proposed by White[230] and modified slightly by Bernstein and Herzberg,[231] as shown in Fig. 8. Each of the three concave mirrors has a radius of curvature equal to the base length of the cell (the distance between A or B and C). The light paths traced represent the central ray only when the mirrors are adjusted for 16 passes. Light is focused into the entrance slit at 0, diverges and fills mirror A, which focuses an image of the entrance slit back on mirror C at point 1. The light path alternates between mirrors A and B, hitting mirror C each time, until it passes through the exit slit at point 8 in the diagram.

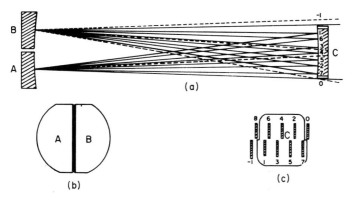

FIG. 8. Multiple-pass absorption cell. Images 0–8 are described in the text. Image −1 results from light entering the cell (at 0) and hitting the wrong mirror (B). This light is wasted, but the observation of −1 image is useful for alignment of the mirror [reproduced from the paper by H. J. Bernstein and G. Herzberg, *J. Chem. Phys.* **16**, 30 (1948), with the kind permission of the authors and publishers].

Although each mirror must be adjusted very carefully, one of the advantages of this system is that the number of traversals can be adjusted (in multiples of four) by a single adjustment on mirror A. The number of traversals may be determined by counting the number of images located between the entrance and exit slit, adding one, and multiplying by four. In the case shown, $(3 + 1) \times 4 = 16$. The main factors that limit the number of traversals are the size of the entrance slit and the reflectivity of the mirrors. (Note that the images between the entrance and exit slit will not be visible except when the scattered light is excessive!) The importance of high reflectivity may be demonstrated by realizing

[230] J. U. White, *J. Opt. Soc. Amer.* **32**, 285 (1942).
[231] H. J. Bernstein and G. Herzberg, *J. Chem. Phys.* **16**, 30 (1948).

that the fraction of light transmitted by the cell is $(R)^{n-1}$, where R is the reflectivity for a single reflection (0.9 is a reasonable value for aluminum surfaces) and n is the number of passes through the cell. It is interesting to note the similarity of this cell design to that which may be used for gathering light *emitted* from within the cell (as with gas phase Raman cells). In such cells, Welsh *et al.*[232] have shown that the intensity of light may be increased by the factor $1/(1 - R)$. Likewise, the cell is one of the possible cavity designs useful for the amplification of laser signals,[233] (pp. 93–103). Recent developments in multiple-pass cells include the low-temperature cells reported by Blickensderfer *et al.*[234] with a base length of 3.8 m (which is commercially available*) and by Horn and Pimentel[235] with a base length of 11 m. Callomon[236] has described an effective imaging system for optimum matching (if possible) of a multiple-pass cell to a spectrograph. Ideally, the exit slit of the cell will be imaged on the entrance slit of the spectrograph and mirror B will be imaged onto the collimator of the spectrograph.

2.3.3. Description of Electronic States

2.3.3.1. Theoretical Models. The analysis and interpretation of molecular spectra depend ultimately on our ability to describe the molecular states involved in the observed transitions. The ability to predict new transitions or properties depends on the quality of the theoretical model.

In principle, any molecular state is fully characterized if we know the total time-dependent wave function. Conversely, knowledge of the behavior of the molecule under all possible conditions is equivalent to a complete knowledge of the wave function. In practice, this goal is approached by two different techniques: (1) through *ab initio* calculations, which depend only on fundamental constants and the principles of quantum mechanics, and (2) through fitting experimental observations of some of the molecular properties to the best existing theoretical model.

[232] H. L. Welsh, C. Cumming, and E. J. Stansbury, *J. Opt. Soc. Amer.* **41**, 712 (1951).
[233] B. Lengyel, "Lasers," 2nd ed. Wiley (Interscience), New York, 1971.
[234] R. P. Blickensderfer, G. E. Ewing, and R. Leonard, *Appl. Opt.* **7**, 2214 (1968).
[235] D. Horn and G. C. Pimentel, *Appl. Opt.* **10**, 1892 (1971).
[236] J. H. Callomon, *Can. J. Phys.* **34**, 1046 (1956).

* Cryogenics Associates, Inc., Indianapolis, Indiana.

The most convenient first approximation (the Born–Oppenheimer approximation* [236a,236b]) for the description of molecular electronic states involves a model that permits the total time-*independent* wave function (ψ_T) to be factored into independent components of electronic, vibrational, and rotational contributions

$$\psi_T = \psi_e \psi_v \psi_r. \qquad (2.3.18)$$

Under these conditions, the total energy of the molecule may be expressed as a simple sum

$$E_T = E_e + E_v + E_r. \qquad (2.3.19)$$

One of the advantages of this factored wave function is that, to a first approximation, the electronic, vibrational, and rotational properties may be treated independently. Indeed, most of the descriptions in preceding chapters for microwave and infrared spectroscopy may be incorporated in the present description of electronic spectroscopy.

2.3.3.1.1. QUANTITATIVE METHODS. The methods and results of *ab initio* calculations are far too extensive for an adequate review in this chapter. Instead the reader is referred to a number of excellent recent texts that discuss the topic in some detail.[237-241] From the viewpoint of the experimental spectroscopist, impressive advances have been made which are yielding accurate descriptions of ground *and excited states*

[236a] M. Born and R. Oppenheimer, *Ann. Phys.* **84**, 457 (1927).

[236b] J. T. Hougen, *in* "Physical Chemistry, an Advanced Treatise" (D. Anderson, ed.), Vol. 4, Chapter 7. Academic Press, New York, 1970.

[237] M. Karplus and R. N. Porter, "Atoms and Molecules: An Introduction for Students of Physical Chemistry." Benjamin, New York, 1970.

[238] F. L. Pilar, "Elementary Quantum Chemistry." McGraw-Hill, New York, 1968.

[239] I. N. Levine, "Quantum Chemistry," Vol. I, Quantum Mechanics and Molecular Electronic Structure. Allyn and Bacon, Boston, Massachusetts, 1970.

[240] I. N. Levine, "Quantum Chemistry," Vol. II, Molecular Spectroscopy, Allyn and Bacon, Boston, Massachusetts, 1970.

[241] N. Mataga and T. Kubota, "Molecular Structure and Electronic Spectra." Dekker, New York, 1970.

* The separation of electronic and nuclear motions, such that $\psi_T = \psi_e \psi_{vr}$, was demonstrated to be a resonable one by Born and Oppenheimer (Ref. 236a) based on the relative masses of the electrons and the nuclei. The additional separation of $\psi_v \psi_r$ was rationalized by them on a similar basis (Ref. 236b) but tends to be more subject to breakdown.

of diatomic and triatomic molecules. An excellent survey of such calculations has been compiled recently by Schaefer.[242] Perhaps two examples will serve as an indication of the present capabilities of such calculations: the dissociation energy of H_2 and the geometry of the ground state of CH_2.

The dissociation energy of H_2 probably provides one of the most interesting examples for comparison between theoretical and experimental results. It also demonstrates the extreme care that must be exercised in both approaches in order to permit realistic comparisons. In 1960 Herzberg and Monfils[243] reported the results of their studies of the far ultraviolet absorption spectra of H_2, HD, and D_2. At that time they felt their dissociation energies were accurate to better than ± 0.5 cm^{-1}. There was very good agreement between these experiments and the high-precision *ab initio* calculations carried out by Kolos and Roothaan[244] about the same time. Subsequent improvements on the calculations by Kolos and Wolniewicz[245-248] disclosed a discrepancy of approximately 4 cm^{-1}—much larger than the estimated uncertainty of either experimental or theoretical results. Furthermore the discrepancy was such that refinements of the theoretical values would be expected to *increase* the discrepancy, because of the variation principle. Recent experiments by Herzberg[249] with a higher-resolution spectrograph and with the H_2 cooled to liquid nitrogen temperatures demonstrated that the problems of overlapping lines and limited resolution were more serious than estimated by Herzberg and Monfils.[243] Table III summarizes the present agreement between experimental and theoretical calculations, as presented by Herzberg.[249] The series of new experiments by Herzberg also led to the determination of dissociation energies of the B $^1\Sigma_u{}^+$, C $^1\Pi_u$, and B $^1\Sigma_u{}^+$ states, some of which could be compared with additional calculations of Kolos and Wolniewicz.[250] A separate investigation of the

[242] H. F. Schaefer III, "The Electronic Structure of Atoms and Molecules. A Survey of Rigorous Quantum Mechanical Results." Addison-Wesley, Reading, Massachusetts, 1972.

[243] G. Herzberg and A. Monfils, *J. Mol. Spectrosc.* **5**, 482 (1960).

[244] W. Kolos and C. C. J. Roothaan, *Rev. Mod. Phys.* **32**, 219 (1960).

[245] W. Kolos and L. Wolniewicz, *J. Chem. Phys.* **41**, 3663 (1964).

[246] W. Kolos and L. Wolniewicz, *J. Chem. Phys.* **43**, 2429 (1965).

[247] W. Kolos and L. Wolniewicz, *J. Chem. Phys.* **48**, 3672 (1968).

[248] W. Kolos and L. Wolniewicz, *J. Chem. Phys.* **49**, 404 (1968).

[249] G. Herzberg, *J. Mol. Spectrosc.* **33**, 147 (1970).

[250] W. Kolos and L. Wolniewicz, *J. Chem. Phys.* **51**, 1417 (1969).

TABLE III. Calculated and Observed Dissociation Energies of Hydrogen[a]

	Theoretical (cm^{-1})		Observed (cm^{-1})	
	Adiabatic[b]	Nonadiabatic[c]	HM[d]	Herzberg[a]
D_0^0 (H$_2$)	36,117.4	36,117.9	36,113.6	<36,118.3 >36,116.3
D_0^0 (HD)	36,405.2	36,405.5	36,400.5	36,406.6 36,405.8
D_0^0 (D$_2$)	36,748.0	36,748.2	36,744.2	36,748.9

[a] G. Herzberg, *J. Mol. Spectrosc.* **33**, 147 (1970).
[b] W. Kolos and L. Wolniewicz, *J. Chem. Phys.* **49**, 404 (1968).
[c] P. Bunker, *J. Mol. Spectrosc.* **42**, 478 (1972).
[d] G. Herzberg and A. Monfils, *J. Mol. Spectrosc.* **5**, 482 (1960).

Rydberg series of H$_2$ by Herzberg and Jungen[251] led to a determination of the ionization potential of H$_2$ and of the dissociation energy of H$_2^+$. The determination of the ionization potential required a detailed analysis of severe rotational and vibrational perturbations between two Rydberg series. In addition, these authors found it necessary to include a correction to the ionization potential which resulted from a pressure shift of all Rydberg lines. The shift amounted to about 1 cm^{-1} at 40 Torr and was found to be independent of the principal quantum number (for n greater than about 10). Their observed ionization potential of 124,417.2 cm^{-1} $\pm$ 0.4 cm^{-1} agrees very well with the theoretical value of 124,417.3 determined by Hunter and Pritchard,[252] with corrections by Jeziorski and Kolos[253] and Bunker.[254]

The absorption spectrum of singlet and triplet CH$_2$, as well as the spectrum of CH$_3$, was reported by Herzberg[255] in 1961. Isotopic substitutions and preliminary rotational analyses left no doubt of the identity of the two absorbing species. In the vacuum-ultraviolet spectrum of CH$_2$, the simple P and R branches and the observed intensity alternations

[251] G. Herzberg and Ch. Jungen, *J. Mol. Spectrosc.* **41**, 425 (1972).
[252] G. Hunter and H. O. Pritchard, *J. Chem. Phys.* **46**, 2153 (1967).
[253] B. Jeziorski and W. Kolos, *Chem. Phys. Lett.* **3**, 677 (1969).
[254] P. Bunker, *J. Mol. Spectrosc.* **42**, 478 (1972).
[255] G. Herzberg, *Proc. Roy. Soc.* (*London*) **A262**, 291 (1961); see also G. Herzberg d.n J. W. C. Johns, *ibid.* **A295**, 107 (1966).

were most easily interpreted in terms of a linear–linear transition $(^3\Sigma_u^- - {}^3\Sigma_g^-)$. The lower state is consistent with the electron configuration for linear CH_2. The alternative interpretation in terms of a bent–bent transition $(A_2 - B_1)$ of CH_2, which was considered, would require that the easily assigned band be a $K' = 0$ to $K'' = 0$ transition. This assignment was rejected, however, because of the absence of the other transitions with $\Delta K = 0$, and $K > 0$, which should have been observed with comparable intensity. Subsequent *ab initio* calculations[256–260] consistently concluded that the lowest triplet state of CH_2 should be bent with a bond angle of about 136°. Observations of the electron-spin resonance spectrum of CH_2 in solid matrices by Bernheim *et al.*[261,262] and by Wasserman *et al.*[263–266] established this structure experimentally. A subsequent paper by Herzberg and Johns[267] restated that their spectrum is consistent with a bent–bent transition, *provided* one of the states is predissociated above the $K = 0$ level. They also pointed out that the HCO spectrum[268,269] does show such a behavior. In that case, the predissociated subbands with $K' \neq 0$ were found later as diffuse features.

These two examples are of very small molecules, but they should demonstrate the important role we may expect rigorous *ab initio* calculations to play in the future. In the search for spectra of unobserved species, information about expected geometries and probable transition energy would be an immense aid to the experimentalist. On the other hand, the

[256] J. F. Harrison and J. C. Allen, *J. Amer. Chem. Soc.* **91**, 807 (1969).

[257] J. F. Harrison, *J. Amer. Chem. Soc.* **93**, 4112 (1971).

[258] W. A. Yeranos, *Z. Naturforsch.* **26A**, 1245 (1971).

[259] C. F. Bender and H. F. Schaefer III, *J. Amer. Chem. Soc.* **92**, 4984 (1970).

[260] V. O'Neil, H. F. Schaefer III, and C. F. Bender, *J. Chem. Phys.* **55**, 162 (1971).

[261] R. A. Bernheim, H. W. Bernard, P. S. Wang, L. S. Wood, and P. S. Skell, *J. Chem. Phys.* **53**, 1280 (1970).

[262] R. A. Bernheim, H. W. Bernard, P. S. Wang, and P. S. Skell, *J. Chem. Phys.* **54**, 3223 (1971).

[263] E. Wasserman, W. A. Yager, and V. J. Kuck, *Chem. Phys. Lett.* **7**, 409 (1970).

[264] E. Wasserman, V. J. Kuck, R. S. Hutton, and W. A. Yager, *J. Amer. Chem. Soc.* **92**, 7491 (1970).

[265] E. Wasserman, V. J. Kuck, R. S. Hutton, E. D. Anderson, and W. A. Yager, *J. Chem. Phys.* **54**, 4120 (1971).

[266] E. Wasserman, R. S. Hutton, V. J. Kuck, and W. H. Yager, *J. Chem. Phys.* **55**, 2593 (1971).

[267] G. Herzberg and J. W. C. Johns, *J. Chem. Phys.* **54**, 2276 (1971).

[268] G. Herzberg and D. A. Ramsay, *Proc. Roy. Soc.* (*London*) **A233**, 34 (1955).

[269] J. W. C. Johns, S. H. Priddle, and D. A. Ramsay, *Discuss. Faraday Soc.* **35**. 90 (1963).

development of experimental data will continue to provide the theoretician with the need for further refinements and developments.

2.3.3.1.2. SEMIQUANTITATIVE METHODS. It is useful to consider some qualitative descriptions of the energies and symmetries of ψ_e for diatomic and polyatomic molecules. These arguments provide a rationale for the selection of efficient basis sets for *ab initio* calculations, as well as a framework for making useful semiempirical correlations of molecular properties.

Perhaps the single most important property of a molecular electronic wave function is its symmetry. This property will be used extensively later to determine if a particular transition is allowed, as well as the rotational selection rules when it is allowed. Furthermore, the symmetry of a particular system places restrictions on how that system may dissociate or react with other systems. This approach may be developed by considering the correlation of a molecular state with its dissociation products. As these fragments are brought together, their individual symmetries will limit the symmetries of the molecule to be formed. In diatomic molecules, this may be treated by the techniques of vector coupling, as described by Herzberg,[270] pp. 315–322,[271] in his discussions of the *separated atom approximation*. In this case, the atomic angular momenta (defined in terms of the symmetry of the atomic state) have quantized projections along the developing internuclear axis. The molecular angular momenta (and hence symmetry properties) are defined then by the resultant of these angular momenta. Identical results may be obtained by applying the techniques of group theory. In this approach, the symmetry of the separated fragments is reduced to that of the molecule to be formed. The group symmetry species of the possible resulting molecular states are simply the "direct products" of those of the individual separated fragments. This procedure offers the advantage of being applicable to any molecule (not just a linear one) and any set of fragments (not just atoms). Herzberg,[272] (pp. 276–296) provides a detailed discussion of this procedure, along with a very convenient set of tables to aid in

[270] G. Herzberg, "Molecular Spectra and Molecular Structure," 2nd ed., Vol. I, Spectra of Diatomic Molecules. Van Nostrand–Reinhold, Princeton, New Jerşey, 1950.

[271] G. Herzberg, "The Spectra and Structures of Simple Free Radicals." Cornell Univ. Press, Ithaca, New York, 1971.

[272] G. Herzberg, "Molecular Spectra and Molecular Structure," Vol. III, Electronic Spectra and Electronic Structure of Polyatomic Molecules. Van Nostrand–Reinhold, Princeton, New Jersey, 1966.

the group theory manipulations. More detailed discussions of the principles of group theory may be found in standard texts (for example, those by Wigner,[273] Cotton,[274] and Tinkham[275]). The determination of molecular vibrational symmetries, as discussed by Wilson, Decius and Cross[276] also involves the use of the direct product procedures. Herzberg,[272] (pp. 277, 278) also discusses the related methods involving the *united atom approximation* (where the molecule is "built" from a single atom, e.g., Ne → CH_4) and the formation of a molecule from a related molecule of different symmetry (e.g., linear CH_2 → bent CH_2).

The above procedures are of considerable importance in developing concepts of bonding and in dealing with dissociative processes. They also suggest methods of describing molecular wave functions as linear combinations of atomic orbitals. Unfortunately, they usually do not provide much information about relative stabilities or geometries of the various states. In many cases the symmetry restrictions are too easily satisfied, so that they do not offer much assistance in eliminating alternate paths or processes.

In view of the wide applicability of the concepts of atomic orbitals, an equivalent *molecular orbital* description is a logical ideal model. Such a model would predict approximate energies, symmetries, and geometries of polyatomic molecules, given only the number of electrons in the various orbitals. This is equivalent to a search for molecular orbitals that are one-electron functions, i.e., independent of interactions with other electrons except through the Pauli exclusion principle.

For diatomic molecules, the molecular orbitals are based on the correlations between united atom and separated atom approximations, and on experimental data such as ionization potentials (see, for example, Herzberg's book,[270] pp. 322–348). The individual molecular orbital is indicated by $\sigma, \pi, \delta, \ldots$, corresponding to an axial component of angular momentum λ_i of $0, 1, 2, \ldots$, respectively. For homonuclear diatomic molecules (point group $D_{\infty h}$) a subscript g or u must be added to specify the symmetry with respect to the center of symmetry. The number

[273] E. P. Wigner, "Group Theory and Its Application to the Quantum Mechanics of Atomic Spectra." Academic Press, New York, 1959.

[274] F. A. Cotton, "Chemical Applications of Group Theory," 2nd ed. Wiley (Interscience), New York, 1971.

[275] M. Tinkham, "Group Theory and Quantum Mechanics." McGraw-Hill, New York, 1964.

[276] E. B. Wilson, Jr., J. C. Decius, and P. C. Cross, "Molecular Vibrations. The Theory of Infrared and Raman Vibrational Spectra." McGraw-Hill, New York, 1955.

of electrons occupying each orbital is indicated by a right superscript, except when a single electron is present, in which case the superscript is omitted. Most first-row homonuclear diatomic molecules may be described in terms of the following order of molecular orbitals

$$K\,K\,1\sigma_g\,1\sigma_u\,1\pi_u\,2\sigma_g\,1\pi_g\,2\sigma_u, \qquad (2.3.20)$$

where K represents the atomic $1s$ orbitals. Numbers preceding the symmetry species of the orbitals are arbitrary labels that indicate the number of times that symmetry species has occurred. They are *not* related to the principal quantum number of atomic orbitals or Rydberg series in any way. The Pauli exclusion principle permits two electrons in σ orbitals and four electrons in π orbitals.

There is an important distinction between electron *configurations* and *molecular states*. The former simply describes the occupancy of molecular orbitals, each of which may yield several molecular states, as shown in Table IV for the ground and excited states of C_2. The symmetry species of the states resulting from a particular electron configuration may be obtained by vector coupling for linear molecules, or by the more general procedure of forming the direct product of the symmetry species of occupied, but unfilled, orbitals. (Closed shells always will yield symmetric species.)

TABLE IV. Predicted Ground and Excited States of C_2 [a,b]

Electron configurations	Resulting states
$KK(\sigma_g2s)^2(\sigma_u2s)^2(\pi_u2p)^4$	$X\,^1\Sigma_g{}^+$
$KK(\sigma_g2s)^2(\sigma_u2s)^2(\pi_u2p)^3(\sigma_g2p)$	$a\,^3\Pi_u\,,\,A^1\Pi_u$
$KK(\sigma_g2s)^2(\sigma_u2s)^2(\pi_u2p)^2(\sigma_g2p)^2$	$b\,^3\Sigma_g{}^-\,,\,^1\Delta_g\,,\,^1\Sigma_g{}^+$
$KK(\sigma_g2s)^2(\sigma_u2s)^2(\pi_u2p)^4(\sigma_g2p)$	$c\,^3\Sigma_u{}^+\,,\,D^1\Sigma_u{}^+$
$KK(\sigma_g2s)^2(\sigma_u2s)(\pi_u2p)^3(\sigma_g2p)^2$	$d\,^3\Pi_g\,,\,C^1\Pi_g$
$KK(\sigma_g2s)^2(\sigma_u2s)^2(\pi_u2p)^2(\sigma_g2p)(\pi_g2p)$	$e\,^3\Pi_g\,,\,^3\Pi_g(3)\,,\,^1\Pi_g(3)\,,\,^5\Pi_g\,,\,^3\Phi_g\,,\,^1\Phi_g$
$KK(\sigma_g2s)^2(\sigma_u2s)^2(\pi_u2p)^3(\sigma_g3s)$	$^3\Pi_u\,,\,F^1\Pi_u$
$KK(\sigma_g2s)^2(\sigma_u2s)^2(\pi_u2p)^2(\sigma_g2p)(\sigma_g3s)$	$f\,^3\Sigma_g{}^-\,,\,g\,^3\Delta_g\,,\,^3\Sigma_g{}^-\,,\,^3\Sigma_g{}^+\,,\,^3\Sigma_g{}^-\,,\,^1\Sigma_g{}^+\,,\,^1\Sigma_g{}^-\,,\,^1\Delta_g$

[a] Reproduced from G. Herzberg, "The Spectra and Structures of Simple Free Radicals." Cornell University Press, Ithaca, New York, 1971, with the kind permission of the author and publishers.

[b] The state designations follow the recommendations of G. Herzberg, A. Lagerqvist, and C. Malmberg, *Can. J. Phys.* **47**, 2735 (1969).

For polyatomic molecules an equivalent description is possible, but it is necessary to specify the type of molecule being considered (e.g., XH_2, XH_3, XY_2, HXY, XY_3, X_2XY, etc.) and the point group that describes the equilibrium symmetry properties of the nuclei [i.e., linear XH_2 ($D_{\infty h}$) or nonlinear XH_2 (C_{2v})]. The molecular orbitals then are indicated by a small symbol indicating the representation that describes the symmetry properties of the orbital within the specified point group.

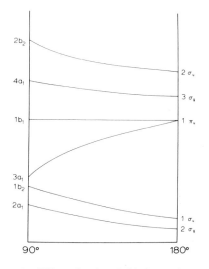

FIG. 9. Walsh diagram for XH_2 molecules. Orbital energies are plotted as a function of the HXH angle [reproduced from the paper by A. D. Walsh, *J. Chem. Soc.* 2260 (1953), with the kind permission of the author and the publishers].

For linear molecules ($D_{\infty h}$) this procedure resulted in labels such as $1\sigma_g$, $2\sigma_g$, $3\sigma_g$, ..., $1\pi_g$, $2\pi_g$, ... For nonlinear XH_2 molecules (point group C_{2v}), the equivalent symbols would be $1a_1$, $2a_1$, $3a_1$, ..., $1b_2$, $2b_2$, ..., etc. The most successful general description of the relative energies of molecular orbitals for small molecules was given in a series of papers by Walsh[277] in 1953. The *Walsh diagrams* consider each class of molecules (such as XH_2) and correlate in graphical form the qualitative behavior of the orbital energies as a function of one geometrical coordinate (for example, the HXH bond angle in the case of XH_2). The Walsh diagram for XH_2 molecules, as modified by Herzberg[272] (p. 319), is given in Fig. 9. The arbitrary numbers that label the orbitals follow the modification of

[277] A. D. Walsh, *J. Chem. Soc.* 2260, 2266, 2288, 2296, 2301, 2306 (1953).

Herzberg and differ slightly from those given by Walsh. (Herzberg has classified the $1s$ orbital of X in the $D_{\infty h}$ and C_{2v} notation, and taken it into account in the numbering of the orbitals.) Perhaps one example will demonstrate the way in which these Walsh diagrams may be used. The ground state of BH_2 (with five valence electrons) would have a linear electron configuration of

$$1\sigma_g^2 2\sigma_g^2 1\sigma_u^2 1\pi_u \Rightarrow {}^2\Pi_u \qquad (2.3.21)$$

or a bent electron configuration of

$$1a_1^2 2a_1^2 1b_2^2 3a_1 \Rightarrow {}^2A_1. \qquad (2.3.22)$$

The $3a_1 - 1\pi_u$ orbital would be expected to favor the bent geometry, and the experimental bond angle of $131°$ for the $\tilde{X}\,{}^2A_1$ state confirms this. The first one-electron excitation would involve promotion of the electron from the $3a_1 - 1\pi_u$ orbital to the $1b_1 - 1\pi_u$ orbital. Since the energy of the latter is independent of bond angle, the occupied $2a_1 - 2\sigma_g$ and $1b_2 - 1\sigma_u$ orbitals would favor the linear geometry. In other words, the first excited state of BH_2 would be expected to have the electron configuration

$$1\sigma_g^2 2\sigma_g^2 1\sigma_u^2 1\pi_u \Rightarrow {}^2\Pi_u. \qquad (2.3.23)$$

The observed state is linear, but subject to Renner–Teller interactions.

The Walsh diagrams for XY_2, XH_3, HXY, XY_3, and H_2XY molecules are given in Figs. 10–14, respectively. The last three are taken directly from Walsh's papers with some updating and numbering of the symmetry labels; the remaining diagrams have been numbered and modified by Herzberg[272] (pp. 319 and 321). Predictions based on these diagrams must be treated with considerable caution since they are based on very qualitative arguments. In fact, the primary justification for using them is that they do provide a surprisingly good basis for the correlation of experimental results. This was established by Walsh in the original papers and in a more recent review[278] of experimental data.

The designation of molecular electronic states throughout this chapter will follow the convention used by Herzberg[272] for polyatomic molecules. As with diatomic molecules, the symmetry of the state will be designated by a large symbol of the symmetry species in the point group appropriate for the molecule. Individual states will be labeled with an arbitrary identification letter *with a tilde* which precedes the symmetry species.

[278] A. D. Walsh, *Ann. Rep. Prog. Chem.* **63**, 44 (1966).

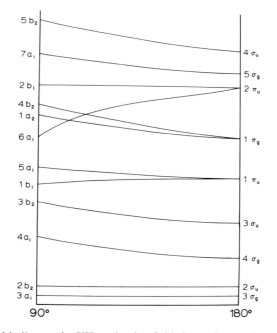

FIG. 10. Walsh diagram for XY$_2$ molecules. Orbital energies are plotted as a function of the YXY bond angle. [Reproduced from the paper by A. D. Walsh, *J. Chem. Soc.* 2266 (1953), with the kind permission of the author and the publishers.]

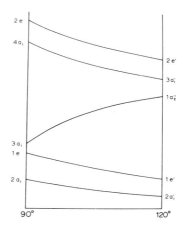

FIG. 11. Walsh diagram for XH$_3$ molecules. Orbital energies plotted as a function of the HXH bond angle (all are assumed equal). [Reproduced from the paper by A. D. Walsh, *J. Chem. Soc.* 2296 (1953), with the kind permission of the author and the publishers.]

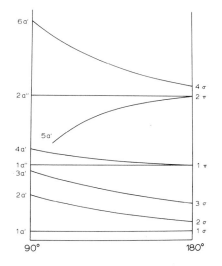

Fig. 12. Walsh diagram for HXY molecules. Orbital energies plotted as a function of HXY bond angle. [Reproduced from the paper by A. D. Walsh, *J. Chem. Soc.* 2290 (1953), with the kind permission of the author and publisher.]

The ground state should be labeled $\tilde{X}$ and higher states of the same multiplicity by $\tilde{A}, \tilde{B}, \ldots$, in order of increasing energies. States of different multiplicities (usually seen with triplets) are labeled with small letters, $\tilde{a}, \tilde{b}, \ldots$, with increasing energies.

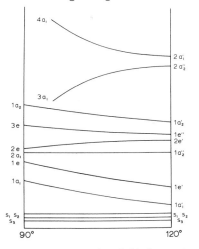

Fig. 13. Walsh diagram for XY_3 molecules. Orbital energies plotted as a function of the YXY bond angle (all are assumed equal). [Reproduced from the paper by A. D. Walsh, *J. Chem. Soc.* 2303 (1953), with the kind permission of the authors and publisher.]

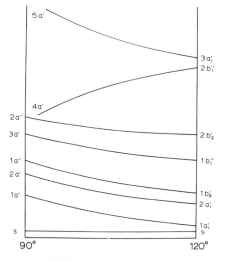

FIG. 14. Walsh diagram for H_2XY molecules. Orbital energies plotted as a function of the HXY = YXY bond angles. [Reproduced from the paper by A. D. Walsh, *J. Chem. Soc.* 2306 (1953), with the kind permission of the author and publisher.]

2.3.3.2. Vibrational Structure. The vibrational energy levels of a polyatomic molecule usually can be described by the expression

$$G = \sum_i \omega_i(v_i + \tfrac{1}{2}d_i) + \sum_i \sum_{j \geq i} x_{ij}(v_i + \tfrac{1}{2}d_i)(v_j + \tfrac{1}{2}d_j) + \sum_i \sum_{j \geq i} g_{ij}l_il_j.$$
(2.3.24)

In this equation v_i are the vibrational quantum numbers, ω_i are the harmonic vibrational frequencies, x_{ij} and g_{ij} are anharmonicity constants, d_i are the degeneracies of the vibrations, and l_i are the quantum numbers for vibrational angular momenta associated with each degenerate vibration. For nondegenerate vibrations $d_i = 1$, $l_i = 0$, and $g_{ij} = 0$. For degenerate vibrations, d_i is an integer greater than one, and the values of l_i are*

$$l_i = \pm v_i, \ \pm(v_i - 2), \ \pm(v_i - 4), \ \ldots, 1 \text{ or } 0. \qquad (2.3.25)$$

These expressions for vibrational energies ignore interactions that lead to Fermi resonance as well as the smaller interactions between degenerate

---

* Hougen (Ref. 236b) discusses some of the conventions and problems associated with the use of signed and unsigned quantum numbers.

vibrations that lead to small splittings in overtone bands. Additional details on these topics are available in standard texts.[272,276,279,280]

The symmetry properties of individual vibrational eigenfunctions for a particular normal mode of vibration may be obtained as described in detail, for example, by Wilson et al.[276] For nondegenerate vibrations the total vibrational symmetry species may be obtained simply by taking the direct product of the species of all vibrational levels that are excited. In the case of H_2CO, for example, if the $v_2(a_1)$, $v_4(b_1)$, and $v_6(b_2)$ modes are each excited by one quantum, the resulting vibrational eigenfunction has the symmetry of an A_2 species (i.e., $^v a_1 \times {}^v b_1 \times {}^v b_2 = {}^v A_2$). As with electrons, small symbols are used to designate the symmetry species of a single element, and large symbols indicate overall symmetry species. Multiple excitations of degenerate vibrations present a bit more of a problem, since symmetric products must be used for multiple excitations of a single vibrational mode. When only double excitations are involved, the resulting direct products may be obtained from appropriate tables (see, for example, Herzberg[272]). When higher excitations of degenerate modes are involved, rules such as the ones summarized by Wilson et al.[276] may be used to determine the proper resulting symmetry species.

The above discussion of vibrational energies and symmetries appears to be totally independent of any consideration of the electronic and rotational states of the molecule. In fact, this approach is a direct result of the extended Born–Oppenheimer approximation. However, within this approximation the expression for the vibrational energy does take into account an interaction of electronic and vibrational motion to the extent that the electronic terms determine the potential energy function of the vibrational problem.

2.3.3.2.1. VIBRONIC STATES. A vibrational level within a specific electronic state may be termed a *vibrational-electronic state*, or *vibronic state*, of the molecule as suggested by Mulliken.[281] Within the Born–Oppenheimer approximation this state is defined by the relations

$$E_{ev} = E_e + E_v, \qquad \psi_{ev} = \psi_e \times \psi_v. \qquad (2.3.26)$$

[279] G. Herzberg, "Molecular Spectra and Molecular Structure," Vol. II, Infrared and Raman Spectra of Polyatomic Molecules. Van Nostrand–Reinhold, Princeton, New Jersey, 1945.

[280] H. C. Allen, Jr. and P. C. Cross, "Molecular Vib-rotors." Wiley, New York, 1963.

[281] R. S. Mulliken, Report on Notation for the Spectra of Polyatomic Molecules, *J. Chem. Phys.* **23**, 1997 (1955).

Even when the interaction between vibrational and electronic terms becomes large and the above approximation is inadequate, the symmetry of the vibronic eigenfunction is retained as the direct product of the electronic and vibrational species. This approximation will break down for very large interactions when additional terms have to be added to the expression as discussed by Hougen.[282] Since the same symmetry species symbols may be used for vibrational and electronic terms as well as for the vibronic terms, a left superscript may be used to identify the eigenfunction associated with the species when it is not clear. If we consider, for example, the benzene molecule (D_{6h} point group) in a vibrational level associated with excitation by one quantum in each of the modes a_{2g}, b_{1u}, and e_{2g}, the resulting vibrational state has the symmetry species $^vE_{1u}$ (i.e., $^va_{2g} \times {}^vb_{1u} \times {}^ve_{2g} = {}^vE_{1u}$). If the electronic state is of symmetry species B_{2u} (i.e., $^eB_{2u}$), then the vibronic state is of species $^vE_{1u} \times {}^eB_{2u} = {}^{ev}E_{2g}$. Obviously, for a totally symmetric electronic state (usually the case for a stable molecule in its ground state), the symmetry of the vibronic state is the same as the symmetry of the vibrational state.

When both the vibrational and electronic states are degenerate, the resulting vibronic levels are of particular interest. In the case of linear molecules, this interaction is referred to as a *Renner–Teller effect*,[283,284] whereas for nonlinear molecules the corresponding (dominant) interaction is referred to as the *Jahn–Teller effect*.[285]

2.3.3.2.2. RENNER-TELLER EFFECT.   In the case of linear molecules it is most convenient to describe the interactions in terms of a coupling of the z component of electronic angular momentum ($\lambda\hbar$) and vibrational angular momentum ($l\hbar$). The total angular momentum (exclusive of electron spin) about the z axis is then $K\hbar$, where

$$K = |\lambda + l|. \tag{2.3.27}$$

For a linear triatomic molecule, $l = l_i$ in Eq. (2.3.25). For larger linear molecules, $l = \sum l_i$. If the electron spin is strongly coupled to the electronic angular momentum [Hund's case (a)], then the resultant total

[282] J. T. Hougen, *in* "Physical Chemistry, an Advanced Treatise," (D. Henderson, ed.), Vol. IV, Chapter 7. Academic Press, New York, 1970.

[283] G. Herzberg and E. Teller, *Z. Phys. Chem.* **B21**, 410 (1933).

[284] R. Renner, *Z. Phys.* **92**, 172 (1934).

[285] H. A. Jahn and E. Teller, *Proc. Roy. Soc. (London)* **A161**, 220 (1937).

vibronic angular momentum including spin is

$$P = |K + \Sigma| = |\lambda + l + \Sigma| = |\Omega + l|. \qquad (2.3.28)$$

The vibronic symmetry of a linear polyatomic molecule is Σ, Π, Δ, ..., corresponding to values of K equal to 0, 1, 2, ... The value of P may be added as a right subscript. Figure 15 demonstrates the energy levels and quantum numbers appropriate for a linear triatomic molecule in a $^2\Pi$ state with large spin–orbit coupling. At this stage, vibronic coupling has not been included and the energy levels may be obtained from the

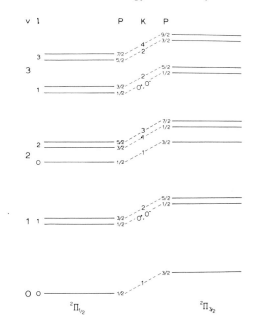

FIG. 15. Vibrational energy levels for the bending vibration of a linear molecule in a $^2\Pi$ electronic state.

expression for vibrational energies plus a spin–orbit term ($\pm \frac{1}{2}A$). Note that for the $v = 3$ level the relative positions for different values of l are governed primarily by the term $g_i l^2$ in Eq. (2.3.24). Within this approximation, the vibronic levels for a given value of l coincide. At some higher approximation, however, we expect to find terms that will permit each vibronic level to assume a slightly different value. The "repulsion" between levels with the same value of K can be evaluated by considering the dependence of the potential energy on the bending coordinate, as was first suggested by Teller[283] and worked out by Renner.[284] When the molecule is bent, the potential energy function must split into two different components, as indicated in Fig. 16, which tends to form two different electronic states of lower symmetry (i.e., in this case, nonlinear). The r axis corresponds to displacement in the degenerate mode of vibration (in this case, a departure of the bond angle from $180°$). Figure 16a corresponds to the potential energy curve when there is no interaction, for example in a $^1\Sigma$ state. This function would be of the form

$$V = ax^2 + bx^4 + \cdots \qquad (2.3.29)$$

and would be appropriate for a linear molecule that is equally free to bend in any plane. Figure 16b depicts the degree of interaction considered by

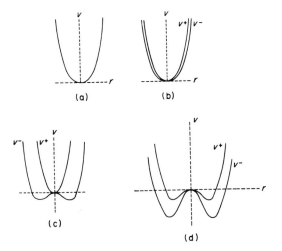

FIG. 16. Potential energy curves for bending of a linear triatomic molecule. See text for details.

Renner.[284] In this case the original curve is split into two components, V^+ and V^-, which indicate the loss of cylindrical symmetry of the molecule. A $^1\Pi$ state, for example, may be "split" into two different electronic states, with the same (e.g., linear) geometry, that have slightly different bending force constants. Figure 16c corresponds to the case (appropriate to the original interpretation of NH_2) in which the lower state is nonlinear and the upper state is linear. Figure 16d indicates the limit in which both upper and lower states are nonlinear as a result of the vibronic interaction. In a sense the Walsh diagrams for XH_2 and XY_2 molecules indicate this kind of splitting of a degenerate π orbital which leads to two different nondegenerate orbitals of different binding energies which favor nonlinear geometries. This comparison also indicates why Fig. 16d is seldom considered a Renner–Teller interaction—the two states are interpreted simply as nonlinear states of the molecule without the necessity of considering the equivalent linear state.

The most frequently observed and analyzed type of Renner–Teller interaction is of the type indicated by Fig. 16b. Renner[284] considered such an interaction for a $^1\Pi$ state of a linear triatomic molecule. He demonstrated that the terms $\omega_2(v_2 + 1) + g_{22}l_2^2$ in Eq. (2.3.24) should be replaced by the following expressions for the vibronic energy levels: For $K = 0$ (Σ vibronic states)

$$G^{\pm}(v_2, K) = \omega_2(1 \pm \varepsilon)^{1/2}(v_2 + 1), \qquad v_2 = 1, 3, 5, \ldots;$$

for $K = v_2 + 1 \neq 0$ (lowest single vibronic state other than Σ)

$$G(v_2, K) = \omega_2[(v_2 + 1) - \tfrac{1}{8}\varepsilon^2 K(K + 1)]; \qquad (2.3.30)$$

for $0 < K < v_2 + 1$ (remaining pairs of vibronic states)

$$G^{\pm}(v_2, K) = \omega_2(1 - \tfrac{1}{8}\varepsilon^2)(v_2 + 1) \pm \tfrac{1}{2}\omega_2\varepsilon[(v_2 + 1)^2 - K^2]^{1/2},$$

where the quantity ε is defined as the *Renner parameter* and may be related simply to equations for the potential energy curves of V^+ and V^-. Occasionally the quantity $\varepsilon\omega_2$ may be referred to as the *Renner parameter*. Hougen and Jesson[286] have extended Renner's treatment to include anharmonic terms x_{ij}. Hougen[287] also has provided a discussion of Fermi resonance in Π electronic states.

Despite the fact that Renner carried out his calculations in 1934, the analysis of the NH_2 spectrum by Dressler and Ramsay[288] provided the first example of a system involving the Renner–Teller interaction. In this case, however, the interaction corresponds to Fig. 16c and the observed transition is *between* the lower bent state and the upper linear state. Pople and Longuet-Higgins[289] derived expressions for the vibronic energy levels of the upper state by including higher-order terms in the perturbation

[286] J. T. Hougen and J. P. Jesson, *J. Chem. Phys.* **38**, 1524 (1963).
[287] J. T. Hougen, *J. Chem. Phys.* **37**, 403 (1962).
[288] K. Dressler and D. A. Ramsay, *Phil. Trans. Roy. Soc.* **251A**, 553 (1959).
[289] J. A. Pople and H. C. Longuet-Higgins, *Mol. Phys.* **1**, 372 (1958).

approach used by Renner. Despite a number of simplifying assumptions, the agreement of their expression with experimental observations was satisfactory. Their expression for the upper state energy levels that depend on the bending vibration may be written in the form

$$G_{vK}^+ = \omega_2\{(v_2 + 1) - \tfrac{1}{2}\bar{a}[(v_2 + 1) - ((v_2 + 1)^2 - K^2)^{1/2}]$$
$$+ \tfrac{1}{2}\bar{b}[3(v_2 + 1)^3 + \tfrac{3}{4}\bar{\beta}(v_2 + 1)((v_2 + 1)^2 - K^2)^{1/2}]\}, \quad (2.3.31)$$

where $\bar{a}$, $\bar{\beta}$, and $\bar{b}$ are coupling parameters that may be related to the form of the two potential energy functions. One interesting property of Eq. (2.3.31) is that for small values of K (i.e., small compared to v_2) the expression reduces to the same form as Eq. (2.3.24). As indicated by Pople and Longuet-Higgins, the difference is that the value of "g_{22}" would be much larger than those expected for a normal linear molecule. In the absence of vibronic coupling and other anharmonic effects, $g_{22} = -\tfrac{1}{3}x_{22}$. Therefore, significant departures from this value can serve as an indication that a Π state is, in fact, the upper component of a state split by Renner–Teller interactions. Dixon[290] extended the theory to permit both upper and lower states to be bent (i.e., Fig. 16d). When he applied his theory to the data of Dressler and Ramsay for NH_2 he concluded that the upper state of NH_2 is bent with an equilibrium bond angle of $144° \pm 5°$ and a barrier height of 777 ± 100 cm^{-1}. Since the lowest vibrational level, which was not observed, would be the only one below the barrier, additional data would be of considerable interest.

The above discussions and results have not included the effects of spin–orbit coupling, despite the fact that NH_2 and many other molecules which exhibit Renner–Teller effect have unpaired electrons. Pople[291] first derived expressions appropriate for a $^2\Pi$ state which result in the energy levels indicated in Fig. 17. For levels with $K = v_2 + 1$, the energy levels are given by

$$G(v_2, K, \pm\tfrac{1}{2}) = \omega_2(v_2 + 1) \pm \tfrac{1}{2}A - \tfrac{1}{8}\varepsilon^2\omega_2 K(K + 1), \quad (2.3.32)$$

where $\pm\tfrac{1}{2}$ represents the z component of the electron spin quantum number. Hougen[292] demonstrated that in a higher approximation A

[290] R. N. Dixon, *Mol. Phys.* **8**, 201 (1964).

[291] J. A. Pople, *Mol. Phys.* **3**, 16 (1960).

[292] J. T. Hougen, *J. Chem. Phys.* **36**, 519 (1962).

should be replaced by the quantity $A[1 + \frac{1}{8}\varepsilon^2 K(K + 1)]$. For levels with $K \neq v_2 + 1$, the energies derived by Pople[291] and corrected by

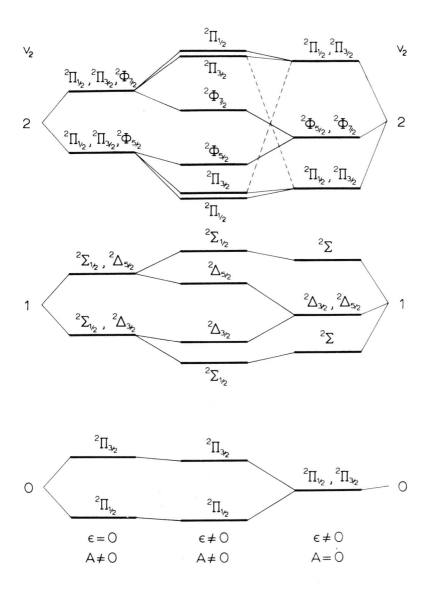

FIG. 17. Correlation of the vibronic levels of a $^2\Pi$ electronic state for zero vibronic (left) and zero spin–orbit (right) interaction.

Hougen[292] are given by the expressions

$$G^{(+)}(v_2, K, \pm \tfrac{1}{2}) = \omega_2(1 - \tfrac{1}{8}\varepsilon^2)(v_2 + 1) + \tfrac{1}{2}A^*_{v_2, K}$$
$$\mp [\varepsilon^2 A \omega_2 K(v_2 + 1)/8A^*_{v_2, K}]$$

$$G^{(-)}(v_2, K, \pm \tfrac{1}{2}) = \omega_2(1 - \tfrac{1}{8}\varepsilon^2)(v_2 + 1) - \tfrac{1}{2}A^*_{v_2, K}$$
$$\pm [\varepsilon^2 A \omega_2 K(v_2 + 1)/8A^*_{v_2, K}],$$

$$(2.3.33)$$

where

$$A^*_{v_2, K} = \{A^2 + \varepsilon^2 \omega_2^2[(v_2 + 1)^2 - K^2]\}^{1/2}$$

is an effective spin–orbit coupling constant. It must be emphasized again that the superscripts plus and minus refer only to the upper and lower vibronic components and not to the symmetry of the levels. Hougen[292] has used the alternative symbols μ and $\varkappa$ in an effort to eliminate this source of possible confusion, but they have not been adopted extensively thus far.

The correlation of states when $\varepsilon \neq 0$, $A = 0$ and $\varepsilon \neq 0$, $A \neq 0$ is somewhat ambiguous when $K \neq 0 \neq v_2 + 1$, as indicated by the dashed lines in Fig. 17. In fact two correlations are possible and the states are interdependent.

Theoretical descriptions have been extended to $^3\Pi$ states with spin–orbit coupling by Hougen,[293] $^2\Delta$ states with spin–orbit coupling by Merer and Travis,[294] and small interactions for $^2\Sigma^+$ states (a "magnetic" interaction) by Chiu and Chen[295] and by Merer and Allegretti.[296]

A relatively large number of triatomic molecules have been observed (most since 1960) that demonstrate the Renner–Teller effect, as is summarized in Section 2.3.4.2.2.

2.3.3.2.3. JAHN–TELLER EFFECT. In nonlinear molecules, degenerate electronic states are subject to interactions with degenerate vibrations, only in this case the interaction usually is referred to as the *Jahn–Teller effect*. Again, the resulting vibronic levels tend to distort the molecule so as to lower the symmetry and remove the degeneracy. Formal theory for the vibronic interactions depends on higher order terms in the potential energy expression for the degenerate vibrations which couple them with electronic terms. In general these functions may be even or

[293] J. T. Hougen, *J. Chem. Phys.* **36**, 1874 (1962).
[294] A. J. Merer and D. N. Travis, *Can. J. Phys.* **43**, 1795 (1965).
[295] C. Chang and Y. Chiu, *J. Chem. Phys.* **53**, 2186 (1970).
[296] A. J. Merer and J. M. Allegretti, *Can. J. Phys.* **49**, 2859 (1971).

odd powers of the nuclear displacement, although the Jahn–Teller coupling usually refers only to the odd-power terms.[272] Thus a nonlinear molecule may undergo Renner–Teller interactions via even-powered terms, as well as Jahn–Teller interactions via odd-powered terms.[297] The symmetry species of the vibronic levels again are simply the direct products of the electronic and vibrational species. Herzberg[272] presents examples of the vibronic species of vibrational levels of nondegenerate and degenerate vibrations in several types of electronic states of D_{3h} and D_{6h} molecules.

Vibronic interactions that alter the potential energy functions of nonlinear molecules in degenerate electronic states are usually termed *static* Jahn–Teller effects. Such effects would be most easily recognized when they resulted in a change in the equilibrium structure of the molecule, comparable to Fig. 16c and d in the case of linear molecules. Considerations of the vibronic energy levels (i.e., with changing coordinates) resulting from such interactions involve the *dynamic* Jahn–Teller effect. Discussions of the problem and solutions of the perturbation equations have been presented by Moffitt and Liehr,[298] Moffitt and Thorson,[299,300] and Longuet-Higgins.[301] Summaries of the results are given by Herzberg,[272] Longuet-Higgins[302] and Child and Longuet-Higgins.[303] In addition, Clinton and Rice[304] have treated the Jahn–Teller effect in terms of a coupling of forces via the Hellman–Feynman theorem rather than a coupling of potential energy functions as mentioned above.

Unfortunately, there are no experimental examples of the Jahn–Teller effect that are as detailed and unambiguous as those for the Renner–Teller effect. Perhaps the best example thus far is the spectrum of CH_3I [305–307] where the excited electronic states of the 2012-Å system are degenerate.[307a] In this case, weak Jahn–Teller effects are inferred from intensity ratios

[297] J. T. Hougen, *J. Mol. Spectrosc.* **13**, 149 (1964).

[298] W. Moffitt and A. D. Liehr, *Phys. Rev.* **106**, 1195 (1957).

[299] W. Moffitt and W. Thorson, *Phys. Rev.* **108**, 1251 (1957).

[300] W. Moffitt and W. Thorson, *Coll. Int. C.N.R.S.* **82**, 141 (1958).

[301] H. C. Longuet-Higgins, U. Öpik, M. H. L. Pryce, and R. A. Sack, *Proc. Roy. Soc. (London)* **A244**, 1 (1958).

[302] H. C. Longuet-Higgins, *Advan. Spectrosc.* **2**, 429 (1961).

[303] M. S. Child and H. C. Longuet-Higgins, *Phil. Trans. Roy. Soc.* **254A**, 259 (1961).

[304] W. L. Clinton and B. Rice, *J. Chem. Phys.* **30**, 542 (1959).

[305] A. Henrici and H. Grieneisen, *Z. Phys. Chem.* **B30**, 1 (1935).

[306] G. Herzberg and G. Scheibe, *Z. Phys. Chem.* **B7**, 390 (1930).

[307] G. Scheibe, F. Povenz, and C. Linström, *Z. Phys. Chem.* **B20**, 283 (1933).

[307a] R. S. Mulliken and E. Teller, *Phys. Rev.* **61**, 283 (1942).

only; the expected splitting of the degenerate vibrational levels was obscured by an overlapping band. On the other hand, the ground state of C_5H_5 is expected to be a degenerate state, and calculations have indicated that the Jahn–Teller interactions should be small but observable.[308–311] The spectra observed by Engleman and Ramsay[312] do present some unusual doublings, but these investigators have been unable to assign these unambiguously to a Jahn–Teller effect thus far.

Jahn–Teller effects do play an important role in studies of the structure and properties of inorganic compounds.* Unfortunately most of the observations rely on indirect evidence or may be obtained only for the condensed phase of the system. The quotation of Van Vleck,[314] "It is a great merit of the Jahn–Teller effect that it disappears when not needed," was repeated by Ballhausen[315] and strengthened by the observation that the effect also is very difficult to find when it should be present.

2.3.3.2.4. VIBRONIC COUPLING BETWEEN DIFFERENT ELECTRONIC STATES. The previous two sections have dealt with vibronic couplings within a single electronic state. For a degenerate state this could be interpreted as the interaction between two states of the same symmetry and energy. In addition, it is possible to have vibronic coupling between different states with significantly different energies. When the two states are of the same symmetry species, the interaction may be essentially electronic (i.e., configuration interaction), although the question of avoided crossings, which seems to be valid for diatomic molecules, is less stringent for polyatomic molecules. Teller[316] demonstrated that states of the same symmetry can cross under some conditions for polyatomic molecules, in which case perturbations between vibrational levels would be quite important.

[308] A. D. Liehr, Z. Phys. Chem. 9, 338 (1956).

[309] A. D. Liehr, Ann. Rev. Phys. Chem. 13, 41 (1962).

[310] L. C. Snyder, J. Chem. Phys. 33, 619 (1960).

[311] W. D. Hobey and A. D. McLachlan, J. Chem. Phys. 33, 1695 (1960).

[312] R. Engleman, Jr. and D. A. Ramsay, Can. J. Phys. 48, 964 (1970).

[313] R. Engleman, "The Jahn–Teller Effect in Molecules and Crystals." John Wiley (Interscience), New York, 1972.

[314] J. H. Van Vleck, J. Chem. Phys. 7, 61 (1939).

[315] C. J. Ballhausen, "Introduction to Ligand Field Theory," p. 205. McGraw-Hill, New York, 1962.

[316] E. Teller, J. Phys. Chem. 41, 109 (1937).

* A recent book (Ref. 313) provides an excellent summary of the theory and experimental methods of interest in evaluating the Jahn–Teller effect in such systems.

Vibronic coupling is of particular importance for polyatomic molecules, since it permits the interaction of electronic states of different symmetry species. In order for two electronic states to perturb each other $\int \psi_e' \psi_e'' \, dt_e$ must be nonzero, or the product

$$\Gamma(\psi_e') \times \Gamma(\psi_e'') \tag{2.3.34}$$

must contain the totally symmetric species. However, when an interaction is not allowed between two electronic states because of this selection rule, the interaction may be permitted through vibronic coupling in either of the states, as discussed by Herzberg and Teller.[283] Such a vibronic interaction (the *Herzberg–Teller interaction*) will be allowed when

$$\Gamma(\psi_e') \times \Gamma(\psi_v) \times \Gamma(\psi_e'') \tag{2.3.35}$$

contains the totally symmetric species. $\Gamma(\psi_v)$ is the species of the normal vibration that "couples" the two electronic states. For example, in a linear molecule, vibronic interaction of a π_u vibration in a Σ_g^+ state results in a vibronic state of symmetry ${}^{ev}\Pi_u$ which may interact with a ${}^e\Pi_u$ electronic state. Likewise, for a molecule of C_{2v} point group, vibronic perturbations between A_2–B_1, A_2–B_2, and B_1–B_2 electronic states will be permitted only through interaction with normal vibrations of type b_2, b_1, and a_2, respectively. Note that the vibrations could be excited in either state, as long as the overall direct product satisfies Eq. (2.3.35). Fulton and Gouterman[317] have presented a recent detailed theoretical treatment of such interactions. These types of interactions will be discussed again in connection with vibronically allowed electronic transitions in Section 2.3.4.1.2.

2.3.3.3. Rotational Structure. As a first approximation, the molecular rotational energies required for a description of electronic transitions can be taken directly from previous discussions regarding microwave and infrared spectroscopy. In practice, this approximation (which is equivalent to saying that $\psi_T = \psi_e \psi_v \psi_r$) is inadequate, and additional interactions must be considered.

Rotational energies of any rigid rotor can be described in terms of its three principal moments of inertia, I_a, I_b, and I_c, or the corresponding rotational constants, A, B, and C, respectively, where

$$A = h/8\pi^2 cI_a, \qquad B = h/8\pi^2 cI_b, \qquad C = h/8\pi^2 cI_c. \tag{2.3.36}$$

[317] R. L. Fulton and M. Gouterman, *J. Chem. Phys.* **35**, 1059 (1961).

By convention,[281] the axes are chosen so that

$$A \geq B \geq C \qquad \text{or} \qquad I_a \leq I_b \leq I_c. \tag{2.3.37}$$

The rotational energies are obtained by finding the eigenvalues of the rotational Hamiltonian

$$\mathscr{H} = AP_a^2 + BP_b^2 + CP_c^2, \tag{2.3.38}$$

where P_a, P_b, and P_c are the operators for angular momentum about the three inertial axes. In special cases, expressions for the energies may be obtained quite readily.

When $A = \infty$ ($I_a = 0$) and $B = C$, i.e., for the case of a linear molecule, the energies are given by the expression

$$F(J) = BJ(J + 1). \tag{2.3.39}$$

For prolate symmetric rotors ($A \neq B = C$) or oblate symmetric rotors ($A = B \neq C$) the energies are given by the following respective equations.

$$F(J, K_a) = BJ(J + 1) + (A - B)K_a^2$$

or $$\tag{2.3.40}$$

$$F(J, K_c) = BJ(J + 1) + (C - B)K_c^2.$$

The quantum number for total angular momentum is again indicated by J, and K_a or K_c is the quantum number for the component of J along the unique axis (a or c for prolate or oblate rotors, respectively). For the more general case of an asymmetric rotor ($A \neq B \neq C$) the energies cannot be obtained from closed-form expressions. It is useful, however, to specify an energy level of an asymmetric rotor by the notation $E(J_{K_a K_c}, \varkappa)$, where K_a and K_c are the symmetric rotor quantum numbers with which the level would correlate in the prolate and oblate limits, respectively. The asymmetry parameter

$$\varkappa = (2B - A - C)/(A - C) \tag{2.3.41}$$

takes values between -1 for the prolate symmetric rotor ($B = C$) and $+1$ for the oblate symmetric rotor ($A = B$).

The calculation of energy levels for an asymmetric rotor may be facilitated by a rearrangement of the general Hamiltonian, Eq. (2.3.38), into a more convenient form such as the one proposed by Ray[317a]

$$\mathcal{H} = \tfrac{1}{2}(A + C)P^2 + \tfrac{1}{2}(A - C)\mathcal{H}(\varkappa),$$
$$\mathcal{H}(\varkappa) = P_a^2 + \varkappa P_b^2 - P_c^2, \tag{2.3.42}$$

where $P^2 = P_a^2 + P_b^2 + P_c^2$. This form has been used extensively, since it offers a convenient method of tabulating solutions of $\mathcal{H}(\varkappa)$, which can be applied to any particular system through the corresponding energy expression

$$E(J_{K_aK_c}, \varkappa) = \tfrac{1}{2}(A + C)J(J + 1) + \tfrac{1}{2}(A - C)F(J_{K_aK_c}, \varkappa). \tag{2.3.43}$$

The quantities $F(J_{K_aK_c}, \varkappa)$ are the eigenvalues of $\mathcal{H}(\varkappa)$ (usually identified as $E_{J_\tau}(\varkappa)$, where $\tau = K_a - K_c$) and have been tabulated for J values up to 30 by King et al.[318] This form has been used extensively in the past, since it offered a convenient method of tabulating solutions to the difficult part of the Hamiltonian, which could then be used to determine the asymmetric rotor energies of any particular molecule. The necessity of such tabulations is virtually eliminated by the convenient availability of computer systems which permit efficient computations of the rotational energies for any desired system. Detailed discussions of these computer techniques and of other convenient formulations of the Hamiltonian are given in several texts on microwave or infrared spectroscopy, such as Gordy and Cook,[319] Allen and Cross,[280] and Townes and Schawlow.[320]

Even with such programs it often is necessary to make judicious choices of the most appropriate approximation technique. This is especially true for spectra that include high J values, as is the case for large polyatomic molecules. Recent reviews of the high-resolution electronic spectra by Ross[321] and by Goodman and Hollas[322] include discussions of some of the more widely used techniques for the efficient computation of rotational energies as well as the calculation of band contours. As emphasized by

[317a] B. S. Ray, Z. Phys. **78**, 74 (1932).

[318] G. W. King, R. M. Hainer, and P. C. Cross, J. Chem. Phys. **11**, 27 (1943).

[319] W. Gordy and R. L. Cook, "Microwave Molecular Spectra." Wiley (Interscience), New York, 1970.

[320] C. H. Townes and A. L. Schawlow, "Microwave Spectroscopy." McGraw-Hill, New York, 1955.

[321] I. G. Ross, Advan. Chem. Phys. **20**, 341 (1971).

[322] L. Goodman and M. Hollas, in "Physical Chemistry, and Advanced Treatise" (D. Henderson, ed.), Vol. III, Chapter 7. Academic Press, New York, 1969.

Ross,[321] the Gora approximation for nearly prolate tops[323] and the extension by Brown[324] for nearly oblate tops are very useful approximations for many molecular systems. Brown[324] also points out the rather surprising fact that the low K_a energy levels of a nearly *prolate* top (say, with $\varkappa = -0.75$) may be described adequately with an *oblate*-top approximation for $J \geq 10$. In this example, the levels with $K_a = 2$–4 would be described poorly by either symmetric rotor approximation and levels with $K_a > 5$ would be described most efficiently by the *prolate* approximation.

2.3.3.3.1. VIBRATIONAL–ROTATIONAL INTERACTIONS. These interactions include primarily vibrational interactions, centrifugal distortion, and Coriolis interaction, all of which are discussed in earlier chapters of this volume (2.1 and 2.2) and elsewhere (see, for example, Gordy and Cook,[321] Allen and Cross,[280] and Herzberg[272,279]). Such interactions must be evaluated if *equilibrium* rotational constants are desired.

For linear molecules with N atoms the rotational energy expressions take a form similar to that for diatomic molecules, namely

$$F_v(J) = B_v J(J+1) - D_v J(J+1) + \cdots,$$

where

$$B_v = B_e + \sum \alpha_1 [v_1 + (d_1/2)]$$

and

$$D_v = D_e + \sum \beta_1 [v_1 + (d_1/2)].$$

(2.3.44)

The sums in these expressions are to be taken over the $3N - 5$ vibrations.

In addition, the presence of one or more degenerate (bending) vibrations leads to the necessity of including the quantum number for vibrational angular momentum l [see Eq. (2.3.25)] in the expression for rotational energies. In the absence of centrifugal distortion, a linear molecule behaves essentially as a symmetric top and

$$F_v(J) = B_v J(J+1) + (A_v - B_v)l^2.$$

(2.3.45)

The term $(A_v - B_v)l^2$ may be considered as another vibrational term $(g_{22}l^2)$, but its presence emphasizes that each rotational level is doubly degenerate when $l \neq 0$. The removal of this degeneracy may be described in a manner similar to that used for Λ-doubling in diatomic

[323] E. K. Gora, *J. Mol. Spectrosc.* **16**, 378 (1965).
[324] J. M. Brown, *J. Mol. Spectrosc.* **31**, 118 (1969).

molecules. For Λ-type doubling the splitting depends on electron-rotational interactions (e.g. Van Vleck's pure precession model), whereas for l-type doubling the splittings depend on vibration–rotation interaction.

For a nonrigid prolate symmetric top the rotational energy expression becomes

$$F_v(J, K_a) = B_v J(J + 1) + (A_v - B_v)K_a{}^2 - D_J J^2(J + 1)^2$$
$$-D_{JK} J(J + 1)K_a{}^2 - D_K K_a{}^4 + \cdots, \qquad (2.3.46)$$

where

$$B_v = B_e - \sum \alpha_i{}^B[v_i + (d_i/2)], \qquad A_v = A_e - \sum \alpha_i{}^A[v_i + (d_i/2)].$$

Similar expressions could be written for the nonrigid oblate symmetric rotor by substituting C everywhere for A in the above equations.

The rotational energies of a symmetric top in a degenerate vibrational level are subjected to a splitting because of the interaction between vibrational and rotational angular momentum. A molecule that is a symmetric rotor because of its symmetry (i.e., it has a threefold or higher symmetry axis) will always have such degenerate vibrations and the vibrational angular momentum will contribute to the total angular momentum about the top axis (K_a or K_c). As a result, the rotational energy formula for the rigid prolate top [Eq. (2.3.40)] multiply excited in a degenerate vibration must be modified by replacing $AK_a{}^2$ by the term

$$A(K_a - l_i\zeta_i)^2, \qquad (2.3.47)$$

where l_i is the previously defined quantum number for vibrational angular momentum. The Coriolis coupling constant ζ_i takes a value between $+1$ and -1 that is characteristic of each degenerate vibration. When more than one degenerate mode is involved, $AK_a{}^2$ is replaced by $A(K_a - \zeta_v)^2$, where

$$\zeta_v = \sum_i l_i\zeta_i \qquad (2.3.48)$$

is the *total* vibrational angular momentum along the symmetry axis.

Considerable attention has been given to the sum rules, selection rules, and theoretical values of ζ_i (see, for example, the following reviews: Herzberg,[272] pp. 63–65, 84–87; Herzberg,[279] pp. 401–406; Allen and Cross,[280] pp. 48–67; and Gordy and Cook,[319] pp. 145–149, 196–199, 528–532).

The theory of vibration–rotation interactions for asymmetric rotors (as well as symmetric rotors) has been summarized in a very informative manner by Gordy and Cook,[319] (Chapter 8). Their discussion includes a development of centrifugal distortion in terms of the constants $\tau_{\alpha\beta\gamma\delta}$, where each of the subscripts may take the values of x, y, or z (i.e., τ_{xxxx}, τ_{xxyy}, ...). In addition, for special cases (planar molecules and symmetric rotors), tables are given that summarize the relations between these τ's and the D_J, D_K, and D_{JK} constants mentioned above for symmetric rotors. Further discussions deal with the development of higher-order distortion effects and with their relation to molecular force fields.

2.3.3.3.2. ROTATIONAL–ELECTRONIC INTERACTIONS. Molecular rotations may interact with electronic properties in two important ways: (1) by coupling with electronic orbital angular momentum, and (2) by coupling with electron-spin angular momentum. The first of these applies only to degenerate states of symmetric rotors where the electronic angular momentum has a value of ζ_e along the symmetry axis. For linear molecules ζ_e takes integral values and is called λ, whereas it may take nonintegral values for nonlinear symmetric rotors. The nature of the interaction is the same as described above for the interaction between vibrational and rotational angular momenta. In other words, AK_a^2 of Eq. (2.3.40) is replaced by $A(K_a \pm \zeta_e)^2$ for a prolate rotor and CK_c^2 is replaced by $C(K_c \pm \zeta_e)^2$ for an oblate rotor. When both electronic and vibrational angular momenta are present, it is the resultant vibronic angular momentum ζ_t, which couples with rotation about the symmetry axis. In this case, the expression for rotational energies (disregarding centrifugal distortion) of a prolate symmetric rotor may be given by the expression

$$F(J, K_a) = BJ(J + 1) + (A - B)K_a^2 \mp 2A\zeta_t K_a + A\zeta_t^2. \quad (2.3.49)$$

So long as the vibronic coupling is small, $|\zeta_t| = |\zeta_v + \zeta_e|$ and the signs may be derived from the discussion by Mills.[325] Child and Longuet-Higgins[303] have derived more detailed expressions for very small coupling for an X_3 molecule, and Child[326,327] has considered the complications introduced by slightly stronger vibronic coupling.

Coupling between electron spin and molecular rotation requires the introduction of the quantum number N, as for diatomic molecules, which

[325] I. M. Mills, *Mol. Phys.* **7**, 549 (1964).
[326] M. S. Child, *Mol. Phys.* **5**, 391 (1962).
[327] M. S. Child, *J. Mol. Spectrosc.* **10**, 357 (1963).

indicates the total angular momentum apart from spin (i.e., it replaces J in the previous discussions). The spin S is then added to N to give the total angular momentum J (i.e., $J = N + S$). As for diatomic molecules, each rotational energy level (of a particular value of N) is split into $2S + 1$ components. In the absence of large spin–orbit coupling [Hund's case (b)], the spin splittings for linear molecules follow the normal equations for diatomic molecules.[270] The equivalent case for nonlinear molecules has been considered in detail by Henderson[328] and by Raynes[329] for doublet and triplet states. In both cases, the spin splitting terms are added as corrections to the rigid-rotor energy expressions.

2.3.4. Analysis of Electronic Transitions

2.3.4.1. Theoretical Descriptions.   The purpose of this section will be to summarize the nomenclature and selection rules for electronic transitions. These will permit the application of the previous descriptions of molecular energy levels to the analysis of observed molecular spectra. Sections 2.3.4.2 and 2.3.4.3 will discuss examples of characteristic systems and indicate some of the progress made in recent years.

In order for an electronic transition to take place, the following integral must be nonzero

$$\mathbf{R} = \int \psi^{*\prime} \mathbf{M} \psi^{\prime\prime} \, d\tau, \qquad (2.3.50)$$

where ψ' and ψ'' refer to the stationary state wave functions of the upper and lower states, respectively. The quantity $\mathbf{M}$ is a multipole expansion term that provides for coupling between the molecule and the electromagnetic field associated with the transition. In this section we will be concerned with arguments based on symmetry which will allow us to say whether the integral must be zero. We will not consider the related question of comparisons between $\mathbf{R}^2$ and experimental measurements of the transition probability.

In order for the above integral to be nonzero, it is necessary that $\psi^{*\prime} \mathbf{M} \psi^{\prime\prime}$ be a totally symmetric function. This condition will be fulfilled if

$$\Gamma(\psi') \times \Gamma(\mathbf{M}) \times \Gamma(\psi'') \qquad (2.3.51)$$

contains the totally symmetric species within a point group appropriate

[328] R. S. Henderson, *Phys. Rev.* **100**, 723 (1955).
[329] W. T. Raynes, *J. Chem. Phys.* **41**, 3020 (1964).

for the system. This statement is equivalent to the requirement that the product $\psi^{*\prime}\psi^{\prime\prime}$ have the same transformation properties as some component of $\mathbf{M}$. In the simplest case, $\mathbf{M}$ is an electric dipole moment operator with components M_x, M_y, or M_z given by $\sum e_i x_i$, $\sum e_i y_i$, and $\sum e_i z_i$, respectively. The transition will be allowed if

$$\Gamma(\psi^{\prime}) \times \Gamma(\psi^{\prime\prime}) = \Gamma(T_x, \quad T_y, \quad \text{or} \quad T_z). \qquad (2.3.52)$$

The problem of discussing selection rules thus simplifies to one of determining the symmetry properties of the three quantities involved. The appropriate point group normally is chosen to be consistent with the equilibrium geometry of the molecule in both states.

2.3.4.1.1. ALLOWED ELECTRONIC TRANSITIONS. An "allowed" electronic transition is one for which

$$\mathbf{R}_e = \int \psi_e^{*\prime} \mathbf{M} \psi_e^{\prime\prime} \, d\tau_e \qquad (2.3.53)$$

is nonzero when $\mathbf{M}$ is an electric dipole operator. In other words,

$$\Gamma(\psi_e^{\prime}) \times \Gamma(\psi_e^{\prime\prime}) = \Gamma(T_x, \quad T_y, \quad \text{or} \quad T_z). \qquad (2.3.54)$$

Thus if we know the electronic symmetries of the upper and lower states, we know immediately whether the transition is "allowed" or "forbidden." For example, a transition between an A_1 and an A_2 state of a molecule possessing symmetry appropriate for the C_{2v} point group would be "forbidden" since the direct product ($A_1 \times A_2 = A_2$) is different from

$$\Gamma(T_x) = B_2, \qquad \Gamma(T_y) = B_1, \qquad \text{or} \qquad \Gamma(T_z) = A_1.$$

For an allowed transition, the *transition moment* would be along the x, y, or z axis depending on whether Eq. (2.3.54) contained the species of T_x, T_y, or T_z, respectively.

The relation between Eq. (2.3.50) and Eq. (2.3.53) provides additional information on the vibrational, rotational, and spin selection rules of allowed transitions, as well as an indication of what kinds of interactions will permit the observation of "forbidden" transitions.

As in Section 2.3.3.1, it is convenient to assume that the total wave function for a state ψ may be factored into terms dependent only on electronic q and nuclear Q coordinates,

$$\psi_{ev} = \psi_e \psi_v,$$

$$\psi \approx \psi_{ev} = \psi_e(q, Q)\psi_v(Q), \tag{2.3.55}$$

where we have ignored the contributions of rotation and spin for the present. If, in addition, we assume the electric dipole moment can be factored into a component due to the electrons and a component due to the nuclei, $\mathbf{M} = \mathbf{M}_e + \mathbf{M}_n$, then

$$\mathbf{R} \simeq \mathbf{R}_{ev} = \int \psi_{ev}^{*\prime} \mathbf{M} \psi_{ev}^{\prime\prime} \, d\tau_{ev}$$

$$\simeq \int \psi_v^{*\prime} \psi_v^{\prime\prime} \, d\tau_v \int \psi_e^{*\prime} \mathbf{M}_e \psi_e^{\prime\prime} \, d\tau_e$$

$$+ \int \psi_v^{*\prime} \mathbf{M}_n \psi_v^{\prime\prime} \, d\tau_v \int \psi_e^{*\prime} \psi_e^{\prime\prime} \, d\tau_e. \tag{2.3.56}$$

The last term vanishes because of the orthogonality of the electronic eigenfunction at a particular value of the nuclear coordinates Q. Therefore,

$$\mathbf{R} \simeq \mathbf{R}_{ev} = \int \psi_v^{*\prime} \psi_v^{\prime\prime} \, d\tau_v \int \psi_e^{*\prime} \mathbf{M}_e \psi_e^{\prime\prime} \, d\tau_e \tag{2.3.57}$$

or, substituting Eq. (2.3.53),

$$\mathbf{R} \simeq \mathbf{R}_{ev} = \mathbf{R}_e(Q) \int \psi_v^{*\prime} \psi_v^{\prime\prime} \, d\tau_v. \tag{2.3.58}$$

This immediately demonstrates that very definite restrictions are placed on the particular vibrational transitions permitted in an allowed electronic transition: namely, $\Gamma(\psi_v{'}) \times \Gamma(\psi_v{''})$ must contain the totally symmetric species. In other words, allowed vibrational components of an allowed electronic transition may consist of $\Delta v_i = 0, \pm 1, \pm 2, \pm 3, \ldots$ for totally symmetric vibrations or $\Delta v_i = 0, \pm 2, \pm 4, \ldots$ for nontotally symmetric vibrations.*

Equation (2.3.58) also provides the basis for applications of the Franck–Condon principle (see Herzberg,[272] for more detailed discussions). If

---

* These results apply strictly only for *nondegenerate* vibrations. For degenerate vibrations each case should be worked out in detail. Note, for example, that $\Delta v = \pm 3$ would be allowed for an e vibration in the C_{3v} point group.

one assumes that

$$R_{\mathrm{e}}(Q) \approx R_{\mathrm{e}}(Q_0) = \text{constant}, \qquad (2.3.59)$$

i.e., that the electronic transition moment varies only slightly with nuclear coordinates, then the intensity distribution among vibrational components is determined by the "overlap integrals," $\int \psi_{\mathrm{v}}^{*\prime} \psi_{\mathrm{v}}^{\prime\prime} \, d\tau_{\mathrm{v}}$. Use of the same nuclear coordinates for upper and lower states conforms to the vertical transitions required by Franck–Condon arguments. The magnitude of this integral is such that one expects to observe *progressions* ($\Delta v_i \neq 0$) in the vibrational modes that favor the change in geometry associated with the electronic transition and only *sequences* ($\Delta v_i = 0$) in all other modes. Thus if the transition results in a change in a bond angle, progressions would be expected in the bending frequency that involves that angle.

It must be emphasized, of course, that the above conclusions depend on the Born–Oppenheimer approximation. The discussion of forbidden transitions (and forbidden vibrational components of allowed electronic transitions) will be discussed in the next section.

Molecular rotation can be added to the above discussion by assuming $\psi = \psi_{\mathrm{e}}\psi_{\mathrm{v}}\psi_{\mathrm{r}}$. In this case it becomes necessary to distinguish between a *space-fixed axis system* (designated $F = X, Y, Z$) and a *molecule-fixed axis system* (designated by $g = x, y, z$). The component of the electric dipole moment (due to electron contributions) along the space-fixed X axis becomes

$$\mathbf{M}_X = M_x \cos xX + M_y \cos yX + M_z \cos zX, \qquad (2.3.60)$$

where the direction cosines between the two coordinate systems are functions of the rotational coordinates alone. Equation (2.3.60) can be denoted more conveniently by

$$\mathbf{M}_X = \sum_g \Phi_{Xg}\mathbf{M}_g, \qquad g = x, y, z, \qquad (2.3.61)$$

where Φ_{Xg} are the appropriate direction cosines. We may now express the F component of Eq. (2.3.50) as

$$\mathbf{R}_F = \sum_g \left[\int \psi_{\mathrm{e}}^{*\prime} \psi_{\mathrm{v}}^{*\prime} \psi_{\mathrm{r}}^{*\prime} \Phi_{Fg}\mathbf{M}_g \psi_{\mathrm{e}}^{\prime\prime} \psi_{\mathrm{v}}^{\prime\prime} \psi_{\mathrm{r}}^{\prime\prime} \, d\tau \right]. \qquad (2.3.62)$$

If we write $\mathbf{M}_g = (\mathbf{M}_{\mathrm{e}})_g + (\mathbf{M}_{\mathrm{n}})_g$ as was done for Eq. (2.3.56), and assume that Φ_{Fg} depends only on the rotational coordinates, then

$$\mathbf{R}_F = \sum_g \left[\int \psi_e^{*\prime} (\mathbf{M}_e)_g \psi_e^{\prime\prime} \, d\tau_e \int \psi_v^{*\prime} \psi_v^{\prime\prime} \, d\tau_v \int \psi_r^{*\prime} \Phi_{Fg} \psi_r^{\prime\prime} \, d\tau_r \right.$$
$$\left. + \int \psi_e^{*\prime} \psi_e^{\prime\prime} \, d\tau_e \int \psi_v^{*\prime} (\mathbf{M}_n)_g \psi_v^{\prime\prime} \, d\tau_v \int \psi_r^{*\prime} \Phi_{Fg} \psi_r^{\prime\prime} \, d\tau_r \right]. \quad (2.3.63)$$

For an electronic transition, the last part of this expression must vanish because of the orthogonality of the electronic wave functions. Since $\mathbf{R} = \sum \mathbf{R}_F$, we can write a more general expression as

$$\mathbf{R} = \sum_F \sum_g \left[\int \psi_e^{*\prime} (\mathbf{M}_e)_g \psi_e^{\prime\prime} \, d\tau_e \int \psi_v^{*\prime} \psi_v^{\prime\prime} \, d\tau_v \int \psi_r^{*\prime} \Phi_{Fg} \psi_r^{\prime\prime} \, d\tau_r \right]. \quad (2.3.64)$$

Thus, for example, if the only nonzero value of the transition moment is along the x axis [i.e., $(\mathbf{M}_e)_x \neq 0$, $(\mathbf{M}_e)_y = (\mathbf{M}_e)_z = 0$], then

$$\mathbf{R} = \int \psi_e^{*\prime} (\mathbf{M}_e)_x \psi_e^{\prime\prime} \, d\tau_e \int \psi_v^{*\prime} \psi_v^{\prime\prime} \, d\tau_v \int \psi_r^{*\prime} \sum_F \Phi_{Fx} \psi_r^{\prime\prime} \, d\tau_r. \quad (2.3.65)$$

In other words, as with vibration–rotation spectra, the rotational selection rules depend only on the direction of the transition moment and the symmetry properties of the direction cosines. Care must be exercised in matching the x, y, and z axes with the a, b, and c inertial axes of the molecule, but this problem is covered thoroughly in standard discussions of microwave and infrared spectroscopy.

The rotational selection rules for symmetric rotors can be summarized as follows. When the transition moment is parallel to the symmetry axis

$\parallel$ transitions: $\Delta K = 0$, $\Delta J = 0, \pm 1$, $J = 0 \not\leftrightarrow J = 0$.

When the transition moment is perpendicular to the symmetry axis,

$\perp$ transitions: $\Delta K = \pm 1$, $\Delta J = 0, \pm 1$, $J = 0 \not\leftrightarrow J = 0$.

For asymmetric rotors the selection rules are best summarized in terms of the evenness or oddness of K_a and K_c indicated, for example, as eo for even K_a and odd K_c. Recall that each level can be specified by $J_{K_a K_c}$. For a transition moment parallel to the a axis of the molecule,

Type A band; ee $\leftrightarrow$ eo, oe $\leftrightarrow$ oo.

For a transition moment parallel to the b axis of the molecule,

$$\text{Type } B \text{ band: } \quad \text{ee} \leftrightarrow \text{oo,} \quad \text{oe} \leftrightarrow \text{eo.}$$

For a transition moment parallel to the c axis of the molecule,

$$\text{Type } C \text{ band: } \quad \text{ee} \leftrightarrow \text{oe,} \quad \text{eo} \leftrightarrow \text{oo.}$$

These selection rules, together with the more general one for J ($\Delta J = 0$, ± 1 but $J = 0 \not\leftrightarrow J = 0$) include the symmetric rotor limits when one considers only K_a or K_c. For strongly asymmetric rotors or for large changes in asymmetry, ΔK may be greater than 1 or 0. Fortunately, initial assignments of rotational transitions can be made within the symmetric rotor approximation in most observed spectra.

The effects of electron spin may be included, as a first approximation, by assuming that spin–orbit interaction is small so that $\psi_{es} = \psi_e \psi_s$. In this case the transition moment integral becomes

$$\int \psi_{es}^{*\prime} \mathbf{M} \psi_{es}^{\prime\prime} \, d\tau = \int \psi_e^{*\prime} \mathbf{M} \psi_e^{\prime\prime} \, d\tau_e \int \psi_s^{*\prime} \psi_s^{\prime\prime} \, d\tau_s. \qquad (2.3.66)$$

In other words, the expected orthogonality of the spin functions requires that there be no change in multiplicity ($\Delta S = 0$) for an allowed electronic transition.

The assumptions made in arriving at the selection rules for allowed transitions obviously are extremely restrictive. For polyatomic molecules in particular, it is essential to consider some of the ways in which these assumptions break down.

2.3.4.1.2. FORBIDDEN ELECTRONIC TRANSITIONS. The most important factors that permit the observation of forbidden transitions involve: (1) higher power electric and magnetic multipole components of $\mathbf{M}$, (2) coupling between the spatial and spin properties of the electron, so that $\Delta S \neq 0$, and (3) vibronic interactions which invalidate the assumption that $\psi_{ev} = \psi_e \psi_v$. Each of these may be considered briefly in order to indicate their effects on the selection rules. In cases (2) and (3) the interaction is regarded as a mixing of one of the states with another state of appropriate symmetry. This concept of "intensity borrowing" from an allowed transition to permit the observation of a forbidden transition merely reflects the inability to describe the states completely independently of one another.

If the dipole moment operator of Eq. (2.3.53) is replaced by an operator appropriate for *higher electric* or *magnetic multipoles*, one may obtain the transition probability for these forbidden transitions. A very thorough quantum-mechanical development of this topic for atomic transitions is given by Shore and Menzel.[330] They include relative transition probabilities up through electric octupole and magnetic quadrupole. At 5000 Å, these estimated ratios for electric dipole : magnetic dipole : electric quadrupole : magnetic quadrupole : electric octupole are approximately $1 : 10^{-5} : 10^{-8} : 10^{-13} : 10^{-15}$. In view of the magnitudes of these ratios, it is not surprising that electric dipole transitions account for most observed spectra. Nevertheless, magnetic dipole and electric quadrupole transitions may be observed in some cases.

The transition probability may be nonzero for a magnetic dipole transition if the product $\Gamma(\psi_e') \times \Gamma(\psi_e'')$ contains the species of one of the components of the magnetic dipole operator. This operator has the symmetry properties of the three rotations R_x, R_y, and R_z. Likewise, an electric quadrupole transition may be observed if the product $\Gamma(\psi_e')$ $\times \Gamma(\psi_e'')$ transforms as one of the components of the polarizability $(\alpha_{xx}, \alpha_{xy}, \ldots)$.* In diatomic molecules, transitions of these types are not uncommon, but there is *only one* example of such a forbidden transition for a polyatomic molecule in the gas phase. The 0–0 band of the 3500-Å system of H_2CO has been established as a magnetic dipole transition as suggested by Sidman.[331] A rotational analysis by Callomon and Innes,[332] as well as a study of electric field splitting of rotational lines by Lombardi et al.[333] confirmed this interpretation. Even in this case, the remainder of this extensive spectrum owes most of its intensity to vibronic interaction.[333a]

Transitions involving a *change of multiplicity* are the most familiar of the "forbidden" transitions. The breakdown of the $\Delta S = 0$ selection rule is normally associated with spin-orbit coupling achieved by "heavy atom effects." To a reasonable approximation, the sym-

[330] B. W. Shore and D. H. Menzel, "Principles of Atomic Spectra," pp. 430–441. Wiley, New York, 1968.

[331] J. W. Sidman, *Chem. Rev.* **58**, 689 (1958).

[332] J. H. Callomon and K. K. Innes, *J. Mol. Spectrosc.* **10**, 166 (1963).

[333] J. R. Lombardi, D. E. Freeman, and W. Klemperer, *J. Chem. Phys.* **46**, 2746 (1967).

[333a] V. A. Job, V. Sethuraman, and K. K. Innes, *J. Mol. Spectrosc.* **30**, 365 (1969); **33**, 189 (1970).

---

* The polarizabilities transform as the products of translations $(T_x T_x, T_x T_y, \ldots)$.

metry properties of ψ_{es} may be obtained as the direct product $\Gamma(\psi_e)$ $\times \Gamma(\psi_s)$, even though $\psi_{es} \neq \psi_e \psi_s$. Herzberg[272] (Table 56) has tabulated the symmetry species of spin functions so that $\Gamma(\psi_{es})$ may be obtained readily from direct product tables. States of even multiplicity do require the use of extended point group tables (the double groups). Aside from linear molecules the best documented examples of $\Delta S \neq 0$ transitions involve singlet–triplet systems of asymmetric rotors. Hougen[334] has worked out the selection rules and line intensities for singlet–triplet transitions of molecules belonging to the D_{2h} point group. The D_{2h} point group represents the highest symmetry an asymmetric rotor may possess, therefore Hougen's tabulations and conclusions may be applied to any asymmetric rotor with only slight modifications. For a triplet state, the spin function transforms as $B_{1g} + B_{2g} + B_{3g}$. Therefore, a triplet–singlet transition may occur if $\Gamma_T \times \Gamma(\psi_s) \times \Gamma_S$ contains the species of T_x, T_y, or T_z [$\Gamma_T = \Gamma$ (triplet ψ_e), $\Gamma_S = \Gamma$ (singlet ψ_e)]. There are nine possible ways this condition can be satisfied, three for each of the possible spin species

$$\mathbf{M}(B_{1g}) = \langle \Gamma_T \times B_{1g} \mid \mathbf{M} \mid \Gamma_S \rangle,$$

$$\mathbf{M}(B_{2g}) = \langle \Gamma_T \times B_{2g} \mid \mathbf{M} \mid \Gamma_S \rangle, \qquad (2.3.67)$$

$$\mathbf{M}(B_{3g}) = \langle \Gamma_T \times B_{3g} \mid \mathbf{M} \mid \Gamma_S \rangle.$$

Since the species of T_x, T_y, and T_z are all different for this point group, only one component in each of the above equations will be nonzero. The most interesting conclusion of Hougen's study for nearly symmetric rotors is that the selection rules for ΔK will be 0, ± 2, or ± 1 depending on whether $\Gamma_T \times \Gamma_S = A_u$, B_{1u} or $\Gamma_T \times \Gamma_S = B_{2u}$, B_{3u}, respectively. In other words, the apparent band type (i.e., parallel or perpendicular) is determined solely by the symmetry species of the electronic portion of the states (i.e., $\psi_e' \psi_e''$). The determination of how the three spin components of Eq. (2.3.67) contribute to the transition (and, consequently, which nearby singlet states are mixing with the triplet state) can only be determined by a detailed evaluation of the line intensities. Hougen does give explicit expressions for the line intensities and explains how his results may be applied to molecules of lower symmetry.

SO_2 and H_2CO provide convenient examples of the application of Hougen's results to molecules of point group C_{2v}. The character table

[334] J. T. Hougen, *Can. J. Phys.* **42**, 433 (1964).

TABLE V. Character Table for D_{2h} Point Group

D_{2h}	E	$C_2(z)$	$C_2(y)$	$C_2(x)$	i	$\sigma(xy)$	$\sigma(xz)$	$\sigma(yz)$	
A_g	1	1	1	1	1	1	1	1	
B_{1g}	1	1	−1	−1	1	1	−1	−1	R_z
B_{2g}	1	−1	1	−1	1	−1	1	−1	R_y
B_{3g}	1	−1	−1	1	1	−1	−1	1	R_x
A_u	1	1	1	1	−1	−1	−1	−1	
B_{1u}	1	1	−1	−1	−1	−1	1	1	T_z
B_{2u}	1	−1	1	−1	−1	1	−1	1	T_y
B_{3u}	1	−1	−1	1	−1	1	1	−1	T_x

for point group D_{2h} is given in Table V for convenience. As may be seen, there are three C_2 axes, only one of which will be retained as a symmetry element in the C_{2v} point group. In order to make use of Hougen's tabulations, it is necessary to correlate the two point groups in such a way as to maintain the z axis as the (near) symmetric-top axis. For H_2CO the C_2 axis coincides with the (near) symmetric-top axis, whereas for SO_2 the C_2 axis is perpendicular to the (near) symmetric-top axis. Table VI summarizes the possible correlations for these two cases, depending on whether the x or y axis is chosen to be perpendicular to the molecular

TABLE VI. Correlations of D_{2h} with C_{2v}

D_{2h}	ΔK^a	$C_2(z) \rightarrow C_2(z)$ $\sigma(xz) \rightarrow \sigma(yz)$	$C_2(z) \rightarrow C_2(z)$ $\sigma(yz) \rightarrow \sigma(yz)$	$C_2(y) \rightarrow C_2(z)$ $\sigma(yz) \rightarrow \sigma(yz)$	$C_2(x) \rightarrow C_2(z)$ $\sigma(xz) \rightarrow \sigma(yz)$
A_g	—	A_1	A_1	A_1	A_1
B_{1g}	—	A_2	A_2	B_1	B_1
B_{2g}	—	B_2	B_1	A_2	B_2
B_{3g}	—	B_1	B_2	B_2	A_2
A_u	$0, \pm 2$	A_2	A_2	A_2	A_2
B_{1u}	$0, \pm 2$	A_1	A_1	B_2	B_2
B_{2u}	± 1	B_1	B_2	A_1	B_1
B_{3u}	± 1	B_2	B_1	B_1	A_1

[a] Selection rules for (near) symmetric-top singlet–triplet transitions [J. T. Hougen, Can. J. Phys. **42**, 433 (1964)].

plane. The first two correlations, in which $C_2(z) \to C_2(z)$, are appropriate for H_2CO; the remaining two are appropriate for SO_2. The selection rules for ΔK are given in the table as well.

In the case of the triplet–singlet absorption spectrum of H_2CO, which has a $^1A_1$ ground state, only transitions with $\Delta K = 0$ were observed by Raynes.[335] Therefore the triplet state, could only be a $^3A_1$ or a $^3A_2$. The state was assigned as $^3A_2$, since it would be expected to be near the lowest excited singlet state that is known to be $^1A_2$. In the triplet–singlet absorption spectrum of SO_2 (with a $^1A_1$ ground state), Brand et al.[336,337] observed only $\Delta K = \pm 1$ transitions. Therefore, from Table VI the triplet state could only be $^3A_1$ or $^3B_1$. In this case, however, missing lines owing to the zero nuclear spin of ^{16}O eliminate $^3A_1$ as a possibility, and the transition is assigned as $^3B_1 - {}^1A_1$.

Neither of the above arguments involved the question of which of the spin functions was operative in making the transition observable through mixing with other singlet states. That information was obtained from both spectra, but only by comparison of the intensities with Hougen's line strength formulas. In the case of H_2CO, Raynes concluded that the major perturbing state is a $^1A_1$ electronic state. Brand et al. concluded that the major perturbing state for SO_2 is a $^1B_2$ electronic state.

Forbidden transitions may become observable through *vibronic interactions*, i.e., a failure of the assumption that $\psi_{ev} = \psi_e \psi_v$. In the presence of such interactions the transition moment integral must be retained in the form

$$\mathbf{R}_{ev} = \int \psi_{ev}^{*\prime} \mathbf{M} \psi_{ev}^{\prime\prime} \, d\tau_{ev} \qquad (2.3.68)$$

rather than in the factored form given in Eq. (2.3.57). Again, the requirement for a nonvanishing integral is that the product of the *vibronic* species of the two states must contain the species of the electric dipole operator $(T_x, T_y, \text{ or } T_z)$. The vibronic species may be obtained from the direct product $\Gamma(\psi_e) \times \Gamma(\psi_v) = \Gamma(\psi_{ev})$, even though $\psi_e \psi_v$ is a very poor approximation to ψ_{ev}. This requirement may be stated in a different way by saying that the product of $\Gamma(\psi_e{}^\prime \psi_e^{\prime\prime}) \times \Gamma(\psi_v{}^\prime \psi_v^{\prime\prime})$ must contain a species of the dipole moment operator.

Since a forbidden transition is one for which Eq. (2.3.53) is zero, vibronic coupling will result in the possibility of observing transitions

[335] W. T. Raynes, *J. Chem. Phys.* **44**, 2755 (1966).
[336] J. C. D. Brand, C. di Lauro, and V. T. Jones, *J. Amer. Chem. Soc.* **92**, 6095 (1970).
[337] J. C. D. Brand, V. T. Jones, and C. di Lauro, *J. Mol. Spectrosc.* **40**, 616 (1971).

that involve $\Delta v_i = 1$ for antisymmetric vibrations. These are the very transitions that should *not* be observed in an allowed transition. For example, the transition $^eA_2 - {}^eA_1$ is forbidden for molecules of point group C_{2v}. The XYZ$_2$ molecule, however, has normal vibrations of symmetry $^vb_1$ and $^vb_2$, either of which may couple with one of the electronic state and lead to a vibronic transition. The examples of formaldehyde and benzene still present the best examples of vibronic transitions, although the upper state of formaldehyde is nonplanar, which complicates the comparison. Herzberg[272] has discussed these assignments in detail and gives the references to the original papers.

Another important consequence of vibronic coupling is that it may permit the observation of *forbidden components of allowed electronic transitions*. According to Eq. (2.3.58), the only vibrational components allowed are those that allow $\psi_v{}'\psi_v''$ to be a totally symmetric function. This requirement would result in all of the observed bands in a system having the same transition moment. In some cases, however, different band types are observed within the same band system of an allowed transition of a linear or nonlinear molecule. The allowed $A_1 - A_1$ transition of a C_{2v} molecule has the transition moment along the z axis. Vibronic mixing in the upper or lower state (for example, with $^vb_1$ or $^vb_2$ modes) would give rise to vibronic transitions with a transition moment along the y or x axes, respectively.

Transitions involving a *change of symmetry* require that only the symmetry elements common to both states be considered. The most common types of change in symmetry are for bent–linear and planar–nonplanar transitions. A few examples of each will be cited in the next section. Another more subtle difficulty arises when the inertial axes of the two states are not parallel, even though both states may be of the same symmetry species. The effects of this "axis switching" have been described theoretically by Hougen and Watson.[338] Experimental observations of the effect are most obvious for HXY molecules, such as HSiCl, HSiBr, HCN, and HNF.

2.3.4.2. Transitions of Linear Molecules

2.3.4.2.1. DIATOMIC MOLECULES. No attempt will be made to summarize the increasingly large number of contributions in this area. Instead, references to existing reviews will be given and a few systems of particular interest will be discussed briefly.

[338] J. T. Hougen and J. K. G. Watson, *Can. J. Phys.* **43**, 298 (1965).

The book by Herzberg[270] serves as the primary reference for this subject. A book by Kovacs[338a] presents a more detailed summary of the equations necessary to describe energy levels and line intensities for multiplet states of diatomic molecules. A short monograph by Hougen[338b] presents a systematic method for the calculation of rotational energy levels and line intensities for diatomic molecules. Herzberg[338c] presents an extensive summary of the observed forbidden transitions in diatomic molecules. Veseth[338d] has published a series of papers dealing with higher order descriptions of multiplet states, especially for doublet states.

Huber[339] has published a tabulation of the ground state constants of diatomic molecules based on a careful evaluation of the existing data. This tabulation is taken from the forthcoming compilation of data for all known diatomic molecules for both upper and lower states. Rosen[340] has edited a compilation of the molecular constants reported in the literature on diatomic molecules. This book includes a large amount of information, such as band heads and relative intensities, which will be of considerable assistance in the identification of known diatomic spectra. Gaydon[341] has included a table of dissociation energies of diatomic molecules in the 1968 edition of his book on the topic. Beckel et al.[342] updated this list in 1971. Several other more specialized reviews include those by Cheetham and Barrow[343] on transition element molecules, by Schofield on Group IIA oxides, by Dunn[344] on nuclear hyperfine

[338a] I. Kovacs, "Rotational Structure in the Spectra of Diatomic Molecules." American Elsevier, New York, 1969.

[338b] J. T. Hougen, "The Calculations of Rotational Energy Levels and Rotational Line Intensities in Diatomic Molecules," NBS Monograph 115, U.S. Govt. Printing Office, Washington, D.C., 1970.

[338c] G. Herzberg, Mem. Soc. Roy. Sc. Liege 16, 121 (1969).

[338d] L. Veseth, Physica 56, 286 (1971); J. Phys. B. 3, 1677 (1971); 4, 20 (1971); 5, 229 (1972); J. Mol. Spectrosc. 38, 228 (1971); Mol. Phys. 20, 1057 (1971); 21, 287 (1971).

[339] K. P. Huber, Constants of Diatomic Molecules, in "American Institute of Physics Handbook," 3rd ed. McGraw-Hill, New York, 1972.

[340] B. Rosen (ed.), "Spectroscopic Data Relative to Diatomic Molecules." Pergamon, Oxford, 1970.

[341] A. G. Gaydon, "Dissociation Energies and Spectra of Diatomic Molecules," 3rd ed. Chapman and Hall, London, 1968.

[342] C. L. Beckel, M. Shafi, and R. Engelke, J. Mol. Spectrosc. 40, 519 (1971).

[343] C. J. Cheetham and R. F. Barrow, Advan. High Temp. Chem. 1, 7 (1967).

[344] T. M. Dunn, in "Molecular Spectroscopy—Modern Research" (K. N. Rao and C. W. Mathews, eds.), pp. 207–257. Academic Press, New York, 1972.

structure in diatomic spectra, and by Brewer and Rosenblatt[345] on dissociation energies and free energy functions of oxides.

A number of individual diatomic spectra do deserve special mention.
The correction to the experimental dissociation energy of H_2 already has
been described in Section 2.3.3.1.1. Comparison with theoretical results
was of particular importance in that case.

Spectroscopic studies of C_2 (and its ions) have provided a wealth of
new data within the past few years. Prior to 1960 a large number of
singlet and triplet states were known for C_2, but no direct information
was available that established the energy of the triplet states relative
to the singlet states. On the basis of indirect evidence the ground state
was considered to be a triplet state. In 1963, Ballik and Ramsay[346,347]
reported the details of their analysis of perturbations between $^3\Sigma_g^-$ and
$^1\Sigma_g^+$ states of C_2, which established firmly that the x $^1\Sigma_g^+$ state is 610
cm^{-1} below the x $^3\Pi_u$ state. This assignment presented some new difficulties since a number of investigators[348–352] reported observing absorption bands in rare-gas matrices that were attributed to the C_2 Swan
system $(A\ ^3\Pi_g - X'\ ^3\Pi_u)$. A simplified diagram of the electronic energy
levels of C_2 is given in Fig. 18, which indicates the systems of importance
to low-temperature studies. Ballik and Ramsay presented a more detailed
diagram, which includes the potential energy curves and extent of
observations for each state. In 1968 Herzberg and Lagerqvist[353] reported
a new absorption spectrum, which they tentatively attributed to C_2^-—the
first molecular negative ion to yield a discrete electronic spectrum. The
spectrum easily established the absorber as a diatomic carbon species,
but the assignment to C_2^- rather than C_2 or C_2^+ was based primarily
on molecular orbital considerations. One of the interesting facets of
this spectrum is that it occurs in the same wavelength region as the
Swan bands of C_2. Subsequent experiments by Milligan and Jacox[354]

[345] L. Brewer and G. M. Rosenblatt, *Advan. High Temp. Chem.* **2**, 1 (1969).

[346] E. A. Ballik and D. A. Ramsay, *Astrophys. J.* **137**, 61 (1963).

[347] E. A. Ballik and D. A. Ramsay, *Astrophys. J.* **137**, 84 (1963).

[348] M. McCarty, Jr., and G. W. Robinson, *J. Chim. Phys.* **56**, 723 (1959).

[349] W. Weltner, P. N. Walsh, and C. L. Angell, *J. Chem. Phys.* **40**, 1299 (1964).

[350] R. L. Barger and H. P. Broida, *J. Chem. Phys.* **43**, 2371 (1965).

[351] W. Weltner and D. McLeod, Jr., *J. Chem. Phys.* **45**, 3096 (1966).

[352] D. E. Milligan, M. E. Jacox, and L. Abouaf-Marquin, *J. Chem. Phys.* **46**, 4562 (1967).

[353] G. Herzberg and A. Lagerqvist, *Can. J. Phys.* **46**, 2363 (1968).

[354] D. E. Milligan and M. E. Jacox, *J. Chem. Phys.* **51**, 1952 (1969).

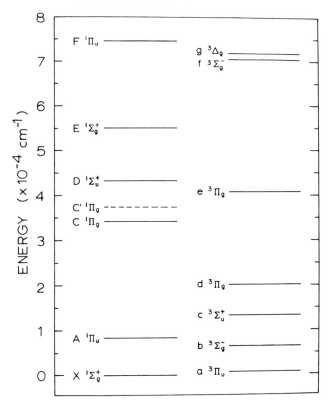

FIG. 18. Partial energy level diagram for C_2.

with cesium (a convenient source of electrons) as well as C_2 in a rare-gas matrix supported the assignment of the Herzberg–Lagerqvist (H–L) bands to C_2^-. Furthermore, they concluded that the bands in rare-gas matrices that had been assigned earlier to the C_2 Swan system should be reassigned to the H–L system. Frosch[355] subsequently confirmed this interpretation by producing C_2 and C_2^- in rare-gas matrices by the x-irradiation of acetylene. He did observe absorption spectra of C_2, but only from the $x\,^1\Sigma_g^+$ state. The Swan bands were observed as well, but only in emission.

Additional experiments by Herzberg *et al.*[356] have resulted in the identification of three new Rydberg states of C_2. These authors also propose a modified identification of the electronic states of C_2 in order to eliminate

[355] R. P. Frosch, *J. Chem. Phys.* **54**, 2660 (1971).
[356] G. Herzberg, A. Lagerqvist, and C. Malmberg, *Can. J. Phys.* **47**, 2735 (1969).

some of the previous difficulties, some of which originated partly through the change in ground state assignment. (The compilation by Rosen[340] uses the modified labels and includes a table for correlation with the old labels.) The continued interest in carbon species produced in a pulsed discharge has led Meinel[357] to the analysis of a new spectrum near 2490 Å, which is assigned to C_2^+. Lineberger and Patterson[357a] have demonstrated unambiguously that the H—L system is associated with C_2^-. They have induced ionization of a beam of C_2^- molecules by a two-photon absorption process with a dye laser. This photodetachment process is greatly enhanced when the laser is tuned to the absorption bands of the H—L system. Their data also indicate the presence of a second electronic absorption of C_2^- near 6800 Å, which has not yet been observed by other techniques.

Recent applications of new techniques to the study of diatomic molecules promise to complement further the conventional spectroscopic investigations. Weltner[358] has reviewed electron-paramagnetic-resonance studies of matrix-isolated molecules, which are preferentially oriented in some cases. Carrington[359] has reviewed microwave and electron resonance studies of radicals in the gas phase. The book by Turner et al.[360] of data derived from photoelectron spectroscopy includes ionization potentials for a large number of diatomic and polyatomic molecules. Hanes et al.[361] have measured the nuclear hyperfine structure of I_2 by saturated absorption ("inverse Lamb dip") spectroscopy. Kroll[362] and Kroll and Innes[363] have used the I_2 absorption spectrum to evaluate the resolution of a Fabry–Perot interferometer–spectrometer, and they have found that nuclear quadrupole interactions contribute significantly to the observed line widths. Resonance fluorescence excited by lasers has been mentioned earlier in Section 2.3.2.1.3.

Significant advances have been made in the study of Rydberg states of diatomic molecules, as indicated by the following papers that have

[357] H. Meinel, Can. J. Phys. 50, 158 (1972).

[357a] W. C. Lineberger and T. A. Patterson, Chem. Phys. Lett. 13, 40 (1972).

[358] W. Weltner, Jr., Advan. High Temp. Chem. 2, 85 (1969).

[359] A. Carrington, in "Molecular Spectroscopy: Modern Research" (K. N. Rao and C. W. Mathews, eds.), pp. 29–48. Academic Press, New York, 1972.

[360] D. W. Turner, C. Baker, A. D. Baker, and C. R. Brundle, "Molecular Photoelectron Spectroscopy." Wiley (Interscience), New York, 1970.

[361] G. R. Hanes, J. Lapierre, P. R. Bunker, and K. C. Shotton, J. Mol. Spectrosc. 39, 506 (1971).

[362] M. Kroll, Phys. Rev. Lett. 23, 631 (1969).

[363] M. Kroll and K. K. Innes, J. Mol. Spectrosc. 36, 295 (1970).

appeared *since* Rosen's compilation[340]: H_2,[251] BH,[364] CH,[365] BF,[366] and NO.[367,368] Many of these new observations have required the development of new theoretical techniques as demonstrated by the evaluation of Rydberg complexes for BH and CH and the theoretical calculations given by Jungen for NO.[368]

TABLE VII. Rotational and Vibrational Constants in the Ground States of Linear Triatomic Nonhydrides[a]

Molecule	State	ΔG (cm^{-1})	B_0 (cm^{-1})	r_0 (Å)	N
C_3	$^1\Sigma_g{}^+$	63.1	0.4305	1.277	12
CCN	$^2\Pi$	(325)	0.3981		
CNC	$^2\Pi_g$	321	0.4535	1.245	13
NCN	$^3\Sigma_g{}^-$	(423)	0.3968	1.232	14
NCO	$^2\Pi$	(539)	0.3894		
N_2O^+	$^2\Pi$	461.2	0.4116	1.155 / 1.185	
N_3	$^2\Pi_g$		0.4312	1.182	15
BO_2	$^2\Pi_g$	464	0.3292	1.265	
CO_2^+	$^2\Pi_g$		0.3804	1.177	
CO_2	$^1\Sigma_g{}^+$	667.4	0.3902	1.162	16

[a] Reproduced from G. Herzberg, "The Spectra and Structures of Simple Free Radicals: "An Introduction to Molecular Spectroscopy." Cornell Univ. Press, Ithaca, New York, 1971, used by permission of the publishers. N refers to the number of valence electrons, and ΔG values enclosed in parentheses refer to values determined by matrix infrared studies or by estimation from Renner–Teller splittings.

2.3.4.2.2. LINEAR POLYATOMIC MOLECULES. The Walsh diagram for XY_2 molecules, Fig. 10, indicates that triatomic molecules with 16 or less valence electrons should be linear in their ground state. Table VII presents a list of the ground-state rotational and vibrational constants for the molecules in this group as given by Herzberg.[271,272] Herzberg[272] discusses in detail the electronic transitions of these molecules and gives references through about 1966. Since that time a number of additions

[364] J. W. C. Johns, F. A. Grimm, and R. F. Porter, *J. Mol. Spectrosc.* **22**, 435 (1967).
[365] G. Herzberg and J. W. C. Johns, *Astrophys. J.* **158**, 399 (1969).
[366] R. B. Caton and A. E. Douglas, *Can. J. Phys.* **48**, 432 (1970).
[367] Ch. Jungen, *J. Mol. Spectrosc.* **33**, 520 (1970).
[368] Ch. Jungen, *J. Chem. Phys.* **53**, 4168 (1970).

have been made. Merer[369] has analyzed additional bands of the C_3 spectrum, which permitted a study of the l-type doubling in the $6\nu_2$ level of the ground state (which is at 452.8 cm^{-1}). Kroto has analyzed the $\tilde{b}\,^1\Pi_u - \tilde{a}\,^1\Delta_g$ electronic transition of NCN[370] and studied the kinetics of the relaxation from the $\tilde{a}\,^1\Delta_g$ state to the $\tilde{X}\,^3\Sigma_g^-$ state.[371,372]

Devillers and Ramsay[373] have analyzed the spectrum of CCO (iso-electronic with NCN), which extends from 5000 to 9000 Å. The spectrum was produced by flash photolysis of C_3O_2, as reported earlier by Devillers.[374] The electronic transition was assigned as $\tilde{A}\,^3\Pi_1 - \tilde{X}\,^3\Sigma^-$ on the basis of a rotational analysis of the 000–000 band near 8580 Å. Renner–Teller interactions in the $\tilde{A}\,^3\Pi_i$ state were evaluated by assignments of the 010–010 bands. This analysis was made much easier by the observation of the 010 ($^2\Sigma^{(-)}$)–000($^2\Sigma^+$) band—a forbidden component of this allowed transition (see Section 2.3.4.1.2). Rotational analysis of this band confirmed Hougen's theory[293] for rotational energies of the $v_2 = 1$ level of a $^3\Pi$ electronic state for a linear molecule. Devillers and Ramsay also suggested a possible revision of the assignments in the analogous NCN transition reported by Herzberg and Travis.[375]

Dixon and Ramsay[376] have assigned absorption bands between 3300 and 4000 Å to two electronic transitions of NCS. Rotational analyses of the 000–000 bands established them as $\tilde{A}\,^2\Pi_i - \tilde{X}\,^2\Pi_i$ and $\tilde{B}\,^2\Sigma^+ - \tilde{X}\,^2\Pi_i$ electronic transitions. Analysis of the 010–010 bands in both transitions permitted an evaluation of the Renner–Teller interaction in terms of $\varepsilon\omega_2$. No 010–000 transitions were observed in these systems; therefore, the evaluation of ω_2 was based on observations of the 020–000 transitions of the A–X system. Since these transitions involve Fermi resonance in the $\tilde{A}\,^2\Pi_2$ state between the 001 and 020 levels, the desired value of ω_2 could only be obtained by assuming a value for the Fermi interaction matrix element by comparison with the observed values for OCS, CO_2, and NCO.

2.3.4.3. Transitions of Nonlinear Molecules.
The discussion of this large group of molecules can be divided most conveniently in terms of XH_2, HXY, XY_2, and larger molecules. Again, the emphasis will be on

[369] A. J. Merer, *Can. J. Phys.* **45**, 4103 (1967).

[370] H. W. Kroto, *Can. J. Phys.* **45**, 1439 (1967).

[371] H. W. Kroto, *J. Chem. Phys.* **44**, 831 (1966).

[372] H. W. Kroto, T. Morgan, and H. Sheena, *Trans. Faraday Soc.* **66**, 2237 (1970).

[373] C. Devillers and D. A. Ramsay, *Can. J. Phys.* **49**, 2839 (1971).

[374] C. Devillers, *C. R. Acad. Sci. Paris* **262C**, 1485 (1966).

[375] G. Herzberg and D. N. Travis, *Can. J. Phys.* **42**, 1658 (1964).

[376] R. N. Dixon and D. A. Ramsay, *Can. J. Phys.* **46**, 2619 (1968).

indicating recent reviews, followed by references to a few of the papers published since those reviews.

2.3.4.3.1. XH_2 MOLECULES. This group of molecules provides a very interesting application of the appropriate Walsh diagram (Fig. 9). Walsh's predictions were made *before* any data were available for these molecules (except the ground states of H_2O, H_2S, and H_2Se). The observed molecules include BH_2, AlH_2, CH_2, SiH_2, NH_2, PH_2, H_2O, H_2S, H_2Se, and H_2Si, as compiled by Herzberg.[272] All of these molecules are bent in their ground state, as expected from Fig. 9. Investigations since the publication of Herzberg's book include reports on the high-resolution spectra of BH_2,[377] SiH_2,[378,379] H_2O,[380] PH_2,[381-384] and AsH_2.[385] Recent reports of the CH_2 system were discussed in Section 2.3.3.1.1. Bender and Schaefer have calculated theoretical potential energy curves for ground and excited states of CH_2^+ and BH_2,[386] as well as the electronic splitting between the $^2B_1$ and $^2A_1$ states of NH_2.[387] Duxbury[388] has treated some of the problems of centrifugal distortion and spin coupling effects in the spectra of NH_2 and CH_2. Earlier papers by Dixon[389] and by Dixon and Duxbury[390] have dealt with similar topics. Additional problems associated with the interpretation of XH_2 spectra include the effects of "quasi-linearity" and large amplitude vibrations (see, for example, Hougen *et al.*,[391] Johns,[392] and Bunker and Stone[392a]).

[377] G. Herzberg and J. W. C. Johns, *Proc. Roy. Soc. (London)* **A298**, 142 (1967).

[378] I. Dubois, G. Herzberg, and R. D. Verma, *J. Chem. Phys.* **47**, 4262 (1967).

[379] I. Dubois, *Can. J. Phys.* **46**, 2485 (1968).

[380] J. W. C. Johns, *Can. J. Phys.* **49**, 944 (1971).

[381] B. Pascat, J. M. Berthou, H. Guenebaut, and D. A. Ramsay, *C. R. Acad. Sci. Paris* **263B**, 1397 (1966).

[382] R. N. Dixon, G. Duxbury, and D. A. Ramsay, *Proc. Roy. Soc. (London)* **296A**, 137 (1967).

[383] B. Pascat, J. M. Berthou, J. C. Prudhomme, H. Guenebaut, and D. A. Ramsay, *J. Chim. Phys.* **65**, 2022 (1968).

[384] B. Pascat and J. M. Berthou, *C. R. Acad. Sci. Paris* **271C**, 799 (1970).

[385] R. N. Dixon, G. Duxbury, and H. M. Lamberton, *Proc. Roy. Soc. (London)* **305A**, 271 (1968).

[386] C. F. Bender and H. F. Schaefer III, *J. Mol. Spectrosc.* **37**, 423 (1971).

[387] C. F. Bender and H. F. Schaefer III, *J. Chem. Phys.* **55**, 4798 (1971).

[388] G. Duxbury, *J. Mol. Spectrosc.* **25**, 1 (1968).

[389] R. N. Dixon, *Mol. Phys.* **10**, 1 (1965).

[390] R. N. Dixon and G. Duxbury, *Chem. Phys. Lett.* **1**, 330 (1967).

[391] J. T. Hougen, P. R. Bunker, and J. W. C. Johns, *J. Mol. Spectrosc.* **34**, 136 (1970).

[392] J. W. C. Johns, *Can. J. Phys.* **45**, 2639 (1967).

[392a] P. R. Bunker and J. M. R. Stone, *J. Mol. Spectrosc.* **41**, 310 (1972).

2.3.4.3.2. HXY MOLECULES. Herzberg[272] lists ground and excited state constants for the molecules HCN, HCP, HCO, HNO, HPO, HCF, HCCl, HSiCl, and HSiBr. Although preliminary constants for HCP were tabulated by Herzberg, the paper by Johns *et al.*[393] should be consulted for the final set of molecular constants, including three new electronic states. This paper is remarkable in characterizing eight different electronic states of HCP, four of which are triplet states. The ground state and all but two of the excited singlet states are linear. Shurvell[394] has computed the force constants and thermodynamic properties of HCP, DCP, and FCN. The "hydrocarbon flame bands" of HCO have been investigated again by Dixon.[395] He presents strong evidence that the lower state in both transitions is the $\tilde{X}\,^2A'$ state and that the $\tilde{B}\,^2A'$ and $\tilde{C}$ states are located 38,691 and 41,270 cm^{-1}, respectively, above the ground state. The bond angle of the $\tilde{B}$ state is shown to be 111° and r_{CO} is 1.36 Å if r_{CH} is assumed to be 1.16 Å. The microwave spectrum of HCO was studied by Bowater *et al.*,[396] who produced it by the reaction of fluorine atoms with formaldehyde.

The spectrum of HNF provides an excellent example of the application of Walsh's diagrams with even low-resolution spectra, as well as the abundance of information that may be available from a detailed rotational analysis. The spectrum of HNF was first observed in 1966 by Goodfriend and Woods[397] at low resolution following the flash photolysis of HNF$_2$. Their tentative identification of the molecule was based primarily on the identity of the parent compound and on an observed deuterium isotope effect. They suggested further that the discrete structure observed in each band of a progression (assigned to an upper state bending frequency) was K-type structure of a perpendicular transition ($\Delta K = \pm 1$). On the basis of Walsh's diagram for HXY molecules (Fig. 12) they concluded the electronic transition should be $^2A' - ^2A''$. A subsequent detailed rotational analysis by Woodman[398] fully confirmed their tentative conclusions. Woodman established that the upper and lower state bond angles are 125 and 105°, respectively, and that the internuclear distances are not appreciably different in the two states. The spectrum provides an excellent example of *axis switching*. [This phenomenon was evaluated

[393] J. W. C. Johns, H. F. Shurvell, and J. K. Tyler, *Can. J. Phys.* **47**, 893 (1969).
[394] H. F. Shurvell, *J. Phys. Chem.* **74**, 4257 (1970).
[395] R. N. Dixon, *Trans. Faraday Soc.* **65**, 3141 (1969).
[396] I. C. Bowater, J. M. Brown, and A. Carrington, *J. Chem. Phys.* **54**, 4957 (1971).
[397] P. L. Goodfriend and H. P. Woods, *J. Mol. Spectrosc.* **20**, 258 (1966).
[398] C. M. Woodman, *J. Mol. Spectrosc.* **33**, 311 (1970).

by Hougen and Watson.[338] They applied their expressions to the 2400-Å systems of C_2H_2[399,399a] and demonstrated that the occurrence of forbidden subbands ($\Delta K = 0, \pm 2$) and their intensities relative to the allowed subbands ($\Delta K = \pm 1$) are consistent with the axis-switching model. This model has been applied to the spectra of HSiCl and HSiBr by Herzberg and Verma.[399b]] Well-resolved spin-doublets also provide an excellent example of the effect of spin–rotation interaction in a nearly symmetric top. These constants also establish that both electronic states correlate with the same $^2\Pi$ state of the equivalent linear molecule (see, for example, the discussion by Dixon[389] on this correlation). The decrease of the upper-state vibrational interval from 1074 to 962 cm^{-1} as v_2 changes from 0 to 6 may be interpreted in terms of the fact first noted by Dixon[400] that the interval goes through a minimum near the top of the barrier which keeps the molecule bent as it vibrates. This interpretation would lead to a barrier height of about 6000 cm^{-1}. In addition, this spectrum provides the first example of a second-order rotational resonance interaction between v_2 and v_3. The development of this theory and the analysis of the interaction permitted an estimate to be made of v_3—which was not available otherwise.

Other reports of HXY molecules since 1966 include a study of a far ultraviolet spectrum of HNO by Callear and Wood,[401] the application of a variable reduced mass model for quasi-linear molecules to HSiBr by Shinkle and Coon,[402] and a study of the emission spectrum of CuOH in the regions 5100–5600 and 6150–6300 Å by Antić-Jovanović and Pesić.[403]

2.3.4.3.3. XY_2 MOLECULES. A large number of studies since 1966 have been reported on XY_2 molecules that are nonlinear in their ground state. Most of the reports discussed here will be limited to those in which gas phase spectra have been observed and analyzed in detail.

The paper by Mathews[404] on the absorption spectrum of CF_2 established firmly that the 2500 Å system is a $\tilde{A}\,^1B_1 - \tilde{X}\,^1A_1$ transition. The

[399] K. K. Innes, J. Chem. Phys. 22, 863 (1954).

[399a] G. Herzberg and R. D. Verma, Can. J. Phys. 42, 395 (1964).

[399b] C. K. Ingold and G. W. King, J. Chem. Soc., 2702 (1953).

[400] R. N. Dixon, Trans. Faraday Soc. 60, 1363 (1964).

[401] A. B. Callear and P. M. Wood, Trans. Faraday Soc. 67, 3399 (1971).

[402] N. L. Shinkle and J. B. Coon, J. Mol. Spectrosc. 40, 217 (1971).

[403] A. M. Antić-Jovanović and D. S. Pesić, Bull. Chem. Soc. (Belgrade) 34, 5 (1969); see also Chem. Abs. 73, 125368g (1970).

[404] C. W. Mathews, Can. J. Phys. 45, 2355 (1967).

ultraviolet data, together with the microwave studies of Powell and Lide,[405] provided data on the ground state, including the geometry, $r_0(CF) = 1.300$ Å and $\angle\ FCF = 104.94°$. The geometry of the $\tilde{A}$ state is $r_0(CF) = 1.32$ Å and $\angle\ FCF = 122.3°$. A search for other electronic transitions of CF_2 yielded spectra of two transient species near 1500 and 1350 Å. Since no rotational structure was resolved, an unambiguous assignment was not possible; however, they were tentatively attributed to the CF_2 molecule. Basco and Hathorn[406] have presented photochemical data that suggest strongly that the 1500-Å system should be assigned to the CF_3 molecule. Modica[407] has studied the electronic oscillator strength of the 2500-Å system, and Marsigny et al.[408] have extended the assignments for the emission spectrum of this system.

The absorption spectrum of SiF_2 near 2300 Å bears a strong resemblance to the 2500-Å system of CF_2. The ground state of SiF_2 was characterized in a series of papers by Timms and co-workers[409-413] through infrared and microwave spectroscopy [$r_0(SiF) = 1.591$ Å, $\angle\ FSiF = 100°59'$]. Khanna et al.[414] recorded the ultraviolet absorption spectrum and revised the earlier vibrational assignments in terms of bending frequencies only, as is the case for CF_2. Dixon and Hallé[415] confirmed their vibrational assignments by analyzing the rotational structure of four bands in the system. The rotational analyses also demonstrated that the transition is $\tilde{A}\ ^1B_1 - \tilde{X}\ ^1A_1$, again analogous to the 2500-Å system of CF_2. The geometry of the upper state is $r(SiF) = 1.601$ Å and $\angle\ FSiF = 115°53'$. Rao[416] has identified a new electronic emission spectrum between 3645 and 4183 Å, which he assigns to SiF_2. His

[405] F. X. Powell and D. R. Lide, Jr., J. Chem. Phys. 45, 1067 (1966).

[406] N. Basco and F. G. M. Hathorn, Chem. Phys. Lett. 8, 291 (1971).

[407] A. P. Modica, J. Phys. Chem. 72, 4594 (1968).

[408] L. Marsigny, J. Ferran, J. Lebreton, and R. Lagrange, C. R. Acad. Sci. Paris 266C, 507 (1968).

[409] P. L. Timms, R. A. Kent, T. C. Ehlert, and J. L. Margrave, J. Amer. Chem. Soc. 87, 2824 (1965).

[410] V. M. Rao, R. F. Curl, Jr., P. L. Timms, and J. L. Margrave, J. Chem. Phys. 43, 2557 (1965).

[411] V. M. Rao and R. F. Curl, Jr., J. Chem. Phys. 45, 2032 (1966).

[412] J. M. Bassler, P. L. Timms, and J. L. Margrave, Inorg. Chem. 5, 729 (1966).

[413] V. M. Khanna, R. Hauge, R. F. Curl, Jr., and J. L. Margrave, J. Chem. Phys. 47, 5031 (1967).

[414] V. M. Khanna, G. Besenbruch, and J. L. Margrave, J. Chem. Phys. 46, 2310 (1967).

[415] R. N. Dixon and M. Hallé, J. Mol. Spectrosc. 36, 192 (1970).

[416] D. R. Rao, J. Mol. Spectrosc. 34, 284 (1970).

vibrational assignments indicate the lower state is the ground state of SiF_2, but no rotational analysis has been reported for this system yet. The 3700–4600-Å system of NO_2* continues to provide very interesting data. Coon et al.[417] observe a large isotope effect upon ^{18}O substitution. They attribute this to a very low vibrational frequency for ν_3', which may be explained in terms of a double-minimum potential. Resonance fluorescence studies have been made by Sakurai and Broida[418] and by Abe et al.[419] The latter article presents evidence for the existence of a $^2B_2$ state near the $^2B_1$ state identified by Douglas and Huber[420] in their analysis of the 3700–4600-Å system. Sackett and Yardley[421,422] have used pulsed dye lasers to study the radiative lifetimes of NO_2. Moraal[423] and Coope[424] have used NO_2 for studies of the Senftleben effect.

Brand and co-workers[425-427] have worked out a detailed analysis of the 4750-Å system of ClO_2. Their analysis includes a very thorough evaluation of the effects of anharmonicity on the structure and spectrum. Intensity "anomalies" among different bands in the system that were difficult to explain previously are accounted for readily with second-order corrected anharmonic vibrational wave functions. The 1969 paper deals primarily with a vibrational analysis of the system when observed under high resolution with ^{35}Cl and ^{37}Cl isotopes. Their revised constants (in reciprocal centimeters) for $^{35}ClO_2$ are: $T_0' = 21,016.3$, $\omega_1'' = 963.5$, $\omega_2'' = 451.7$, $\omega_3'' = 1133.0$, $\omega_1' = 722.4$, $\omega_2' = 296.3$, and $\omega_3' = 780.1$. The observations of 2–0 and 4–0 transitions in ν_3 are interpreted in terms of anharmonic effects rather than in terms of a double-minimum (i.e.,

[416a] N. M. Atherton, R. N. Dixon, and G. H. Kirby, Trans. Far. Soc. 60, 1688 (1964).

[417] J. B. Coon, F. A. Cesani, and F. P. Huberman, J. Chem. Phys. 52, 1647 (1970).

[418] K. Sakurai and H. P. Broida, J. Chem. Phys. 50, 2404 (1969).

[419] K. Abe, F. Meyers, T. K. McCubbin, and S. R. Polo, J. Mol. Spectrosc. 38, 552 (1971).

[420] A. E. Douglas and K. P. Huber, Can. J. Phys. 43, 74 (1965).

[421] P. B. Sackett and J. T. Yardley, Chem. Phys. Lett. 6, 323 (1970).

[422] P. B. Sackett and J. T. Yardley, Chem. Phys. Lett. 9, 612 (1971).

[423] H. Moraal, Chem. Phys. Lett. 7, 205 (1970).

[424] J. A. R. Coope, Mol. Phys. 21, 217 (1971).

[425] A. W. Richardson, R. W. Redding, and J. C. D. Brand, J. Mol. Spectrosc. 29, 93 (1969).

[426] J. C. D. Brand, R. W. Redding, and A. W. Richardson, J. Chem. Soc. (London) D11, 618 (1969).

[427] J. C. D. Brand, R. W. Redding, and A. W. Richardson, J. Mol. Spectrosc. 34, 399 (1970).

* For additional discussions of this system, see Ref. 416a and pp. 507–509, Ref. 272.

unequal Cl—O bonds). The 1970 paper deals primarily with the rotational analysis of the 100–000 band from which the electronic transition is established as $^2A_2 - {}^2B_1$. The geometry of the excited state $[r(\text{ClO}) = 1.619 \text{ Å}, \angle \text{OClO} = 107°0']$ required the derivation of a force field since only the 100 level was analyzed. In addition, this paper presents the best comparison of calculated and observed intensities for the vibrational structure.

Barrow et al.[428] have reported the vibrational analysis of two band systems of NSF near 4050 and 2150 Å. Although no rotational assignments were reported, they did use the observed discrete structure for qualitative comparisons with SiF_2, SO_2, and CF_2. Craig and Fischer[429] also report their observations of the 4050-Å system, as well as another system near 5300 Å, which is assigned by Fischer[430] to a singlet–triplet transition of the molecule.

Additional reports include the following molecules: BCl_2,[431] $GeCl_2$,[432] GeF_2,[433] $HgCl_2$,[434] $HgBr_2$,[435,436] SeO_2,[437] and TeO_2.[438] Hastie et al.[439] have summarized the ultraviolet spectra and electronic structure of Group IVA dihalides. Extensive discussions of the analyses of the SO_2 spectra have been summarized in Section 2.3.4.1.2.

2.3.4.3.4. LARGER MOLECULES. In addition to the tables in Herzberg's book, a number of reviews are available which discuss the spectra of larger molecules (see, for example, those by Innes,[440] Goodman and Hollas,[322] and Ross[322]). Most of the molecules that fall in this class have spectra that cannot be completely resolved. Consequently "band contour fitting" techniques have been developed, as discussed in each of the above reviews. The procedure simply amounts to a comparison of a calculated

[428] T. Barrow, R. N. Dixon, O. Glemser, and R. Mews, Trans. Faraday Soc. 65, 2295 (1969).

[429] D. P. Craig and G. Fischer, Chem. Phys. Lett. 4, 227 (1969).

[430] G. Fischer, Chem. Phys. Lett. 11, 356 (1971).

[431] O. Dessaux, P. Goudman, and G. Pannetier, Bull. Soc. Chim. 5, 447 (1969).

[432] C. M. Pathak and H. B. Palmer, J. Mol. Spectrosc. 31, 170 (1969).

[433] R. Hauge, V. M. Khanna, and J. L. Margrave, J. Mol. Spectrosc. 27, 143 (1968).

[434] S. Bell, R. D. McKenzie, and J. B. Coon, J. Mol. Spectrosc. 20, 217 (1966).

[435] A. Gedanken, B. Raz, U. Even, and I. Eliezer, J. Mol. Spectrosc. 32, 287 (1969).

[436] S. Bell, J. Mol. Spectrosc. 23, 98 (1967).

[437] I. Dubois, Bull. Soc. Roy. Sci. Liege 37, 562 (1968).

[438] I. Dubois, Bull. Soc. Roy. Sci. Liege 39, 63 (1970).

[439] J. W. Hastie, R. H. Hauge, and J. L. Margrave, J. Mol. Spectrosc. 29, 152 (1969).

[440] K. K. Innes, in "Molecular Spectroscopy: Modern Research" (K. N. Rao and C. W. Mathews, eds.), pp. 172–205. Academic Press, New York, 1972.

trial spectrum with the experimental spectrum. The energy and intensity of each rotational line is calculated, given a "width" related to the experimental resolution, and then all transitions within a band are superimposed in order to obtain the calculated spectrum. Complications arise because the molecules usually are asymmetric rotors and because of the large number of transitions within a single band (50,000 lines would not be uncommon for a molecule such as benzene). The review by Ross[321] contains detailed information and leading references on the techniques used in these computations. He also provides a very convenient table of the molecules that have been studied by this technique. The table includes the constants that were determined and a *reproduction of a typical band contour for each molecule*. The data available from this band contour procedure obviously is much more limited than that available from a complete rotational analysis. Kidd and King[441] have offered some comments on this technique, as well as a useful summary of the individual computer programs presently in use for these calculations.

There are a few examples of larger (nonlinear) molecules whose electronic spectra can be analyzed in a reasonably complete fashion. Two of them deserve special mention: formaldehyde (H_2CO) and glyoxal ($C_2H_2O_2$). Innes[440] provides a concise summary of the formaldehyde literature, which will not be repeated here. In the case of glyoxal, Brand[442] reported a detailed vibrational analysis of the 4550-Å system in 1954 for $C_2H_2O_2$ and for $C_2D_2O_2$. He also assigned the bands to a $^1A_u - {}^1A_g$ transition. In 1957 King[443] reported a partial rotational analysis of the O—O bands and showed they were consistent with Brand's assignment. In 1967 Paldus and Ramsay[444] reported rotational analyses of the O—O bands of $C_2H_2O_2$ and $C_2D_2O_2$ in the 4550-Å system. Their precision in the determination of the rotational constants A, B, and C were comparable to those of smaller molecules (i.e., better than one part in 10^4). In 1970 Birss et al.[445] reported similar analyses of 12 bands of $C_2H_2O_2$ and in 1971 Agar et al.[446] reported analyses of 11 bands of $C_2D_2O_2$.

[441] K. G. Kidd and G. W. King, *J. Mol. Spectrosc.* **40**, 461 (1971).

[442] J. C. D. Brand, *Trans. Faraday Soc.* **50**, 431 (1954).

[443] G. W. King, *J. Chem. Soc. (London)* 5054 (1957).

[444] J. Paldus and D. A. Ramsay, *Can. J. Phys.* **45**, 1389 (1967).

[445] F. W. Birss, J. M. Brown, A. R. H. Cole, A. Lofthus, S. L. N. G. Krishnamachari, G. A. Osborne, J. Paldus, D. A. Ramsay, and L. Watman, *Can. J. Phys.* **48**, 1230 (1970).

[446] D. M. Agar, E. J. Bair, F. W. Birss, P. Borrell, P. C. Chen, G. N. Currie, A. J. McHugh, B. J. Orr, D. A. Ramsay, and J.-Y. Roncin, *Can. J. Phys.* **49**, 323 (1971).

The constants contained in these three papers represent a monumental task accomplished by D. A. Ramsay and his co-workers. The volume of data would have been virtually impossible to handle without the development of computer programs in cooperation with F. W. Birss. Holzer and Ramsay[447] studied the emission spectrum of glyoxal which was induced by irradiation with five lines from an argon ion laser. Most of the observed bands were easily assigned in terms of the 4550-Å system of glyoxal, but seven bands could not be correlated with glyoxal. The assignment of these bands to a different system was confirmed subsequently by Currie and Ramsay[448] in their analysis of the 4875-Å band of *cis* glyoxal (a geometrical isomer of the more stable *trans* glyoxal). In addition, Goetz *et al.*[449] studied the $^3A_u - {}^1A_g$ system of *trans* glyoxal (near 5200 Å) by observing its magnetic rotation spectrum. This growing accumulation of data on glyoxal undoubtedly will attract other kinds of studies of its properties, such as the theoretical calculations by Pincelli *et al.*[450] and the determination of its radiative lifetime by Yardley *et al.*[451]

ACKNOWLEDGMENTS

I want to express my appreciation to the large number of associates who have provided valuable suggestions for various portions of this section. I especially want to thank Dr. K. K. Innes, Dr. J. K. G. Watson, and Dr. C. J. Dymek for their critical comments on the manuscript, which helped eliminate a number of errors. I also wish to apologize to those authors whose work has not been discussed because of the choice of topics or because of an oversight. .

[447] W. Holzer and D. A. Ramsay, *Can. J. Phys.* **48**, 1759 (1970).
[448] G. N. Currie and D. A. Ramsay, *Can. J. Phys.* **49**, 317 (1971).
[449] W. Goetz, A. J. McHugh, and D. A. Ramsay, *Can. J. Phys.* **48**, 1 (1970).
[450] U. Pincelli, B. Cadioli, and D. J. David, *J. Mol. Structure* **9**, 173 (1971).
[451] J. T. Yardley, G. W. Holleman, and J. I. Steinfeld, *Chem. Phys. Lett.* **10**, 266 (1971).

2.4. Molecular Lasing Systems*

Stimulated emission from molecules in optical maser devices has been known now for ten years. The original observations were made using the Fabry–Perot cavity of a pulsed xenon laser but replacing the xenon by a flowing stream of carbon monoxide or nitrogen. Intense radiation in the Angstrom system $(B\ ^1\Sigma \rightarrow A\ ^1\Pi)$ of CO and the first positive system $(B\ ^3\Pi_g \rightarrow A\ ^3\Sigma_u^+)$ of N_2, respectively, was obtained.[1] These two band systems lie in the visible and near infrared regions of the electromagnetic spectrum. Later on, stimulated emission involving transitions between vibrational states was observed, generally in the mid-infrared region. The best known example of this is the CO_2 laser, which operates[2] on the P branch of the transition (00^01)–(10^00) in CO_2. A similar line spectrum has been observed[3] from the isoelectronic N_2O molecule, both systems lying near 10.9 μm. Stimulated emission in the far infrared beyond 100 μm must involve rotational transitions since the energy changes are very small. Although some of the transitions observed appear formally to be vibrational transitions between close lying exited states, it is found that for all cases of this kind discovered so far, the mechanism involves resonance perturbation, as a result of which, the transitions take on part of the character of pure rotational transitions. This chapter will be principally concerned with the origin, generation, and use of submillimeter waves produced by stimulated emission in optical laser cavities. We understand the submillimeter region to cover the range suggested by Martin,[3a] namely 3–0.05 mm, i.e., 3.3–200 cm^{-1} or 100–6000 GHz.

[1] L. E. S. Mathias and J. T. Parker, *Phys. Lett.* **7**, 194 (1963); *Appl. Phys. Lett.* **3**, 16 (1963).

[2] C. K. N. Patel, *Phys. Rev.* **136**, A1187 (1964).

[3] L. E. S. Mathias, A. Crocker, and M. S. Wills, *Phys. Lett.* **13**, 303 (1964).

[3a] D. H. Martin, "Spectroscopic Techniques." North Holland Publ., Amsterdam, 1967.

* Chapter 2.4 is by George W. Chantry and Geoffrey Duxbury.

2.4.1. Molecular Beam Masers

Although the devices that produce stimulated emission near 0.5 mm are often known as *masers*, because of the similarity of the resonant cavity techniques to those used in the microwave region, the experimental arrangements have much more in common with lasers and bear little resemblance to the beam masers of Marcuse[4] and De Lucia and Gordy.[5] These latter devices operate by using electrostatic focusing to create a population inversion between molecules in different rotational energy levels of the ground electronic state of the molecule. The HCN beam maser has been operated on the $J = 1 \rightarrow J = 0$ transition at 88.63 GHz and also on the $J = 2 \rightarrow J = 1$ transition at 177.2 GHz. One of the most important submillimeter lasers also involves transitions in HCN (the 337-μm maser described at length later), and because of this the operation of the HCN beam maser will be discussed in some detail so that the performance and spectral output of the two types of devices may be compared. The beam maser works by producing a beam of molecules in a specified rotational state travelling normally to the axis of a Fabry–Perot resonator.[6] It will be seen, therefore, that in ideal operation not only does one have a population inversion (since molecules in the lower state are absent) but also that Doppler broadening is eliminated since the molecules are traveling normally to the axis of propagation of the radiation. A molecular beam apparatus features ultrahigh vacua and because of this, pressure broadening is virtually eliminated and the natural line width ($\sim 10^{-3}$ Hz) is quite negligible. Extremely sharp lines are therefore expected. However, in practice there is a finite Doppler contribution to the line width, since the molecular trajectories do have a small spread about the direction perpendicular to the axis of the resonator and there is always present the effect of thermal noise, which also contributes to the line width. Frequency pulling is another trouble, for the frequency of oscillation can be varied if the resonator tuning is altered. Even if the resonator is constructed of Invar, a change of frequency of ~ 100 Hz for a 1 °C temperature change would result. A thermostatically controlled resonator is mandatory for optimum stability. Nevertheless even with these practical shortcomings extremely high resolving power is available and lines of the order of 10 kHz apart but with absolute frequencies of hundreds of gigahertz may be readily

[4] D. Marcuse, *Proc. Inst. Radio Eng.* **49**, 1706 (1961).
[5] F. De Lucia and W. Gordy, *Phys. Rev.* **187**, 58 (1969).
[6] V. M. Fain and Ya. J. Khanin, "Quantum Electronics." Pergamon, Oxford, 1969.

resolved. With a beam maser, hyperfine splitting due to nuclear quadrupoles can be easily resolved and the much smaller effects due to the
magnetic moments of the nuclei can be observed. The rotational energy
levels for HCN in its ground state have been given by De Lucia and
Gordy in the form

$$E = E_R + E_Q + E_M, \qquad (2.4.1)$$

where

$$E_R = B_0 J(J + 1) - D_0 J^2(J + 1)^2,$$

$$E_Q = -eQq\left[\frac{\frac{3}{4}C(C + 1) - I(I + 1) \cdot J(J + 1)}{2I(2I - 1) \cdot (2J - 1) \cdot (2J + 3)}\right],$$

$$E_M = \tfrac{1}{2}C_N C,$$

$$C = F(F + 1) - J(J + 1) - I(I + 1).$$

In these equations B_0 and D_0 are the usual rotational constants
(44.3159757 GHz and 87.24 kHz, respectively), J is the rotational quantum number, F is the total angular momentum quantum number which
runs in value from $(J + I)$ to $|J - I|$ in integral steps, I is the nuclear
spin of the quadrupolar or magnetic nucleus which for ^{14}N has the value
unity, eQq is the nuclear quadrupole coupling constant $(-4.7091$ MHz),
and C_N is the nuclear magnetic coupling constant of ^{14}N (10.4 kHz).
The selection rules are

$$\Delta J = \pm 1, \qquad \Delta F = 0, \pm 1, \qquad (2.4.2)$$

so that the $J = 1 \rightarrow J = 0$ line will be a triplet and the $J = 2 \rightarrow J = 1$
line will be a sextet. The intensities of the components are complicated
functions of the quantum numbers but are listed in numerical form in
"Microwave Spectroscopy" by Townes and Schawlow.[7]
    The output power of the maser used as an oscillator is usually very
low (10^{-9} W) because there are so few molecules in the beam and the
device is used mainly as an amplifier. A beam of microwave radiation in
the appropriate frequency band is produced by harmonic multiplication
of the output of a phase-locked klystron. The phase locking is achieved
by means of several harmonic multiplications from a fundamental time
standard broadcast in the U.S. from the Bureau of Standards. When the
microwave power is transmitted through the Fabry–Perot cavity, it will

[7] C. H. Townes and A. L. Schawlow, "Microwave Spectroscopy." McGraw-Hill,
New York, 1955.

stimulate emission when the frequency matches, within a very narrow interval, that of one of the hyperfine components of the line. The plot of the detector signal versus frequency of the stimulating field will therefore show peaks that coincide very accurately with the lines of the spectrum. Of the expected nine components of the $J = 1 \rightarrow J = 0$ and the $J = 2 \rightarrow J = 1$ lines, eight have been measured by De Lucia and Gordy and their frequencies agree to within 1 kHz with those calculated using (2.4.1) and the listed parameters. The corresponding transitions in DCN have also been observed near 72.41 and 144.83 GHz, respectively. These spectra are much more complex than are those of HCN because the deuterium nucleus has also a quadrupole moment. Two further parameters $(eQq)_D$ and C_D are required (194.4 and -0.6 kHz, respectively) to account for the spectrum, and the calculation of line frequencies involves the adoption of a suitable coupling scheme for the two nuclear quadrupoles followed by the solution of a rather large secular determinant (19×19). Nevertheless all the observed lines of DCN have been fitted into this scheme to better than 1 kHz using these two parameters and $B_0 = 36.2074627$ GHz, $D_0 = 57.83$ kHz, $(eQq)_N = -4.703$ MHz, and $C_N = 8.4$ kHz.

The observation of this type of spectrum in the submillimeter region complements the molecular beam spectral studies that have been made in the microwave region. From the combined results, coupling constants can be derived and these can be compared with those calculated from molecular wave functions. This provides a good test of the wave function since the coupling constants are very sensitive to the behavior of the wave function near the atomic nucleus. To further this valuable program and to provide amplifiers and oscillators for the 10^{11}-Hz region it is desirable that other beam masers be constructed using different molecules as the emitting species. Already emission has been observed[8] from the $1_{10} \rightarrow 1_{01}$ transition of D_2O at 316 GHz. $HC^{15}N$ should be an interesting species to study, since it would have no hyperfine structure.

2.4.2. Submillimeter Lasers

The first molecular submillimeter laser was discovered late in 1963 at SERL (Services Electronics Research Laboratory, Baldock, England) by Crocker *et al.*[9] These workers took the very long pulsed lasers that

---

[8] F. De Lucia and W. Gordy (to be published).

[9] A. Crocker, H. A. Gebbie, M. F. Kimmitt, and L. E. S. Mathias, *Nature (London)* **201**, 250 (1964).

had been developed for the CO and N_2 experiments and modified them for operation in the far infrared. Silicon windows were used together with sensitive cooled detectors and either Michelson interferometers for broad-band surveys or grating spectrometers for quick wavelength determinations. Typical operating conditions were tube length 4.8 m, tube internal diameter 2.5 cm, gas pressure 1 Torr, pulse length 1 sec, pulse height 46 kV, and peak current 13 A. Nine lines were detected in the wavelength region 23–79 μm when water vapor was present in the cavity. On removal of one end mirror, stimulated emission was still observed at 28 μm. The intense stimulated emission from a cavity that had apparently only one Fabry–Perot mirror was at first interpreted as evidence that the gain of the system was so great that "superradiance" had been achieved. Subsequently it was shown that any reflector in the analyzing spectrometer could act as the second mirror and, in particular, the window of the detector. This exemplifies the feedback problems introduced by the coupling of the analyzer system to the oscillator. If possible, it is most desirable to put some form of isolator between the laser and the detection system.

Very soon after this discovery, Faust et al.[10] obtained far infrared radiation with neon gas, using the higher analogs of the transitions responsible for the well-known visible and near infrared laser emission. Several lines between 31.928 and 57.355 μm were observed arising from $7p$–$6d$ (Racah) transitions. Further work[11] both at SERL and at NPL[12] (National Physical Laboratory, Teddington, Middlesex, England) led to the discovery of more stimulated emission lines both from water and heavy water, D_2O. In 1964 Gebbie and his group[13] at NPL obtained emission from cyanides and organic nitriles at 337 μm. This emission is very strong and in a useful spectral region at one third of a millimeter where the atmosphere is relatively transparent. Initially the emitter was identified[14] as the CN radical, since the visible radiation from the plasma was almost entirely from the violet $B\ ^2\Sigma \rightarrow X\ ^2\Sigma$ and the red $A\ ^2\Pi \rightarrow X\ ^2\Sigma$ band systems of CN. The transition was postulated to be between the

---

[10] W. L. Faust, R. A. McFarlane, C. K. N. Patel, and C. G. B. Garrett, *Phys. Rev.* **133A**, 1476 (1964).

[11] L. E. S. Mathias and A. Crocker, *Phys. Lett.* **13**, 35 (1964).

[12] H. A. Gebbie, F. D. Findlay, N. W. B. Stone, and J. A. Robb, *Nature (London)* **202**, 169 (1964).

[13] H. A. Gebbie, N. W. B. Stone, and F. D. Findlay, *Nature (London)* **202**, 685 (1964).

[14] G. W. Chantry, H. A. Gebbie, and J. E. Chamberlain, *Nature (London)* **205**, 377 (1965).

levels $X\,^2\Sigma$, $v = 2$, $J = 8\frac{1}{2}$, $N = 8$ and the $X\,^2\Sigma$, $v = 2$, $J = 7\frac{1}{2}$, $N = 7$ of the CN radical, which is calculated from optical data to occur at 29.69 cm^{-1}. Subsequent studies[15] of the effects of isotopic substitution on the emission frequencies and calculations[16] of the radiative equilibria in the plasma cast doubt on this assignment. Finally, from a careful analysis of some high-resolution infrared spectra, the emitter was identified[17] as HCN. Similarly[18] H_2O, and not[19] OH, was shown to be the emitter in the water vapor laser.

Since this time many more laser lines have been discovered in the sub-millimeter region, and many of these have been obtained in continuous oscillation. Very few intense stimulated emission lines have been discovered at longer wavelengths than the 300 μm group of emission lines and none have as yet been operated continuously. A list of the sub-millimeter laser lines and their relative strengths is given in the Appendix.

Recently the development of the electrode design of the lasers and the use of stabilized cavities and current stabilized power supplies has led to lasers which provide an output which begins to approach that of backward wave oscillators or klystrons in stability. With these improved lasers use can be made of superheterodyne detection and it has been possible[20] to phase lock an HCN laser to a microwave frequency standard to obtain stabilities of one part in 10^8 or better.

2.4.3. Submillimeter Laser Mechanisms

When a beam of electromagnetic radiation of frequency $\bar{\nu}_0$ is traversing a medium that has a resonant absorption at this frequency, the intensity of the beam as a function of distance into the medium is governed by three effects; (1) spontaneous emission from the upper level to the lower level, (2) stimulated absorption from the lower to the upper level, and (3) stimulated emission from the upper to the lower level. The first of these contributes very little at submillimeter wavelengths since spontane-

---

[15] L. E. S. Mathias, A. Crocker, and M. S. Wills, *Electron. Lett.* **1**, 45 (1965).

[16] H. P. Broida, K. M. Evenson, and T. T. Kikuchi, *J. Appl. Phys.* **36**, 3355 (1965).

[17] D. R. Lide and A. G. Maki, *Appl. Phys. Lett.* **11**, 62 (1967).

[18] B. Hartmann and B. Kleman, *Appl. Phys. Lett.* **12**, 168 (1968); W. S. Benedict, *ibid.* **12**, 170 (1968); M. A. Pollack and W. J. Tomlinson, *ibid.* **12**, 173 (1968).

[19] W. J. Witteman and R. Bleekrode, *Phys. Lett.* **13**, 126 (1964).

[20] V. J. Corcoran, R. E. Cupp, and J. J. Gallagher, *IEEE J. Quantum Electron.* **QE 6**, 724 (1969).

ous emission is always isotropic and incoherent and is additionally very feeble in the far infrared because of the dependence of the intensity on the fourth power of the frequency. Emission into the narrow forward direction of the beam can therefore usually be neglected and we need consider only stimulated absorption and emission. These are governed by identical matrix elements and we may write the integrated absorption coefficient over the band as

$$\int \alpha(\bar{\nu}) \, d\bar{\nu} = h\bar{\nu}_0 B(n_l - n_u), \tag{2.4.3}$$

where B is the Einstein coefficient for stimulated absorption and n_l and n_u are the number of molecules per cubic centimeter in the lower and the upper states, respectively. Under normal circumstances n_l is greater than n_u, the two being related by the Boltzmann distribution function

$$n_u = n_l e^{-h\bar{\nu}_0 c/kT}. \tag{2.4.4}$$

As a result the right-hand side of Eq. (2.4.3) is positive and the beam is attenuated progressively with distance according to Lambert's law

$$I(x) = I_0 e^{-\alpha x}. \tag{2.4.5}$$

If however, n_u is larger than n_l, a wave propagating through the medium will grow in amplitude; α will be negative and, provided the medium is long enough, an initial photon produced by spontaneous emission will be multiplied to such an extent that coherent stimulated emission will be observed. This is the principle on which all masers and lasers operate and a variety of methods have been devised to bring about the necessary population inversion.

Despite early unfavorable prognoses, it has proved possible to achieve population inversions in many molecular and atomic systems. The gas masers work, as pointed out earlier, by using an electrostatic state selector. In some solid-state lasers such as the ruby laser, very short pulses of ultraviolet and visible radiation from a xenon flash tube populate an upper level from which transitions occur to an intermediate level. Stimulated emission from the intermediate to the depopulated ground state can then occur. The flashes have to be extremely intense to achieve a significant depopulation and four level lasers in which the stimulated emission occurs between two intermediate states (for example, the uranium laser) are less restrictive for this reason. Gas lasers such as the helium–neon laser achieve population inversion by means of collisional transfer of energy from metastable atoms to atoms that possess lower states connected to still lower states by allowed transitions. Laser action in pure inert gases arises from the phenomenon of radiation trapping.[21] Atoms in the ground (^1S) state can be transferred to excited (^1P) states by electron collision. Radiative transitions in the ultraviolet back to the ground state are almost immediate because of the large transition probability. However, this radiation is immediately absorbed by other ground state atoms and the radiation can leave the plasma only very slowly. For this reason a significant population of the ^1P states is built up and stimulated emission to the lower lying and unpopulated ^1D states can therefore occur. However, as far as is known at the moment, these processes play little part in the operation of most glow discharge submillimeter molecular lasers where the predominant process is a resonance interaction (or perturbation) between ro-vibrational energy levels. This phenomenon will now be described in some detail.

To the harmonic approximation, the potential energy of a molecule is a diagonal function of the normal coordinates, i.e.,

$$2V = \sum_{i=1}^{3N-6} \lambda_i Q_i^2, \qquad (2.4.6)$$

where $\lambda_i = 4\pi v_i^2$ and Q_i is the ith normal coordinate. Two normal modes may have closely similar energies and the absorption bands corresponding to them may lie in the same region of the spectrum but nevertheless no interaction will occur. This is still true for excited states involving the excitation of several normal modes to any number of vibrational quanta. When, however, cubic and higher terms are included in the potential energy function, the wave equation is no longer separable in terms of the normal coordinates and interactions occur leading to a displacement of levels. This is especially marked for levels which, in the absence of the perturbation, would be very close together. The phenomenon is usually called *Fermi resonance* and like all slight perturbations can be treated by writing a new set of wave functions using the unperturbed wave functions as a basis set. Suppose that we have two states whose wave functions (unperturbed) are ψ_0^i and ψ_0^j and that these are sensibly remote in energy from other states so that we may discuss their mutual perturbation in isolation. Under the influence of the resonance the wave functions to

[21] C. G. B. Garrett, "Gas Lasers." McGraw-Hill, New York, 1967.

first order may be written

$$\psi_1{}^j = \cos\theta \, \psi_0{}^i + \sin\theta \, \psi_0{}^j,$$
$$\psi_1{}^j = -\sin\theta \, \psi_0{}^i + \cos\theta \, \psi_0{}^j,$$

(2.4.7)

where θ is a parameter that is zero for no interaction and can range up to $\pi/4$ when the interaction is complete. The net effect of this quantum-mechanical resonance is that the energy levels are shifted one up and the other down from their unperturbed positions and that each of the observed levels takes on more or less of the character of each of the unperturbed states. A good example of this phenomenon occurs in the infrared spectrum of nitrous oxide.[22] The levels (100), that is the pseudo-symmetric stretching mode, and (02^00), the overtone of the bending mode, have the same symmetry Σ^+ (or A_1) and lie in roughly the same region of the spectrum. Fermi resonance is thus possible and the level formally written (02^00) takes on some of the character of (100) and in particular becomes, like (100), strongly connected to the ground state. A medium-strong band therefore appears in the infrared spectrum at 1167 cm^{-1} near to the intense

$$(000) \rightarrow (100)$$

band at 1285 cm^{-1}. The other component of the overtone, namely (02^20), has species Δ and is therefore unable to interact and there is no evidence of a transition to this level in the infrared spectrum. The corresponding (100)(02^00) resonance in CO_2 is virtually exact and two levels result which are essentially 50/50 mixtures of the unperturbed states.[23] The frequency shifts are large (-50 and $+42.7$ cm^{-1}, respectively) and in the Raman spectrum two intense bands occur at 1285.5 and 1388.3 cm^{-1} whereas one would expect only one at 1336 cm^{-1}. The CO_2 laser operates[2] by collisional transfer of energy from metastable nitrogen molecules in their first vibrationally excited state (2352 cm^{-1}) to ground state CO_2 molecules, thereby raising these to the (00^01) state (2439.3 cm^{-1}). The close match of energy is of course essential for effective transfer. The vibrationally excited CO_2 molecules can then undergo stimulated emission to *both* the perturbed (10^00)–(02^00) levels. The result is two bands

[22] G. Herzberg, "Infrared and Raman Spectra," p. 277. Van Nostrand–Reinhold, Princeton, New Jersey, 1945.

[23] G. Herzberg, "Infrared and Raman Spectra," p. 275. Van Nostrand–Reinhold, Princeton, New Jersey, 1945.

in stimulated emission at 961 and 1064 cm^{-1}. The resonance contributes only marginally to the laser mechanism in so far as the state (10^00) would be metastable without it; with it however, the mixing ensures that the state (10^00) can be rapidly depopulated by taking part in the radiative cascade

$$(02^00) \rightarrow (01^10) \rightarrow (00^00).$$

Another source of departure from simple behavior–known as *Coriolis perturbation*—is a form of vibration—rotation interaction. The simple theory of molecular spectroscopy is developed within a framework in which rotation and vibration are considered to be separable. For most purposes this is quite satisfactory, but there are cases where the simultaneous excitation of two vibrational modes is equivalent to an overall rotation and conversely where the effect of rotation in a molecule excited to one vibrational state is equivalent to exciting another. As before, this leads to a mixing of wave functions, but whereas the Fermi resonance applies to the entire vibration–rotation band, the Coriolis interaction is a sensitive function of J and the effect varies markedly over the band. Like all forms of perturbation the extent of the interaction depends on how close the two unperturbed states are to each other in energy, because the matrix elements involve terms like $(\bar{\nu}_i - \bar{\nu}_j)^{-1}$—the so-called *resonance denominators*. Not surprisingly, therefore, the effect is most marked when the two interacting states would be degenerate without it. The best known examples arise in the case of linear molecules excited to one or more quanta of the degenerate bending mode where the effect of the interaction is to split each rotational level into two close components, which leads to the so-called *l-type doubling*. Because of this it is essential to specify the value of the quantum number l (the vibrational angular momentum) when states of linear molecules are being discussed and this is in fact the superscript applied to the middle quantum number in the previous discussion of N_2O and CO_2. States with $l = 0$ have singlet rotational levels because the vibrational angular momentum by definition is zero. This type of Coriolis perturbation between degenerate levels is called *first order*, but second-order effects are often noticed. An excellent example occurs in the spectrum of methane,[24] where the ν_2 fundamental would be formally forbidden. However, Coriolis perturbation between ν_2 and ν_4 is allowed and ν_2 therefore appears weakly in the spectrum. The

[24] J. Lecomte, "Encyclopaedia of Physics" (S. Flügge, ed.), Vol. 26, Light and Matter II. Springer-Verlag, Berlin, 1958, "Spectroscopie dans l'infrarouge"; J. S. Burgess, E. E. Bell, and H. H. Nielson, *J. Opt. Soc. Amer.* **43**, 1058 (1953).

selection rule for Coriolis perturbation is that the direct product of the two symmetry species should contain a rotation. In the present case for v_2 and v_4 of an XY_4 T_D molecule

$$E \times F_2 = F_1 + F_2 \qquad (2.4.8)$$

and F_1 contains the three rotations R_x, R_y, and R_z. For states with J not zero a mixing of wave functions occurs and v_2 takes on partly the character of v_4, which is strongly allowed. The v_2 band becomes also allowed and appears in the infrared spectrum of gaseous methane, and the origin of the forbidden intensity is evident from the very anomalous distribution of line strength as a function of J. In particular, the $J' = 0$ lines are missing.

In 1967, Lide and Maki[17] suggested that the stimulated emission lines at 337 and 311 μm were due to vibration–rotation transitions of HCN. The population inversion and transition moment both arise from Coriolis perturbation between excited states of this molecule. In fact the two states $(11^1 0)$ and $(04^0 0)$ are very close together and Coriolis interaction is allowed by the selection rules. The rotational origin of $(04^0 0)$ lies some 2.51 cm^{-1} below that of $(11^1 0)$ but the former state has the larger B value (1.4916 cm^{-1} as compared with 1.468 cm^{-1}) and so the rotational levels eventually cross. The crossing occurs near $J = 10$ and as a result an intense perturbation occurs here. The relevant levels are shown schematically in Fig. 1. The infrared spectrum of HCN in the 2800–2900-cm^{-1} region[25] is dominated by the $(00^0 0) \rightarrow (11^1 0)$ absorption band, transitions to the level $(04^0 0)$ from the ground state having very small probabilities. Nearly all the lines in the observed band fall where they would be expected using the parameters $v_0 = 2805.58$ cm^{-1} and $B(00^0 0) = 1.4784$ cm^{-1} together with the appropriate centrifugal distortion constants $D' = 4.4 \times 10^{-6}$ cm^{-1} and $D'' = 3.3 \times 10^{-6}$ cm^{-1}. There are, however, some exceptions and the lines $R(8)$, $R(9)$, $P(10)$, $P(11)$, and $P(12)$ are slightly shifted from their expected positions and furthermore the line $R(9)$ appears as a doublet under high-resolution observation. These phenomena arise from the strong perturbation at $J = 10$ and the somewhat weaker interaction at the adjacent levels $J = 9$ and $J = 11$. The two states $J = 10$ of $(11^1 0)$ and $(04^0 0)$ are strongly mixed and both share the character of the hypothetical unperturbed $J = 10$ level of $(11^1 0)$; in particular, they both have similar transition probabilities to the ground

[25] A. G. Maki and L. R. Blaine, *J. Mol. Spectrosc.* **12**, 45 (1964).

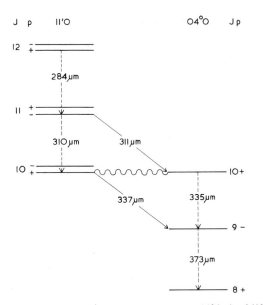

Fig. 1. Coriolis perturbations (wavy line) between the 11^10 the 04^00 levels of HCN and the origin of some stimulated emission lines of the HCN laser (—: primary lines; – –: secondary lines) [D. R. Lide and A. G. Maki, *Appl. Phys. Lett.* **11**, 62 (1967); L. O. Hocker and A. Javan, *Phys. Lett.* **25A**, 489 (1967)].

state. This is why just one line [one of the components of the $R(9)$ doublet] of the $(00^00) \rightarrow (04^00)$ band appears in the spectrum. For identical reasons, the transition

$$(11^10)J = 11 \rightarrow (04^00)J = 10$$

takes on some of the character of the pure rotation line $J = 11 \rightarrow J = 10$ within (11^10) and since HCN has a very large dipole moment $(2.986D)$ this is strongly allowed. In a similar way the apparently vibration–rotation line

$$(11^10)J = 10 \rightarrow (04^00)J = 9$$

takes on some of the character of a pure rotation line within (04^00) and becomes optically active.

Pollack[26] has recently considered the mechanism of the far-infrared laser gain resulting from rotational perturbations of both the Fermi and Coriolis kind, assuming an initial Boltzman distribution of rotational

---

[26] M. A. Pollack, *IEEE J. Quantum Electron.* **QE 5**, 558 (1969).

energy level population. The inversion is then produced mainly by the rotational perturbation of the above situation. The laser gains are calculated on this basis in terms of the perturbation parameters and the vibrational populations. His study shows that laser action can take place due to the perturbation of one ro-vibrational state by another, even when there is little or no vibrational population inversion between the two states involved. This near equality of the vibrational populations is the result of the rapid collisional cross-relaxation caused by their near degeneracy, and this interstate relaxation may be considerably faster than that of either component to the ground vibrational state.

The transition frequencies calculated from infrared data are 29.712 and 32.172 cm^{-1}, which are in very good agreement with those observed[15] for the intense lines of the HCN laser near 337 and 311 μm. Under vigorous excitation with high gain conditions many more lines are observed. The origin of some of these is shown in Fig. 1. The lines at 284, 310, 335, and 373 μm are secondary lines due to the strong depopulating or populating action of the intense primary lines.

The correctness of the assignment in Fig. 1 has been partially proved by the work of Frenkel et al.[27] who measured the frequency of the 337 and 311 μm lines and showed that the frequencies were not affected by the application of a magnetic field, and therefore unlikely to arise from the CN radical. Further proof came from the work of Hocker and Javan,[28] who succeeded in measuring the frequencies of the four lines at 310, 311, 335, and 337 μm. Their method was to beat the radiation from the laser with a high harmonic of a klystron in a suitable silicon crystal video detector. With appropriate choice of fundamental klystron frequency (Javan and Hocker used a V band klystron near 52.4 GHz) and harmonic number, the two sources of radiation can be brought close enough in frequency for the difference frequency to be measured. From this the absolute frequency of the laser radiation can be found by addition. Javan and Hocker's results are

$$310 \quad \mu m = 967.9658 \quad GHz,$$
$$311 \quad \mu m = 964.3134 \quad GHz,$$
$$335 \quad \mu m = 894.4142 \quad GHz,$$
$$337 \quad \mu m = 890.7607 \quad GHz,$$

[27] L. O. Hocker, A. Javan, D. Ramachandra Rao, L. Frenkel, and T. Sullivan, *Appl. Phys. Lett.* **10**, 147 (1967).

[28] L. O. Hocker and A. Javan, *Phys. Lett.* **25A**, 489 (1967).

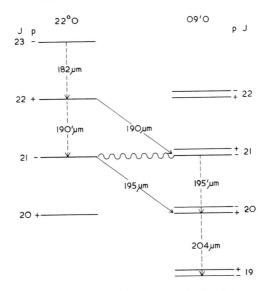

FIG. 2. Schematic diagram of the rotational energy levels of the states 22^00 and 09^10 of DCN near the Coriolis perturbation (wavy line) at $J = 21$ (—: primary lines; – –: secondary lines). The origin of laser action in DCN near 190 μm is shown [after A. G. Maki, *Appl. Phys. Lett.* **12**, 122 (1968); L. O. Hocker and A. Javan, *Appl. Phys. Lett.* **12**, 124 (1968)].

with an estimated error of 0.001 GHz. From these figures, the total energy difference between $J(11)$ of (11^10) and $J(9)$ of (04^00) by the two paths is found to be 1858.7276 ± 0.001 GHz and 1858.7265 ± 0.001 GHz, which is a beautiful illustration of the Ritz combination principle.

Experiments by Mathias *et al.*[15,29] at SERL showed that when hydrogen was replaced by deuterium in compounds containing $H^{12}C$ and ^{14}N, no emission was observed at 337 and 311 μm, but instead new emission lines were observed at 190 and 195 μm. Maki[30] has explained these lines as arising from perturbations between the (22^00) and (09^10) vibrational levels of $D^{12}C^{14}N$ in the vicinity of $J = 21$. The energy level diagram is shown in Fig. 2 and a listing of the observed lines is included in the Appendix. Maki[30] has identified a series of lines in the $HC^{14}N$ system between 135 and 125 μm as arising from transitions in the $(12^00)(05^10)$ (12^20) system in the vicinity of the resonance near $J = 26$. Owing to the lack of infrared data on higher vibrational states of HCN no further

[29] L. E. S. Mathias, A. Crocker, and M. S. Wills, *IEEE J. Quantum Electron.* **QE 4**, 205 (1968).
[30] A. G. Maki, *Appl. Phys. Lett.* **12**, 122 (1968).

positive assignments have yet been made, but it is possible that the group between 200 and 223 μm may arise from a resonance in the $(13^10)(06^20)$ system.

During early experiments[14] at NPL no emission was detected when ^{14}N was replaced by ^{15}N in the $H^{12}C^{14}N$ laser system, but subsequently, under much higher gain conditions, Mathias et al.[29] showed that weak emission could be obtained at 110, 113, 139 and 165 μm, which is attributed to $H^{12}C^{15}N$.

Initially the idea that the HCN molecule was the emitter was in doubt, since it was claimed that emission at 337 and 311 μm was obtained when there was no hydrogen-containing compound present in the system.[31] However, it soon became clear that it was difficult to ensure that there was no hydrogen occluded on the walls of the laser. Further spectroscopic evidence that HCN was present in this type of electrical discharge came from the observation[32] of the microwave spectrum at 88.63 GHz, where the characteristic triplet of HCN (see Section 2.4.1) was observed. The inability to detect HCN by normal infrared methods means that the concentration in the discharge must be very low. This implies that a high proportion of the HCN molecules formed must take part in the laser process to account for the relatively large output power of about 10 mW obtained[33] in continuous operation with a 2-m laser.

Two other emission lines have been obtained at much longer wavelengths than the 337-μm group. Kneubuhl and his collaborators at Zurich[31] have obtained laser action at 538 and 774 μm from pulsed discharges through iodine cyanide, ICN. In both cases Kneubuhl has shown that iodine is necessary for laser action at these wavelengths. This was confirmed for the 538-μm line by Kon and his colleagues[34] in Japan. They postulated that the laser action is due to ICN itself. However, since the vibration–rotation spectrum of ICN has not yet been studied under high resolution the assignment is at present only provisional. It is interesting, in retrospect, to observe that one pre-

[31] H. Steffen, J. Steffen, J. F. Moser, and F. K. Kneubuhl, Phys. Lett. 20, 20 (1966); 21, 425 (1966).

[32] M. Lichtenstein, V. J. Corcoran, and J. J. Gallagher, IEEE J. Quantum Electron. QE 3, 696 (1967).

[33] H. A. Gebbie, N. W. B. Stone, J. E. Chamberlain, W. Slough, and W. A. Sheraton, J. Chim. Phys. No. 1, 80 (1967); Nature (London) 211, 62 (1966). See also W. M. Müller and G. T. Flesher, Appl. Phys. Lett. 8, 217 (1966).

[34] M. Yamanaka, S. Kon, J. Yamamoto, and H. Yoshinaga, Jap. J. Appl. Phys. 7, 554 (1968).

diction[14,16] of the CN theory was that there should be emission at 18.56 cm^{-1} (538 μm) if there was at 29.70 cm^{-1} (337 μm). It is a remarkable coincidence that radiation at 18.56 cm^{-1} has been observed from discharges through CN containing vapors and yet one knows that it does not come from the CN radical. Rather intriguingly emission, which indubitably comes from CN, has been observed by Pollack[35] in the products of the flash photolysis of cyanogen.

The water vapor laser provides the richest source of stimulated emission lines in the far infrared. More than 100 lines are known distributed throughout the wavelength range from 7 to 220 μm. The identification of the emitter was achieved almost simultaneously by Hartmann and Kleman, by Benedict and by Pollack and Tomlinson.[18] A more detailed description of the laser mechanism has been given in a comprehensive paper by Benedict et al.[36] The vibration frequencies of the water molecule are determined mostly by the very small mass of the two hydrogen atoms and for this reason, the symmetric stretching mode v_1 and the antisymmetric stretching mode v_3 are very close in frequency, 3651.7 and 3755.8 cm^{-1}, respectively. The bending mode v_2 has a frequency which is roughly half that of the stretching modes (1595 cm^{-1}), and therefore its overtone (020) lies in the same spectral region as the two stretching fundamentals. The rotational constants A, B, and C are very large ($A = 27.77$, $B = 13.39$, and $C = 9.95$ cm^{-1} for the ground state) and as a result the energy levels even for moderate J values can be thousands of reciprocal centimeters above the rotational ground state. Many close coincidences are therefore possible and when it is borne in mind that H_2O is a highly anharmonic molecule, strong perturbations are to be expected. It is these perturbations between (100) and (020) and between (020) and (100) that have been invoked by Benedict to explain the operation of the H_2O submillimeter laser. The rotational energy levels of an asymmetric rotor such as H_2O cannot be written in closed form as a function of J. The water molecule is classed as a type b asymmetric rotor in that its intermediate principal axis of inertia coincides with the twofold symmetry axis. Although only J is a "good" quantum number for an asymmetric rotor it is usual to define two other quantum numbers—in this case K_a and K_c—which are the values of the quantum number K which the state would have if the molecule were distorted into a limiting

[35] M. A. Pollack, *Appl. Phys. Lett.* **9**, 230 (1966).

[36] W. S. Benedict, M. A. Pollack, and W. J. Tomlinson, *IEEE J. Quantum Electron.* **QE 5**, 108 (1968).

prolate and oblate symmetric top, respectively. These numbers are given as right lower subscripts to J in specifying a given rotational level. Because H_2O is a case b asymmetric rotor, it follows that the symmetry axis does not coincide with the unique axis of inertia in either extreme, and because of this the selection rules bear little resemblance to those for a symmetric rotor. As an example, two of the strongest lines in the pure rotation spectrum of water at low frequencies are

$$1_{01} \rightarrow 1_{10} = 18.56 \quad \text{cm}^{-1} \quad \text{and} \quad 2_{02} \rightarrow 2_{11} = 25.08 \quad \text{cm}^{-1}$$

which for the limiting cases would be $\Delta J = 0$, $\Delta K = \pm 1$ type. The dipole selection rules can be deduced from the overall parity of the levels. For the absorption of infrared radiation this must change because of the negative parity of the dipole moment operator. The parity depends on whether K_a and K_c are even or odd and four cases arise, namely, $(++)$, $(--)$, $(+-)$, and $(-+)$. The selection rules are then

$$(++) \leftrightarrow (--) \quad \text{and} \quad (+-) \leftrightarrow (-+).$$

This definition of parity follows the original suggestion of Herzberg,[37] but in their paper Benedict et al. prefer a variant in which the parity of K_a and the magnitude of the old quantum number τ are given. In terms of this $(++)$ and $(+-)$ remain unchanged but $(--)$ becomes $(-+)$ and vice versa. The parity is affected also by the symmetry of the vibrational wave function and this will be negative if an odd number of quanta of the antisymmetric stretching mode ν_3 are excited. When this is the case the parity is given not by the odd/even quality of (K_a, K_c) but rather by that of (K_{a+1}, K_c). The perturbations between the rotational levels of (100), (020), and (001) have a very complicated nature. Those between (100) and (020) are of the Fermi type and those between (020) and (001) are of the Coriolis type. Some of these perturbations have been discussed by Pollack[26] and the treatment he uses follows that outlined for HCN. Once again it is found that, even with little or no vibrational population inversion, the perturbations give sufficient gain for laser action.

A simplified energy level diagram showing the origin of the main lines of the water vapor laser is given in Fig. 3. Pollack and Tomlinson[18] have measured the Zeeman effect on some of the water vapor laser lines, and using the theory of the Zeeman laser have shown that the lines arise from

[37] G. Herzberg, "Infrared and Raman Spectra," p. 52. Van Nostrand–Reinhold, Princeton, New Jersey, 1945.

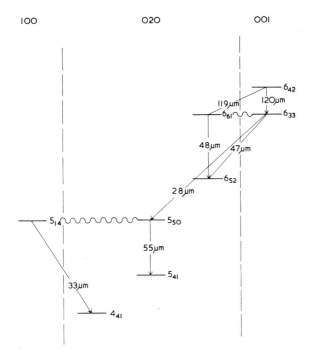

FIG. 3. Simplified energy level diagram for the electronic ground state of H_2O showing the origin of some of the far-infrared laser lines [W. S. Benedict, M. A. Pollack, and W. J. Tomlinson, *IEEE J. Quantum Electron.* **QE 5**, 108 (1968)].

$\Delta J = 0$ and $\Delta J = \pm 1$ transitions, which is additional evidence for the assignments. A full list of the laser lines in H_2O, D_2O, and $H_2{}^{18}O$ is given in the Appendix.

2.4.4. Submillimeter Stimulated Emission Devices—Experimental Details

2.4.4.1. Introduction. This section describes the practical aspects of the two types of stimulated emission device—the molecular beam maser and the molecular laser—which were discussed in theoretical detail in the preceding three sections.

2.4.4.2. The Molecular Beam Maser. The experimental arrangements used by De Lucia and Gordy[5] for their observation of maser action on the $0 \rightarrow 1$ and $1 \rightarrow 2$ pure rotational lines of HCN are shown in Figs. 4 and 5. The beam collimator, state selector, and Fabry–Perot resonator are all inside a vacuum chamber that can be evacuated to a pressure of approximately $2\ \mu$Torr by means of a large diffusion pump fitted with

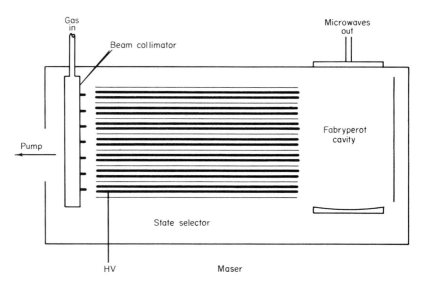

FIG. 4. Molecular beam maser of F. De Lucia and W. Gordy [*Phys. Rev.* **187**, 58 (1969)].

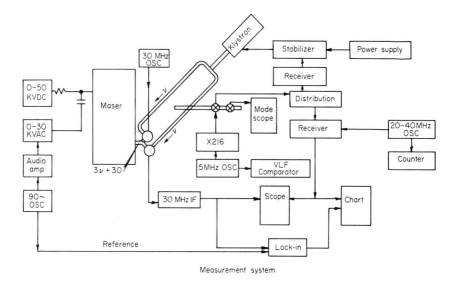

Measurement system

FIG. 5. Electronic supply and measurement system used by De Lucia and Gordy for their molecular beam maser.

liquid nitrogen cooled cold traps. The beam collimator is made from a number of sections of needle tubing embedded in one side of a flat rectangular box. The state selector consists of a set of stacked quadrupole electrodes separated by cooled copper fins, the purpose of which is to trap scattered or deflected molecules and thus prevent them from entering the resonator. In a quadrupolar field, a molecule in a state that increases in energy in an electric field (i.e., a positive Stark effect) will be focused into the beam, and one that decreases in energy will be deflected out of the beam. In general, the upper and lower states of any transition are unlikely to have exactly the same Stark behavior, so some degree of perturbation of the molecular Boltzmann distribution will occur as the result of the passage of the beam through the state selector. The ideal case, so far as stimulated emission is concerned is where the upper state is focused entirely into the beam and the lower state entirely out. This is found for the

$$J = 2, \quad F = 1 \rightarrow J = 1, \quad F = 1$$

line of HCN and with this favorable occurrence, De Lucia and Gordy were able to make the maser oscillate on this line. All three of the $J = 1 \rightarrow J = 0$ lines show measurable gain and at least one component has been made to oscillate, but not all of the $J = 2 \rightarrow J = 1$ lines show gain and in fact, with some of them, the operation of the state selector is to increase the population of the ground state relative to the upper state. Such lines show an enhanced absorption, but in the context of molecular investigation this can be just as valuable as an observed amplification.

The principal difference between the centimeter wavelength masers and that used by De Lucia and Gordy is that the latter features a Fabry–Perot resonator in place of the usual microwave cavity. This is partly because of the extremely difficult problems encountered in attempting to scale down the cavities to match the millimetric wavelengths, but is also indicated by the need to maintain a high Q and an acceptably narrow bandwidth. The Fabry–Perot resonator provides both these but there is the difficulty of filling all its volume with active gas and for this reason, the performance of the maser has not yet reached that expected from theory. The resonator is of the semiconfocal type with hole coupling to a microwave guide. The flat mirror which had the coupling hole bored in it was 10 cm in diameter, and the spherical mirror was the same size and mounted on a slide arrangement so that the cavity could be tuned. This part of the maser is entirely analogous to the cavities used in molecular lasers and the two devices differ only in the means used to bring about the necessary population inversion. The operation of, and mode patterns in, Fabry–Perot resonators will be discussed in the following subsection.

The millimeter wave measurement and detection system is shown in Fig. 5. The high stability 5-MHz oscillator is constantly compared with the standard broadcast frequency (WWVB) to ensure a frequency stability quoted as one part in 10^{10}. The output of the oscillator is multiplied 216 times to give a 1080-MHz signal and is then mixed along with a phase-related 180-MHz signal in a silicon crystal with part of the klystron output. The resulting rf signal is used both to drive a feedback loop providing automatic klystron frequency control and also to give the necessary frequency markers. The output of the klystron is split in two, and one half goes to a crystal harmonic generator where it is multiplied either twice or three times and simultaneously mixed with a 30-MHz signal from an oscillator. This mixed signal is used to stimulate the maser and the output of the maser is itself mixed in another crystal with the other half of the klystron output. The detector is therefore a super-heterodyne receiver and the 30-MHz beat can, in the usual way, be amplified in an i.f. strip, demodulated, and displayed on a suitable recorder. The signal-to-noise ratio can be improved, as usual, by employing an audio frequency modulation of the maser output followed by "lock-in" or phase-sensitive second detection of the demodulated rf signal. The necessary modulation is readily achieved by applying a 90-Hz high voltage to the state selector plates in addition to the constant dc potential: The alternating potential leads to an oscillating efficiency of the state selector and gives amplitude modulation of the maser output. Enhanced absorption, which is really the converse of stimulated emission, can also be detected readily by the system since an enhanced absorption when the state selector is on will look just the same as an apparent amplification when it is off. The two sorts of signal will therefore be in quadrature. It should be emphasized that stimulated emission is a highly nonlinear phenomenon and the relative intensities observed in amplification by the maser may bear little resemblance to those observed in the conventional absorption spectrum.

2.4.4.3. Constructional Details of Submillimeter Lasers. As we have remarked earlier, the submillimeter stimulated emission devices that are principally discussed in this chapter bear a close resemblance to the lasers that operate in the visible and near-infrared regions. However, because of the longer wavelengths, and because of the absence of any really rigid transparent optical material in the submillimeter region, there are some marked differences in practical construction.

Since the wavelength of the radiation is becoming comparable with the diameter of the cavity, it is necessary to consider the diffraction losses. These losses are related to the Fresnel number N so that a high Fresnel number implies a low diffraction loss. The Fresnel number is given by

$$N = a^2 L^{-1} \lambda^{-1},$$

where a is the radius of the mirrors and L is the tube length. The diameter must be chosen so that for a given tube length and wavelength, $N \geq 1$. Thus a is seldom less than 2 cm.

Since it is difficult to construct good Brewster-angle windows with the materials available in the submillimeter region, it is usual to mount both mirrors inside the discharge tube, in order to avoid serious absorption and reflection losses. Brewster windows have, however, been used on occasion with some success.[38] The cavities used for submillimeter lasers are usually of the nearly confocal or nearly semiconfocal type described by Fain and Khanin[6] and Garrett.[21]

Five types of output coupling have been used—hole coupling, beam splitter coupling, diffraction round the edge of one of the end mirrors, the zero-order reflection from a diffraction grating used as one end mirror, and residual transmission over the whole of one end mirror. The final method was the last to be developed, since there are no dielectric materials that can be used for this purpose at submillimeter wavelengths, as germanium has been used for the 10.6-μm CO_2 laser. However, it has been shown by Ulrich[39] that metal-mesh filters can approximate the function of artificial dielectric films, and so residual transmission is now available for the submillimeter region as well.

The two principal lasers in this region are the HCN laser and the H_2O laser and for both, the molecules responsible for the emission are unstable with respect to electron bombardment. Sealed-off discharge tubes are not therefore ordinarily used and the common practice is to flow an appropriate fuel through the laser at a suitable rate to maintain a pressure of the order of 0.1 Torr in the tube. The best fuel for the HCN laser is a mixture of methane and nitrogen, though almost any compound

[38] R. G. Jones, C. C. Bradley, J. Chamberlain, H. A. Gebbie, N. W. B. Stone, and H. Sixsmith, *Appl. Opt.* **8**, 701 (1969).

[39] R. Ulrich, *Appl. Opt.* **7**, 1987 (1968); see also R. Ulrich, T. J. Bridges, and M. A. Pollack, *ibid.* **9**, 2511 (1970); A. F. Wickersham, *J. Appl. Phys.* **29**, 1537 (1958).

or mixture containing hydrogen, carbon and nitrogen will do; for the H_2O laser the simplest fuel is just water vapor, but once again, provided the two elements are present many other alternatives are possible. As an example of this, intense emission is obtained at 28 μm with nitromethane CH_3NO_2 as the fuel. The operation of the HCN laser is complicated by the formation of a brown solid polymer from the discharge products. This polymer coats the walls of the tube and eventually the rest of the inside of the laser. For this reason, after a running period of the order of 100 hr, the instrument has to be dismantled and thoroughly cleaned in order to bring the performance back to an acceptable level. The nature of the polymer is as yet unknown but it presumably contains carbon, hydrogen, and nitrogen for, if the walls of the laser are allowed to get hot, the polymer decomposes to form products which are themselves suitable fuels for the laser. Interesting experiments have been carried out with a sealed-off HCN laser that successfully operated for several hours when the tube walls were allowed to get much hotter than they do in normal operation. Fuller at NPL has operated a sealed-off CW HCN laser for some tens of hours and Murai has shown[40] that the presence of bromine in the discharge is very beneficial for long time sealed-off operation. He has operated a sealed-off pulsed laser fueled with BrCN and H_2 for up to 70 hr.

Submillimeter lasers are operated in two ways, either pulsed or continuous. The gain per meter is less in continuous operation and therefore only the stronger lines can usually be obtained in this form of operation. In pulsed operation, the electrodes of the laser are connected in series with a thyratron or ignitron and a capacitor (~0.1 μF) which is charged up to a voltage of the order of 25 kV from a dc power supply. When a triggering pulse is delivered to the thyratron the condenser discharges through the tube. The length of the discharge pulse is typically a few microseconds and the peak current of the order 100 A. More powerful pulses can be obtained from delay-line type pulse generators followed by a step-up (5:1) pulse transformer. The early experiments were carried out with such equipment delivering a peak voltage of 46 kV and a peak current of 300 A. High repetition rates are not possible because of the time required to charge the capacitor, but values up to 10 Hz can be achieved and this matched the response of the Golay cell, which is optimal for frequencies of this order. The output of the detector can be fed to a gating circuit that is switched by the triggering pulse to the thyratron.

[40] A. Murai, *Phys. Lett.* **28A**, 540 (1969).

In this way a dc signal suitable for display on a chart recorder can be obtained. It has been found that by careful design of the delay lines, pulses reproducible to 5% can be obtained.[41] Although pulsed lasers have been used in the past as sources in various experiments, they are really most valuable for their inherent high gain, and are best used in searches for new stimulated emission lines, or in experiments requiring time resolution, such as some pulsed plasma studies.

CW lasers are much more stable and with good design of laser and especially of the electrodes a continuous power output varying in the short term by as little as 1% is possible. The original electrodes were of solid brass and the resulting arc discharge was rather unstable. A considerable improvement came with the development of hot filament (barium zirconate) cathodes and lasers fitted with these ran very stably. Unfortunately, new filaments had to be fitted daily and this was very time consuming. The modern design[42] features a hollow cathode, and this gives a stable discharge without the inconveniences of the filament type. For the most refined work, i.e., where high frequency stability is desired, a hollow anode is also used, but in general the composition or configuration of this electrode has little effect on the discharge characteristics. The simplest type of power supply used to run this type of laser consists of a power pack capable of delivering 3 kV at 1.5 A, with electric fire elements used as ballast resistors. Once the discharge is established, the voltage across the electrodes drops to 1.5 kV. It is useful to have available a high voltage supply as well, to be used for initially breaking down the discharge path: this is especially so when long lasers are being used since the voltage necessary to maintain the discharge is much less than that required to initiate it. For high-precision work it is necessary to use current stabilized power supplies. At NPL, the type manufactured by K. S. M. Brookmans Park, Herts, England, is used. Unless the current flowing through the plasma is constant the power output of the laser will vary and in fact with some badly stabilized power supplies, nearly 90% mains frequency modulation of the laser output has been observed.

The mirrors used in the laser are normally made of optical quality glass, front surfaced with aluminum or gold by evaporation. Reflectivities

---

[41] Unpublished observations from NPL, but see commercial literature produced by Molectron Corp., Far Infrared Div.

[42] See for example, M. Yamanaka, H. Yoshinaga, and S. Kon, *Jap. J. Appl. Phys.* **7**, 250 (1968).

of metals are very high in the far infrared and it is not necessary to employ mirrors coated with multilayer dielectric films of the type used for lasers operating in the visible and near-infrared regions. This is very convenient since multilayer techniques are not yet available for the far infrared and the mirrors have to be capable of withstanding the arc discharge. The early forms of pulsed laser used a plane mirror cavity but this is tending toward obsolescence except in applications where a plane wave front is desired. CW lasers, where the inherent gain per meter is substantially less, are almost invariably constructed with the confocal or folded confocal configurations since the higher diffraction losses associated with the plane mirror arrangement make it unattractive. The alignment tolerances are also much more critical with the plane mirror system. In order to maximize the power output of a submillimeter laser, it is essential to be able to vary the length of the cavity (see later) and this is achieved in the usual arrangement by mounting one mirror on the spindle of a micrometer. The spindle passes into the discharge tube through on O-ring vacuum seal. Good optical quality of the mirrors aids in the visual alignment of the cavity which can be achieved using small torch bulbs, a low power He–Ne laser or else an autocollimator.

A diagram of a typical experimental laser is shown in Fig. 6. The discharge tube is constructed from standard borosilicate glass tube supplied (in England) by QVF and more normally used as chemical pipeline. Because of its large scale application, the tubing is cheap, available in many lengths and diameters and ends in flanges which may be readily vacuum sealed using flat washers. The two metal end plates are separated by four Invar rods which minimize the length changes of

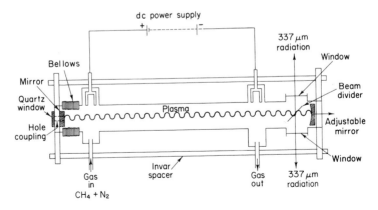

FIG. 6. Basic construction of HCN laser.

the cavity when the temperature of the laser varies. Fans may be mounted on these rods to cool the electrodes by forced convection.

Output coupling may be achieved using any of the five methods mentioned earlier. If the second method is adopted, the beam divider may be either a thin polyethylene terephthalate (melinex) film or else a metal mesh film. A variable degree of output power coupling can be readily realized by replacing one window by a mirror for then the combination of two mirrors and a beam divider forms a Michelson interferometer and the output power will go from a maximum to zero as the path difference changes by $\lambda/2$. The beam divider does introduce some losses into the cavity but its use tends to give a nearly plane wave front for the emerging radiation and because the angle of incidence $45°$ is not too far away from the Brewster angle $59°$ the radiation is strongly polarized in the plane of Fig. 6. This is of course contrary to the usual result of Brewster polarization which produces an enhancement of the electric vector component in the perpendicular direction. The answer to the paradox is that the losses by reflection of vertically polarized waves in the laser are so high that the laser will not oscillate. The oscillations which do occur are necessarily, therefore, horizontally polarized.

Output coupling via a hole requires us to introduce the concept of cavity modes. A Fabry–Perot resonator does not possess modes in the same sense as does a microwave cavity, since the boundary conditions are indeterminate. Nevertheless Fox and Li[43] and Boyd and Kogelnik[44] have shown that mode labels can be assigned to the radiation patterns inside the cavity. The lowest loss mode is the even symmetric TEM_{00} mode, which is consequently often called the *dominant mode*. This mode is most intense at the mirror center, and falls off to zero intensity at the mirror edge. If the coupling hole is small, the losses in the higher modes are higher than in the dominant mode, and power may be coupled out in this mode. However, if the hole is large, oscillation tends to occur in one of the modes with a node at the center of the mirror, such as the TEM_{01} mode, which has the next lowest loss. In general, in a submillimeter laser, one observes output from several different modes at the appropriate mirror separations. This is because the longitudinal mode spacing and the nonaxial mode spacing ($\sim c/2L$) are greater than the Doppler width of the gain curve, whereas in a visible or near-infrared

[43] A. G. Fox and T. Li, *Bell Syst. Tech. J.* **40**, 453 (1961).

[44] G. D. Boyd and H. Kogelnik, *Bell, Syst. Tech. J.* **41**, 347 (1962); see also D. C. Sinclair, *Appl. Opt.* **3**, 1067 (1964).

laser the reverse is true. It is for this reason that the cavity length adjustment is needed, since it may be necessary to alter the separation between the mirrors to achieve oscillation or else to select the desired mode. The disadvantage of hole coupling is that the beam obtained is very divergent, because the wavelength of the radiation is comparable with the hole diameter, which is usually 1 or 2 mm. The topic of hole coupling has been discussed comprehensively by McNice and Derr.[45] Output coupling by diffraction round the edge of the laser mirrors is useful for the observation of very weak lines, since this form of coupling does not introduce any additional losses into the cavity. For the TEM_{00} mode in particular, the absence of any coupling away of power at the mirror center where the mode is most intense gives the system the best chance to oscillate on marginal lines. This method has been used by Mathias et al.[29] to study some of the very weak lines of HCN.

The fourth method, i.e., zero-order coupling from a diffraction grating used as one end mirror, is used mostly to provide a beam to act as a monitor of the laser performance while the main output is taken from the mirror at the other end of the laser.

The fifth method, employing metal meshes as the laser mirrors, has recently been described by Ulrich et al.[39] In the usual arrangement one mirror is replaced by an infrared interference filter consisting of two metal-mesh reflector grids. By adjusting the spacing between the two parallel grids, the reflectance of the filter can be varied in rather the same way as can that of a Michelson interferometer. Ulrich has presented the theory of these filters in a series of papers that are listed in the paper by Ulrich et al.[39] The filter can be thought of as an artificial dielectric and its operation can be discussed in terms of microwave circuit theory. The output beam from a mesh coupled laser has similar properties to that from a conventional beam-splitter coupled laser. However, it is possible to induce a particular linear polarization by replacing the filters of square section by ones that are of rectangular section or even, in the limit, strips. In the latter case additional losses occur due to the necessity of supporting the strips on a substrate.

If a laser with a low Fresnel number is being used, the walls of the cavity can act as a dielectric waveguide. Kneubuhl and his colleagues have set up the theory of this effect,[46] and McCaul has verified some of

[45] G. T. McNice and V. E. Derr, *IEEE J. Quantum Electron.* **QE 5**, 569 (1969).
[46] P. Schwaller, H. Steffen, J. F. Moser, and F. K. Kneubuhl, *Appl. Opt.* **6**, 827 (1967).

the theory by the demonstration of oscillations in a laser that had a convex mirror at one end.[47] In general it is good practice to choose lasers in which the wall effects are unimportant, especially when working with the HCN system. The deposits of polymer that form on the walls cause attenuation of the dielectrically guided waves and so the output from the laser will soon deteriorate if the wall effects are significant.

The power levels available in CW HCN lasers vary from 10 mW with a tube length of 2 m to some 600 mW, which was obtained by Kotthaus.[48] This worker used a water-cooled laser 6.5 m long and inside diameter 10 cm; the cavity was plano-concave and the fuel ether vapor plus nitrogen. The power output in the TEM_{00} mode reached a peak of 600 mW when the output coupling hole was between 21 and 25 mm in diameter.

Because the mode separation is greater than the Doppler gain width, the power output of a laser as a function of cavity length shows strong, very sharp, peaks whose maxima correspond to cavity lengths that are even multiples of $\lambda/2$. A cavity fringe pattern observed with a water vapor laser is shown in Fig. 7 and some very fine records of both HCN and H_2O lasers have been shown by Müller and Flesher.[49] At high power levels several sets of peaks are often observed corresponding to the various modes of the cavity. The spacing of the peaks is strictly $\lambda/2$ for the TEM_{00} modes, but is very slightly different from this value for the other modes so that under normal conditions where one is moving the mirror through a range of cavity lengths all much greater than λ, the sets will have become displaced. One must of course be particularly careful not to misidentify these subsidiary peaks as arising from other transitions. When several lines are present, several sets of peaks will be observed, but since the spacings will vary significantly, the relative shifts of peaks over a reasonable travel of the mirror will enable a positive identification to be made. The best procedure, however, is to analyze the radiation with a separate passive interferometer, and the Mach–Zehnder type is preferred since there is no risk of radiation being returned to the cavity. The wavelength of the radiation may be calculated by counting the number of fringes observed in a known travel of the interferometer mirror. The use of grating spectrometers is not advisable because of the possibility of mistaking a strong line observed in a high order for a

[47] B. W. McCaul, *Appl. Opt.* **9**, 653 (1970).
[48] J. P. Kotthaus, *Appl. Opt.* **7**, 2422 (1968).
[49] W. M. Müller and G. T. Flesher, *Appl. Phys. Lett.* **8**, 217 (1966).

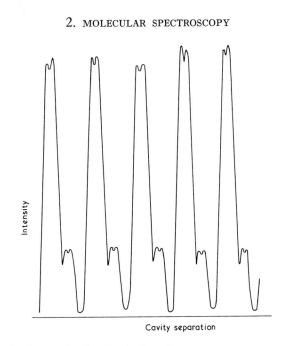

Cavity separation

FIG. 7. Cavity fringes showing Lamb dips of a water vapor laser operating at 118.6 μm (C. C. Bradley, unpublished). The stronger "doublets" arise from the TEM_{00} mode and the weaker from the TEM_{01} mode.

weak line observed in the first order. As an example of this, Hard and his colleagues reported[50] the observation of a line at 469 μm which they afterward realized was due to the 118-μm line of H_2O appearing in the fourth order of their grating.

This observation[50] by Hard et al. was most interesting, for they were observing HCN emission from a diethyl ether plus nitrogen fuel. They correctly deduced that the appearance of H_2O lines was due to a slight leak resulting in the presence of molecular oxygen in the tube. Following this up, they succeeded in finding mixtures of gases which were such that both HCN and H_2O lines could be obtained simultaneously. This confirmed the earlier observation of simultaneous emission by Müller and Flesher[49] when operating an HCN laser with water vapor in the tube. The composition of the gas mixture to give both HCN and H_2O lines is very critical, but when it is achieved there is no competition or cascade interaction between the lines, which is as expected for emission by chemically distinct species.

[50] T. M. Hard, F. A. Haak, and O. M. Stafsudd, *IEEE J. Quantum Electron.* **QE 5**, 132 (1969).

HCN and H_2O lasers can be forced into single-frequency operation by replacing one of the cavity mirrors with a diffraction grating. This technique was first used by Moeller and Dane Rigden[51] for the CO_2 laser, which normally gives only the P lines of the $00^\circ1 \rightarrow 10^\circ0$ band, but which, when fitted with the grating, gives also the R branch lines. As the grating is rotated, the cavity becomes resonant for a particular frequency, given by the grating parameters and the angle of rotation. In this way competing lines (the P branch in this case) are suppressed and much weaker lines may oscillate. The technique is very valuable for studying competitive processes. Jeffers[52] subsequently employed a grating for a water vapor laser operating between 23 and 57 μm, and the technique was extended further into the submillimeter region by Brannen and his colleagues,[53] who constructed a laser giving only the 118.6-μm line of H_2O. A diffraction grating has considerably different reflectivities for radiation polarized perpendicular and parallel to the grooves, and Brannen et al. found that the output of their laser was 99.8% polarized with the electric vector perpendicular to the grooves. A monochromatic and virtually 100% linearly polarized source is clearly of great value for the solid-state studies described in Section 2.4.7.

Polarization of the laser output can also be induced by the application of a magnetic field, and this proved to be important in the researches which led to the successful identification of the emitter in the water vapor laser. In the absence of a field, the laser emission corresponding to a $\Delta J = 0$ transition is bistable circularly polarized, and whether the observed emission is right-handed or left-handed depends on the initial starting conditions.[18,54] In the presence of a field the degeneracy is removed, and the tuning curve of the laser shows a sharp switch from the one to the other state as the cavity goes through resonance.[54] This was observed for the 118.6-μm line of H_2O, when a solenoid wound on the laser tube was supplying an axial magnetic field. From this observation it was concluded that the transition was of the $\Delta J = 0$ type, and that J was probably greater than 2. From the magnitude of the intensity change when the polarization switched, the effective magnetic moment of the molecule may be inferred and for the water vapor laser μ_{eff} for the emitter was found to be only $\frac{3}{4}$ of a nuclear magneton. This strongly

[51] G. Moeller and J. Dane Rigden, Appl. Phys. Lett. **8**, 69 (1966).

[52] W. Q. Jeffers, Appl. Phys. Lett. **11**, 178 (1967).

[53] E. Brannen, M. Hoeksema, and W. J. Sarjeant, Can. J. Phys. **47**, 597 (1969).

[54] W. J. Tomlinson, M. A. Pollack, and R. L. Fork, Appl. Phys. Lett. **11**, 150 (1967).

indicated that the H_2O molecule was the emitter, since it was unlikely that a free radical could have such a low magnetic moment. The 28-μm line was later shown to be a $\Delta J = \pm 1$ transition and all the effective magnetic moments measured lay between 0.5 and 1.0 nuclear magnetons. For a $\Delta J = \pm 1$ line the output is linearly polarized in the presence of an axial magnetic field. The direction of the electric vector for such linearly polarized radiation is determined by anisotropies of the cavity, and Yamanaka et al. have shown[42] that a set of parallel scratches on the plane mirror of the cavity will ensure a polarization parallel to these scratches for HCN lasers, where, of course, all the lines are of the $\Delta J = \pm 1$ type. In this way, a polarized output can be obtained from a hole coupled laser.

2.4.5. Dynamics of the Active Plasma

2.4.5.1. Introduction. The instantaneous population distribution inside the plasma is governed by three processes: (1) there is the combination of electron bombardment plus chemical reaction plus spontaneous radiation emission which produces the molecules in the necessary states and with the required population inversion; (2) there is the effect of molecular collision which tends to rapidly restore the Boltzmann distribution; and (3) there is the effect of the stimulated radiation field itself which can produce further inversions leading to cascade transitions and which can depopulate levels in more than one way giving the phenomenon of competitive emission.

2.4.5.2. Time Behavior of the Output. The principal topic investigated experimentally has been the correlation in time between the laser output pulse and the driving current pulse. For this work, fast detectors are necessary since the times to be measured are of the order of microseconds. Two types have been used with submillimeter lasers, the copper-doped germanium detector for the shorter wavelengths and the indium antimonide detector of Putley[55] for the longer wavelengths. Both detectors have to be operated at temperatures of 4.2 K or less. These detectors and their associated circuitry can have response times of the order 0.5 μsec. Mathias and his colleagues[15] reported that for the HCN laser the emission occurred in the afterglow, starting between 1 and 17 μsec after the current had ceased and lasting between 5 and 44 μsec. These times varied from fuel to fuel. In general each stimulated

---

[55] E. H. Putley, Phys. Status Solidi 6, 571 (1964); Appl. Opt. 4, 649 (1965).

emission line shows a different delay and pulse length, but certain similarities are noticed. Thus Mathias *et al.*[29] state that the 116, 126, 129, 131, and 135 μm lines of HCN behave similarly (delay 2 μsec, pulse length 5 μsec), but that they were all quite different from the 201, 211, and 223 μm set which showed a longer (~ 10 μsec) initiation time and a longer pulse length (~ 15 μsec). These patterns of common behavior agree well with the suggestion mentioned earlier, that the members of each set do in fact belong together: the shorter wavelength group to transitions in the $(12^00)(05^10)(12^20)$ system and the longer to transitions in the $(13^10)(06^20)$ system. Yamanaka *et al.* have studied[34] the time dependence of the output at 538 μm of the pulsed ICN laser. Again a delay of about 10 μsec is observed with a long emission pulse (~ 15 μsec), in fact quite reminiscent of the 200-μm group of HCN. Two sorts of explanation have been proposed to explain the time behavior of the output. The first type of explanation is concerned with the effect of the plasma electrons on the output delay. The second type of explanation is in terms of the chemical processes that can occur in the laser.

2.4.5.2.1. PLASMA ELECTRON AND RESONANT CAVITY EFFECTS.   McCaul has carried out some interesting experiments[47] concerning the effect of the plasma electrons in pulsed submillimeter lasers upon the delay of the output pulse. He has shown that the plasma forms a temporarily divergent refractive element. The divergence introduced can be strong enough to prohibit stable laser modes. As electron–ion recombination proceeds, the cavity rapidly becomes stable and Q-switching of the laser occurs. This effect has been referred to as *self Q-switching*.

A further effect occurs for a uniform radial distribution in the plasma. As the cavity length is increased near cavity resonance, the delay becomes shorter. This has been attributed by Steffen and Kneubuhl[56] to the effects of a time-dependent refractive index. McCaul[47] has extended this idea and shown that it can be correlated with the electronic contribution to the index of refraction as deduced from electron density measurements. Because the electron density is declining in the plasma afterglow, the index of refraction at the tube center is increasing toward unity [see Section 2.4.7.3]. This means that the apparent cavity length nL is increasing toward its vacuum value L. Thus we can picture the cavity mode being swept over the gain profile from high to low frequency.

---

[56] H. Steffen, B. Keller and F. K. Kneubuhl, *Electron. Lett.* **3**, 562 (1967); H. Steffen and F. K. Kneubuhl, *IEEE J. Quantum Electron.* **QE 4**, 992 (1968).

Measurements by microwave techniques of the electron density change during a laser pulse, showed that the change of refractive index was sufficient, for a typical laser, to ensure that the apparent cavity length changed by one longitudinal mode spacing.

For the HCN 337-μm emission, Yamanaka and his colleagues[42] report structure to the output pulse and that this structure was pressure dependent. Fine structure of a nonrepetitive nature (i.e., "spiking") was reported for the water vapor laser by Jeffers and Coleman[57] who also noted, as have several others, that the pulse shape varied in an erratic pattern from pulse to pulse. With the incorporation of a diffraction grating in the laser cavity which restricted the laser to oscillation on only one line at a time, they found that, with optimum tuning of the cavity length, pulses were obtained which were completely reproducible and which showed no "spiking." Jeffers[52] has concluded that the spiking arises from mode–mode interactions on a single emission line, whereas the erratic pulse envelopes are mainly due to line–line interactions in the active medium.

2.4.5.2.2. CHEMICAL AND MOLECULAR PROCESSES INVOLVED IN THE INVERSION MECHANISM. Although the identity of several of the molecular species that are responsible for many of the stimulated emission lines in the submillimeter region has now been established, the mechanism by which population of the energy levels occurs is not very well understood. The laser about which most is known is the H_2O laser. Jeffers and Coleman[58] have studied the relaxation behavior of 17 lines using a split discharge laser. Two discharges were incorporated in the same resonator and pulsed at varying times. The laser signal produced by the first discharge then probed the decaying second discharge at subsequent time intervals. They came to the conclusion that after the current pulse excitation, the rotational levels of ν_1, $2\nu_2$, and ν_3 came to rapid equilibrium with a time constant of about 0.5 μsec. A laser line involving a pure rotational transition would be strongly attenuated if passed through the gas at this time. Next, the vibrational states ν_1, $2\nu_2$, and ν_3 relax with a much longer time constant in the region of 100 μsec, causing the absorption to decay. Also the ν_1, $2\nu_2$, and ν_3 states may be fed from higher-lying, vibrationally excited levels and/or by the OH radical. The evidence for the participation of some process other than relaxation of the low-lying vibrational levels is the observation that the

[57] W. Q. Jeffers and P. D. Coleman, *Appl. Phys. Lett.* **10**, 7 (1967).
[58] W. Q. Jeffers and P. D. Coleman, *Appl. Phys. Lett.* **13**, 250 (1968).

relaxation time constant increases as the water vapor pressure increases. This is contrary to what would normally be expected of collisional relaxational processes in a gas, but is consistent with the decay constant behavior of the OH radical observed using the OH ultraviolet emission from the laser. Further evidence that OH may play a part in the lasing process are the experiments of Duxbury,[59] which showed that intense water vapor laser emission could be observed from lasers fueled by a mixture of ammonia and nitrogen dioxide. This mixture is known from the work of Del Greco and Kaufman[60] to be a very good source of OH radicals and with it as the fuel, direct excitation of water vapor molecules is excluded.

A rate equation approach to the study of oscillations in a water vapor laser has been developed by Takeuchi et al.[61] They have extended the work of Pollack[26] on rotation–vibration interactions of both the Fermi and the Coriolis type. On the basis of the assumption that the rotational relaxation time is much shorter than the duration of the exciting pulse, they have derived expressions for the output power in the steady state as a function of pumping rate and the initial oscillation condition. The effect of multiwavelength oscillation has also been obtained. The theory fits the experimental results quite well.

In the case of the HCN laser much less is known about the processes involved. However, it is known that the best sources of the emission are not discharges in HCN itself, but rather in mixtures of methane and either ammonia or nitrogen, and hence direct excitation of HCN is eliminated. Since it is known that CN is produced in these discharges, and that CN readily abstracts hydrogen from ammonia or methane to form HCN, it is possible that CN plays the same role in the HCN laser that OH may in the H_2O laser. The delayed pulses observed by Jones and Bradley and their collaborators may be evidence for participation of CN in the laser mechanism, although an alternative explanation has been proposed in Section 2.4.5.2.1. In the case of the ICN laser, the delay and pulse shape are unaffected in going from pure ICN to a mixture of I_2 and C_2N_2 as fuel. Thus it is possible that in this laser also, chemical recombination is involved.

The SO_2 and H_2S lasers (see Section 2.4.9) operate under much more stringent conditions than do the HCN, H_2O, or ICN lasers. It is possible

[59] G. Duxbury, unpublished work (1968).

[60] F. P. Del Greco and F. Kaufman, Discuss. Faraday Soc. 33, 139 (1962).

[61] V. Takeuchi, T. Kasuya, and K. Shimoda, Jap. J. Appl. Phys. 9, 1119 (1970).

that in the SO_2 and H_2S lasers direct excitation takes place, since lasing is only observed with very gentle excitation. Furthermore, H_2S once dissociated into SH and H does not easily reform, but tends to decompose still more forming sulfur which coats the walls and the electrodes of the laser.

2.4.5.3. Cascade and Competitive Emission. Cascade emissions are well illustrated in the HCN and DCN systems (see Figs. 1 and 2). They arise because the intense primary radiation produces further population inversions either by overpopulating an upper level (e.g., the 373-μm line of HCN) or else depopulating a lower level (e.g., the 284-μm line of HCN). It follows that they will only be observed when the cavity length is such that the cavity is resonant for both the primary and the cascade lines. Competitive emissions are commonplace in the H_2O system. Pollack et al.[62] reported competitive effects in the CW H_2O laser in 1967. Their experimental arrangement featured a normal, nonfrequency selective laser, which they scanned by translating one mirror of the cavity. Cavity mode patterns were observed which showed periodic variations in the intensity of various peaks. A spectacular example was for the 79.106-μm line and the 78.455-μm line. Because of the closeness of the wavelengths, the entire mode patterns effectively coincide around certain mirror positions. This overlap was sufficient that oscillation on the 79.106-μm line was entirely suppressed over five orders on each side of the best coincidence. Similar observations laid the groundwork[63,64] for the final successful identification of the lasing transitions. The 79.106-μm and 79.106-μm competition arises because the two lines go to a common ground state $(020\ 8_{35})$ of the $H_2^{16}O$ molecule. With the use of a frequency selective cavity,[52] competition can be eliminated and many lines that are normally weak because of competition can be enhanced often by several orders of magnitude. As extreme examples of this, the lines at 34.60 and 48.19 μm, which are ordinarily hardly observable, become quite strong in frequency selective operation. The latter line is in competition with 47.244 (common lower level) and 23.365 μm (common upper level).

2.4.5.4. Q-Switching of Submillimeter Lasers. Q-switching or Q-spoiling are names used for processes that prevent the laser cavity from being

[62] M. A. Pollack, T. J. Bridges, and W. J. Tomlinson, Appl. Phys. Lett. 10, 253 (1967).
[63] D. P. Akitt and W. Q. Jeffers, J. Appl. Phys. 40, 429 (1969).
[64] T. Kasuya, N. Takeuchi, A. Minoh, and K. Shimoda, J. Phys. Soc. (Japan) 26, 148 (1969).

resonant for all but a very short time. The laser is therefore constrained to radiate in very short (<1 μsec) pulses and these are consequently extremely intense and are for this reason sometimes called "giant" pulses. Two methods are in general use, mechanical and passive (or absorptive) Q-switching. In the former, one mirror of the cavity is rotated about a vertical axis at a high speed and will therefore only be in the correct position for resonance for a short time. In the latter an absorbing medium is placed in the cavity and will become saturated due to the intense fields. The mechanical method only relies on the lasing levels having a reasonably long relaxation time and mechanical Q-switching of CO_2 lasers is readily achieved with a rotating mirror since the relaxation time is some hundreds of microseconds. Patel,[65] for example, has rotated the mirror at 120 Hz and observed "giant" pulses each 250 nsec wide and with a peak power of 3 kW. The passive method depends, in addition to a relaxation time, on the initial radiation intensity. The phenomenon of saturation,[66] which is the observation of intensity-dependent absorption coefficients, arises because the incident radiation may be transferring molecules by stimulated absorption from the lower to the upper state at a rate comparable to that at which they are relaxing by collisional or radiative processes. The ultimate state would be when the populations became equal, for then α would be zero. The proper line shape equations, taking account of saturation, feature a saturation parameter β given by

$$\beta = I\tau\alpha_{max}/nh\nu_0, \qquad (2.4.9)$$

where I is the incident power in ergs per square centimeter per second, τ is the relaxation time, α_{max} is the peak absorption coefficient of the line in question, n is the number of molecules per cubic centimeter, and ν_0 is the line-center frequency. If β is comparable to, or greater than, unity, saturation effects will be observed. Wood and Schwarz calculated[67] that for SF_6, whose ν_3 band overlaps the CO_2 laser frequency, β should be 32 for a pressure of 1 Torr and an intensity of 5 W, focused down into an area of 0.03 cm². Passive Q-switching should therefore occur and was indeed observed when SF_6 gas was included in the cavity of a frequency-selective laser. Pulses were observed having a half-width of the order 1 μsec, a peak power of 1 kW, and a repetition frequency between 10^3 and

[65] C. K. N. Patel and E. D. Shaw, *Phys. Rev. Lett.* **24**, 451 (1970).
[66] See for example C. H. Townes, and A. L. Schawlow, "Microwave Spectroscopy," p. 371. McGraw-Hill, New York, 1955.
[67] O. R. Wood and S. E. Schwarz, *Appl. Phys. Lett.* **11**, 88 (1967).

10^4 Hz. It was necessary to use a frequency selective cavity, for otherwise the presence of the absorbing gas merely transferred the laser to CW operation on the R branch lines since the absorption coefficient is less for these. It is now known that a large number of vapors will passively Q-switch the CO_2 laser. CH_3F is very good (see also Section 2.4.9.2) and PF_5 remarkably so, since it will Q-switch virtually all the lines of the 10.6 μm P branch.[68]

Attempts to obtain higher peak intensities from the HCN and H_2O lasers by Q-switching them have proved disappointing. Rotating mirror methods were tried[38] by the group at NPL but when CW lasers were Q-switched, peak powers only twice as large as those readily obtainable from pulsed lasers were observed. The much more difficult experiment of Q-switching pulsed lasers gave an improvement by a factor up to 10, but the complexity of the experiment was not justified by the final outcome. These results must mean that the relaxation time is quite short, in fact of the order 10 μsec. This is perhaps reasonable when one considers the small energy differences between rotational levels as compared with that between the vibrational levels of the CO_2 laser.

2.4.6. Velocity of Light Determinations

2.4.6.1. Introduction.   The direct method by which the velocity of light was inferred from the time taken by a radiation pulse to travel a known distance is now only of historical interest. Used in reverse, however, as a means of distance measurement *knowing* the velocity of light, this type of technique has recently been rejuvenated not only for determining cosmic distances but also for very accurate metrology on the surface of the earth. The modern methods for determining c all rely on the equation

$$\nu\lambda = c, \qquad (2.4.10)$$

where ν is the time frequency and λ is the wavelength of a suitable highly monochromatic source. An alternative way of writing (2.4.10) is

$$\nu = c\bar{\nu}, \qquad (2.4.11)$$

where $\bar{\nu}$ is the wave number of the radiation. Wave numbers are the observed quantities in infrared spectroscopy, whereas time frequencies are usually observed in microwave spectroscopy. It follows that if a

---

[68] T. Y. Chang, C. H. Wang, and P. K. Cheo, *Appl. Phys. Lett.* **15**, 157 (1969).

quantity such as B_0 has been observed in both, the velocity of light may be inferred. As an example of this, B_0 for $HC^{14}N$ has been found to be[69]

$$B_0 = 1.478218 \pm (8 \times 10^{-7}) \quad cm^{-1}$$

which, when combined with the value in time frequency given earlier, yields

$$c = (2.997932 \pm 0.000016) \times 10^{10} \quad cm/sec.$$

Nearly all the error in this determination comes from the wave number measurement since time frequencies are now readily available to an accuracy of one part in 10^8. Unfortunately there does not seem to be any prospect for a significant improvement in wave number determinations at present, so this *band spectrum method*, as it is called, has fallen into abeyance.

The development of powerful, highly monochromatic, and coherent stimulated emission devices throughout the infrared has renewed interest in the direct method [Eq. (2.4.10)]. Basically the position is that we have a microwave standard of time and a visible region standard of length. The time standard defines a transition between two hyperfine levels of cesium to occur at 9.192631770 GHz and the length standard defines the meter to be that length which contains 1,650,763.73 wavelengths of the unperturbed transition $2p_{10} \to 5d_5$ in ^{86}Kr. Immediate determinations of the velocity of light by measuring the wavelength of the time standard or the frequency of the length standard are not possible at the moment; the former because of diffraction difficulties and the latter because of the absence of direct frequency measuring methods for the visible region. Froome[70] in 1958, worked at high (4 mm) microwave frequencies where the diffraction problems are not too severe and yet where frequency measurements are still relatively straightforward and obtained the result

$$c = (2.997925 \pm 0.000001) \times 10^{10} \quad cm/sec$$

which agrees with the accepted result[71]

$$c = (2.9979266 \pm 0.0000009) \times 10^{10} \quad cm/sec.$$

[69] D. H. Rank, A. H. Guenther, J. N. Shearer, and T. A. Wiggins, *J. Opt. Soc. Amer.* **47**, 148 (1957); D. H. Rank, *J. Mol. Spectrosc.* **17**, 50 (1965).

[70] K. D. Froome, *Nature (London)* **181**, 258 (1958); *Proc. Roy. Soc. (London)* **A246**, 109 (1959).

[71] E. R. Cohen and J. W. M. Dumond, *Rev. Mod. Phys.* **37**, 537 (1965).

It would be most desirable to have the value of c accurate to one part in 10^8, for then it would be possible to dispense with the length standard altogether. To achieve accuracy of this order, still shorter wavelengths must be chosen, so that the diffraction uncertainties will be acceptably small. The molecular lasers radiating between 337 and 9.6 µm are attractive candidates, the CO_2 laser is particularly so since its short wavelength guarantees small diffraction errors and the high power levels available augur well for high-precision frequency measurements.

2.4.6.2. The Determination of the Velocity of Light Using Submillimeter Lasers.   The technique of frequency measurement in the infrared depends on the construction of a chain of substandards extending from the accurately known low-frequency primary standards to higher and higher frequencies. The first measurement of a submillimeter laser frequency came in 1967 from Hocker and his colleagues.[27] An experimental arrangement, similar to theirs, used at NPL by Bradley and Knight is shown in Fig. 8. The 1-MHz standard frequency is successively multiplied to 18.6 GHz and this is mixed with the output of the 74.4-GHz klystron in a crystal diode. The beat note between the two is fed via a phase-lock loop back to the klystron. In the phase-lock loop, an input signal from a 30-MHz crystal oscillator serves to provide a correcting dc voltage to the klystron if the beat note drifts away from 30 MHz. In this way a

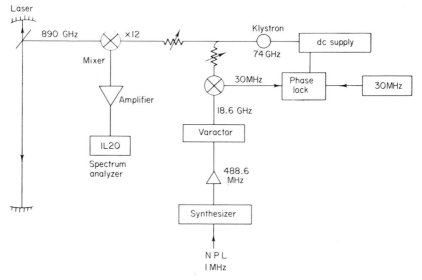

FIG. 8. Schematic of the apparatus used at NPL by Bradley and Knight to measure the frequency of the HCN laser emission.

highly stable output of known frequency is obtained from the klystron. The 74.4-GHz signal is mixed with the laser radiation in another crystal and the beat notes analyzed with a spectrum analyzer. Bradley and Knight[72] find that the frequency of the 337-μm radiation from HCN is 890.7602 ± 0.0002 GHz. The HCN laser is a most important link in the chain of frequency measurement because of its power and wavelength, there being no other suitable source in the 300-μm region. However, no Lamb dip (see later) has yet been observed with this laser and the problem of stabilizing its frequency output is severe. Two possibilities arise: (1) to lock it to an artificial narrow dip provided by a suitably absorbing gas inside the cavity (see Section 2.4.7.6), and (2) to stabilize the output frequency by very careful construction and operation of the laser. Fuller at NPL has shown[73] that if the discharge conditions are set such that the striations are very stable, the output frequency is also stable. Water-cooled electrodes are valuable in this connection as is the use of carefully controlled flow rates of fuel and good quality electrical supplies. Fuller[73] has mixed the radiation from two separate lasers and Bradley[74] has mixed two different modes from a single laser and in both cases the line widths of the difference notes were less than 2 kHz. Strauch[75] has analyzed the widths of the difference notes produced by apparatus similar to that of Fig. 8 and again finds a line width of less than 5 kHz. Unfortunately, although the emission line width is amply narrow enough for accurate frequency measurement, the actual emission can occur anywhere within the Doppler-broadened envelope where the gain is sufficient. This means that without special precautions the laser frequency can vary by some megahertz. The best experimental practice is to use servo-control of either the cavity length or else of the laser current to hold the frequency constant, and at the moment this gives a frequency accuracy of 0.2 MHz, which is of the order one part in 5×10^6. With the H_2O laser it is possible to do rather better. Pollack et al. observed[76] a Lamb dip on the 118.6-μm line of H_2O, and Frenkel et al.[77] followed this up with a direct frequency measurement of the laser emission when locked to the Lamb dip. They find the frequency to be 2527.9528 GHz

[72] C. C. Bradley and D. J. E. Knight, Phys. Lett. **32A**, 59 (1970).
[73] D. W. E. Fuller, unpublished work.
[74] C. C. Bradley, unpublished work.
[75] R. G. Strauch, Electron. Lett. **5**, 246 (1969).
[76] M. A. Pollack, T. J. Bridges, and A. R. Strand, Appl. Phys. Lett. **10**, 182 (1967).
[77] L. Frenkel, T. Sullivan, M. A. Pollack, and T. J. Bridges, Appl. Phys. Lett. **11**, 344 (1967).

± 0.1 MHz, which represents an accuracy of one in 4×10^8. Since 1967 the frequency range has been extended still further. Hocker and co-workers[78] have measured the frequency of the 84-μm line of D_2O by detecting the beat note between this radiation and the fourth harmonic of the 337-μm HCN laser radiation. They find $v = 3557.143$ GHz ± 2 MHz. Unfortunately since the HCN laser is involved in the measurement chain, high accuracy is not yet possible for this line.

The range of direct frequency measurement has been enormously extended by the development[79] of the metal–metal diode that has been shown to give video detection at the CO_2 laser frequency. Since frequency multiplication at still higher frequencies is possible using bulk non-linearities, we now have a real prospect of a frequency multiplication chain extending from radio to optical frequencies. Evenson and his colleagues[80] have measured the frequencies of the 28- and 78-μm lines of H_2O by beating these with harmonics of the 337- and 373-μm lines of HCN. The frequencies were 10.718073 THz ± 2 MHz and 3.821775 THz ± 3 MHz, respectively. Daneu et al.[79] have continued this chain building by mixing the 28-μm line of H_2O with the radiation from a CO_2 laser. The $R(10)$ line of the laser lies below and the $R(12)$ line lies above the third harmonic of the 28-μm line, the separations being -19.950 and ± 21.866 GHz, respectively. When the triple mixing experiment involving the CO_2 and H_2O laser radiation plus additional radiation from a K band (29 GHz) klystron was performed, strong beat notes were observed for klystron frequencies of the above values. Combining these results with those of Evenson et al.,[80] we have

$$R(12) = 32.176085 \quad \text{THz} \pm 6 \quad \text{MHz},$$

$$R(10) = 32.134269 \quad \text{THz} \pm 6 \quad \text{MHz}.$$

Wavelength measurements have not been performed as extensively nor as accurately as have frequency measurements. The wavelength of the 337-μm line of HCN has been determined to no better than a few parts in 10^5 though undoubtedly, better accuracy is possible. Daneu and his colleagues[81] have measured the wavelengths of three lines of

[78] L. O. Hocker, J. G. Small, and A. Javan, Phys. Lett. **29A**, 321 (1969).

[79] V. Daneu, D. Sokoloff, A. Sanchez, and A. Javan, Appl. Phys. Lett. **15**, 398 (1969).

[80] K. M. Evenson, J. S. Wells, L. M. Matarrese, and L. B. Elwell, Appl. Phys. Lett. **16**, 159 (1970).

[81] V. Daneu, L. O. Hocker, A. Javan, D. Ramachandra Rao, and A. Szoke, Phys. Lett. **29A**, 319 (1969).

H_2O and one line of D_2O to high accuracy taking care to reduce experimental error to a minimum. They used a Michelson interferometer with helium–neon laser fringes as the markers. This, as is well known, eliminates periodic errors in the determination of the displacement of the moving mirror. The helium–neon laser itself was stabilized on the Lamb dip and had a wavelength of 6329.915 Å in vacuo determined in a separate experiment against the [86]Kr standard. The experimentalists took especial precautions to ensure that the far-infrared and visible laser beams were accurately parallel and employed wide apertures to reduce the diffraction errors to less than one part in 10^8. Their final results were accurate to one part in 10^6 but higher accuracy is possible in principle. The wavelengths and values for c are

H_2O: 118.59085 ± 0.0002 µm, $c = (2.997922 \pm 0.000006) \times 10^{10}$ cm/sec,

D_2O: 84.27907 ± 0.0004 µm, $c = (2.997927 \pm 0.00002) \times 10^{10}$ cm/sec,

H_2O: 47.46316 ± 0.0001 µm,

H_2O: 27.97080 ± 0.00004 µm, $c = (2.997931 \pm 0.000005) \times 10^{10}$ cm/sec.

The CO_2 laser offers the best prospect for c determinations to an accuracy of one part in 10^8 but so far the wavelengths have not been measured to this accuracy and the frequencies of the lines are only known to one part in 10^6. It is most interesting to note, however, that McCubbin[82] has given the wave numbers of the $R(12)$ and $R(10)$ lines of CO_2 as 1073.271 and 1071.877, respectively, both ± 0.005 cm^{-1}. From these wave numbers and the absolute frequencies given earlier, we find

$$c = (2.997945 \pm 0.00002) \times 10^{10} \text{ cm/sec.}$$

This is not particularly accurate, but there is little reason to doubt that far more accurate wavelengths and frequencies will be measured for the CO_2 laser in the near future and that these will yield a value of c having the desired one part in 10^8 accuracy.

2.4.7. Spectroscopy with Submillimeter Lasers

2.4.7.1. Introduction. From Sections 2.4.3, 2.4.9, and the Appendix it can be seen that there now exists a range of emission lines from submillimeter lasers distributed throughout the far-infrared region. Un-

---

[82] T. K. McCubbin, unpublished work, but quoted in Daneu et al.[73]

fortunately, as discussed in Section 2.4.8, it is not possible at present to tune these lasers over more than a few megahertz on either side of the line-center frequency, and this limitation restricts the range of experiments that can be carried out to those in which a wide tuning range is unnecessary.

Lasers have been used to make spot measurements of both the absorption coefficients and the refractive index of liquids[83] and solids.[84] This data can then be used to place the broad-band results from interferometric techniques on an absolute scale. Although this application is very valuable, it must be realized that laser spectroscopy is in competition with conventional far-infrared spectroscopy and because of this the type of experiment which best utilizes the laser sources is that which requires a much higher resolution than is practicable using standard Michelson interferometers. The electric and magnetic resonance experiments fall into this category, and here the tuning is provided by varying the electric or magnetic field applied to the specimen as in conventional ESR and NMR spectroscopy. An alternative type of experiment is the use of "pressure scanning" to vary the line shape of a pure rotational transition, and hence the absorption coefficient at the laser frequency. The diagnosis of plasmas provides yet another good example and in this instance the variation in the optical parameters is provided by varying the electron density and temperature in the plasma. Examples of all these types of experiment will be discussed in this section.

2.4.7.2. The Determination of the Complex Refractive Index of Materials and Plasmas Using Submillimeter Lasers. The complex refractive index $\hat{n}$ of a medium is defined[85] by the relation

$$\hat{n} = n - i(\alpha/4\pi\bar{\nu}), \tag{2.4.12}$$

where n is the refractive index and α is the absorption coefficient at the wavenumber $\bar{\nu}$. This concept enables a theory of the propagation of an electromagnetic wave in a dispersive and absorptive medium to be developed. The real part n and the imaginary part α are connected by the Kramers–Kronig integral transforms. The absorption coefficient is measured by determining the attenuation of beam intensity after tra-

[83] J. Chamberlain, E. G. C. Werner, H. A. Gebbie, and W. Slough, *Trans. Faraday Soc.* **63**, 2605 (1967).

[84] J. Chamberlain and H. A. Gebbie, *Nature (London)* **206**, 602 (1965).

[85] J. Chamberlain, J. E. Gibbs, and H. A. Gebbie, *Infrared Phys.* **9**, 185 (1969).

versing a known distance in the medium, and the refractive index by comparing the apparent path length in the medium with the true path length.

The experimental layout of a typical system[83] used for the measurement of absorption coefficients using the null method is shown in Fig. 9. Radiation from the laser is modulated and divided into two beams, one of which passes, after focusing, through the sample to the Golay detector D_1, while the other constitutes a reference beam and is focused directly on to a second Golay detector D_2. The signal from D_2 is fed via a variable

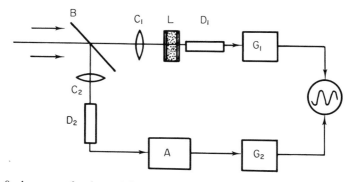

FIG. 9. Apparatus for determining the absorption coefficient of a liquid using a laser source [J. Chamberlain, E. G. C. Werner, H. A. Gebbie, and W. Slough, *Trans. Faraday Soc.* **63**, 2605 (1967)].

attenuator and the preamplifier G_2 on to one input of a differential amplifier, and the output from D_1 via a second preamplifier G_1 on to the second input. For a given thickness of sample the setting of the attenuator is adjusted, and also, if necessary, the gains G_1 and G_2 of the amplifiers altered so that a null reading is obtained. When this occurs, the intensities I_1 and I_2 leaving the beam divider are proportional to G_1/τ and G_2/A, where τ is the transmission factor of the sample and A is the "transmission factor" of the attenuator. Since $I_1/I_2 =$ constant and $\tau = \exp(-\alpha d)$, the absorption coefficient α (neper reciprocal centimeters) of the sample may be calculated by applying the least-squares method to the equation

$$\ln(AG_1/G_2) = \text{constant} - \alpha d, \qquad (2.4.13)$$

where d is the thickness of the sample. The use of several thicknesses of the sample is necessary to compensate for the back reflection in solid samples, and the effect of the windows of the absorption cells used for the studies of liquids and gases. The thickness of the absorbing layer is usually between 0.05 and 5 mm in the case of solids and liquids, but is 50 cm–100 m in the cases of gases. In order to achieve these long path lengths multiple traversal cells of the type described by White[86] are frequently employed. Fabry–Perot interferometers may also be used as long path cells for gases. They do, however, have high insertion losses and their principal value is in applications, such as magnetic resonance experiments, where it is desirable to keep the cell volume as small as possible.

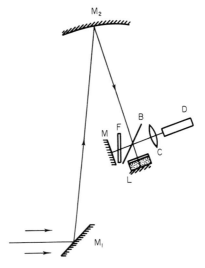

FIG. 10. Apparatus for determining the refractive index of a liquid using a laser source [J. Chamberlain, E. G. C. Werner, H. A. Gebbie, and W. Slough, *Trans. Faraday Soc.* **63**, 2605 (1967)].

The layout for refractive index experiments[83] is similar to that for the absorption measurements, and is shown in a frequently used form in Fig. 10. Radiation from the laser is modulated by a mechanical chopper and is then collimated by the mirrors M_1 and M_2 to form a convergent* beam which is split by the Melinex beam-divider into two beams passing, respectively, to the plane mirror M and to the similar mirror mounted

---

[86] J. U. White, *J. Opt. Soc. Amer.* **32**, 285 (1942); E. R. Stephens, *Infrared Phys.* **1**, 187 (1961).

---

* Corrections to take account of beam convergence have been discussed by Chamberlain, *Appl. Optics* **6**, 980 (1967).

at the rear of the sample. The radiation is recombined and the part not returning toward the laser is focused by the lens C on to the Golay detector D. When the thickness of the sample is changed, the path difference within the interferometer is changed and intensity variations (fringes) are observed at the detector D and displayed either on an oscilloscope or on a chart recorder. The transition from a maximum to a minimum intensity corresponds to a change in optical path length of one-half wavelength, so that the (arbitrarily numbered) Zth fringe is related to sample thickness d (in centimeters) by

$$Z = \text{constant} + 2n\bar{\nu}d, \qquad (2.4.14)$$

where, as before, $\bar{\nu}$ is the wave number of the radiation. n can be obtained by a least-squares fit to this equation. Since the contrast in the fringes is greatest when the loss in each arm of the interferometer is comparable, compensating thicknesses of the attenuator are introduced at F to balance the loss due to the sample.

Recently it has been shown,[87] that phase modulation of the radiation is superior to amplitude modulation and gives signal-to-noise improvements by factors as large as four. In this technique the modulation of the beam is achieved by oscillating the fixed mirror of the interferometer at a suitable low audio frequency through an amplitude A. The degree of modulation is wave-number dependent and is maximal when

$$\bar{\nu} = [8A]^{-1}. \qquad (2.4.15)$$

One can therefore choose A to be appropriate to the laser wave number.

In order to achieve a changing path difference with a solid specimen, Chamberlain rotated a disk of material very much larger than the beam, about an axis perpendicular to the beam. This is equivalent to continuously varying the thickness of a stationary plate. Chamberlain and Gebbie[84] determined the refractive indices of a number of solids at 337 μm in this way.

With liquid cells, it is possible to vary the spacing of the windows, and hence the path difference, continuously. The liquid specimen has to be plane parallel and to ensure this, rigid windows such as those made from crystal quartz are necessary. The rear window of the liquid cell is commonly front surface aluminized to serve as the fixed mirror of the interferometer as in Fig. 10. A more advanced design featuring the

[87] J. Chamberlain and others, *Infrared Phys.* **11**, 25 (1971).

Mach–Zehnder form of interferometer and phase-modulation[87] is shown in Fig. 11.

The change in path difference arising from the introduction of a gaseous sample is usually determined by allowing the gas to gradually fill a cell of known length up to a given pressure while counting the resulting fringes. This procedure may be carried out with a number of different cell lengths to eliminate end corrections.

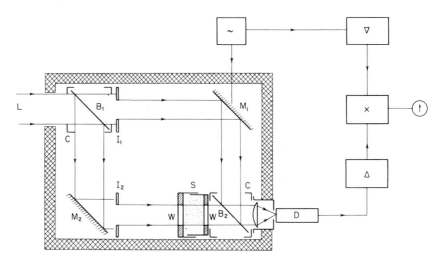

Fig. 11. Schematic layout of a phase-modulated Mach–Zehnder interferometer suitable for the determination of the optical constants of a liquid at a laser frequency.

The Mach–Zehnder form of interferometer has also proved particularly suitable for the study of plasmas.[88,89] The experimental arrangement used by Chamberlain and his colleagues[88] is shown in Fig. 12. The attenuation by the plasma is measured by blocking off one of the split beams with a shutter at S and then observing the decrease in the detected signal when the discharge is switched on. Transmissions varying from 0.95 to 0.08 were observed with a potassium seeded argon plasma 100 mm long. The phase drifts in the plasma (from which the refractive index follows) are measured in two stages. Firstly the amount of movement of mirror M_2 necessary to restore constructive interference at the detector when

[88] J. Chamberlain, H. A. Gebbie, A. George, and J. D. E. Beynon, *J. Plasma Phys.* **3**, 75 (1969).

[89] G. J. Parkinson, A. E. Dangor, and J. Chamberlain, *Appl. Phys. Lett.* **13**, 233 (1968).

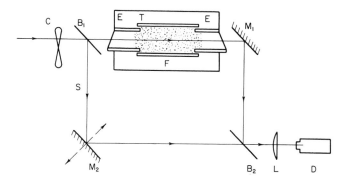

FIG. 12. Apparatus for determining the optical properties of a plasma at a laser frequency.

the discharge is switched on is measured. This gives the part of the phase shift that is not an exact number of wavelengths. To determine the remainder, phase shifts of identical plasmas varying only in length are observed, whence the integer follows from application of the method of exact fractions. What one does is to number the fringes so that the phase shifts are directly proportional to plasma length. Phase shifts as small as one tenth of a fringe can be detected, which gives the minimum concentration (see Section 2.4.7.3) of electrons that can be observed as 7×10^{12} cm^{-3}. Electron densities higher than 10^{16} cm^{-3} cannot be studied with a 337-μm laser since such a density leads to cutoff, i.e., the plasma acts like a metal and reflects back most of the incident power. The dc plasmas can be well studied with CW lasers and the relatively slow response Golay cell, but when we come to consider the technically interesting plasmas produced in nuclear fusion research, much faster response is necessary and the use of pulsed lasers and the Putley indium antimonide detector has been suggested.[89]

2.4.7.3. Some Experimental Results for $\bar{n}$ at Submillimeter Laser Wavelengths and Their Theoretical Interpretation. One of the first experiments performed[90] with an HCN laser was the determination of the absorption coefficient of the liquid halobenzenes at 29.712 cm^{-1}. The absorption and refraction of polar liquids at microwave frequencies had been subject of intense interest since experimental determinations could be compared with the theoretical predictions of Debye.[91] The Debye

[90] H. A. Gebbie, N. W. B. Stone, F. D. Findlay, and E. C. Pyatt, *Nature* (*London*) **205**, 377 (1965).

[91] P. Debye, "Polar Molecules." Chem. Catalogue Co., New York, 1929.

theory indicated that the complex dielectric constant

$$\hat{\varepsilon} = \varepsilon' - i\varepsilon'', \qquad (2.4.16)$$

where

$$\hat{\varepsilon} = (\hat{n})^2$$

for rigid dipoles attempting to rotate in a viscous medium, was given by the expression

$$(\hat{\varepsilon} - \varepsilon_\infty)/(\varepsilon_0 - \varepsilon_\infty) = 1/(1 + i\omega\tau), \qquad (2.4.17)$$

where ε_0 and ε_∞ are the static and high (i.e., infrared) frequency dielectric constants and τ is the relaxation time. Equation (2.4.17) can be rearranged to show that a plot of ε'' against ε' should be semicircular (the so-called *Cole–Cole plot*), and much effort was spent in testing the validity of this plot. It was found to be a good approximation for simple non-hydrogen-bonded liquids, but there was some discussion as to whether the points departed from the semicircle for very high frequencies. Poley suggested[92] that this was so and that the data indicated a high frequency (i.e., submillimeter) dispersion additional to the Debye dispersion. The absorption coefficient α calculated from Eq. (2.4.17) together with Eqs. (2.4.16) and (2.4.12) is found to rise monotonically and to eventually reach an asymptotically limiting maximum value α_∞. This result is not physically reasonable, since polar liquids are not opaque throughout the infrared, visible, and ultraviolet regions of the spectrum. The reason for the breakdown of Eq. (2.4.17) at high frequencies lies in the approximations made by Debye in his model, principally the neglect of inertial effects. Debye was aware of the shortcomings and specifically required that his equation should not be used for $\omega > \tau^{-1}$. Attempts have been made to introduce analytical terms to take account of the molecular moment of inertia but these have had only qualitative success. The best that one can deduce from a theoretical point of view is that α will reach a maximum near τ^{-1} and will then steadily decline to zero. For the halobenzenes τ is around 10^{-10} sec, so that, at 29.712 cm^{-1}, α should be somewhat below α_∞ calculated from microwave data. In fact $\alpha_{29.7}$ for chlorobenzene is found[90] to be 15.2 cm^{-1} whereas α_∞ is only 7 cm^{-1}. This observation established the validity of Poley's hypothesis and it was soon followed by broad-band interferometric spectroscopic studies which characterized the bands.

[92] J. Ph. Poley, *J. Appl. Sci.* **B4**, 377 (1955).

The absorption in chlorobenzene takes the form[93] of a broad continuous band which reaches an ill defined peak at 45 cm^{-1}. All other polar liquids show similar bands and so do nonpolar liquids except that in this case they are very much weaker.[94] Behavior of this type had been predicted[95] by Hill and following her model we call these bands "liquid lattice" bands. Chamberlain[96] has shown, on the basis of some work of Hill,[97] that the broad features can be thought of as made up from several heavily damped resonances. This picture is especially evident from the refraction spectrum, which itself is derived from dispersive Fourier transform spectroscopy coupled with accurate measurement of n at 29.712 cm^{-1}. The technique of using accurate absolute measurements at a laser wavelength to put relative data for a spectral region including this wavelength on an absolute scale has wide applicability for refraction spectra. Less obviously it is also occasionally useful for absorption spectroscopy. In the case of liquid water, the submillimeter absorption coefficients are so high that the thicknesses of specimens that give measurable transmission are too small to be determined with any accuracy. However, one can measure $\alpha_{29.7}$ with a relatively thick specimen and then, knowing this ($\alpha_{29.7}$ = 220 cm^{-1}), one can measure the thickness of the spectroscopic specimen by measuring the attenuation of the HCN laser radiation in passing through it. In this way absorption and refraction spectra throughout the submillimeter region have been obtained for liquid water.[98] It is interesting to note that the refractive index for water at 29.712 cm^{-1} and 20°C is 2.132, showing that the microwave dispersion, which is intense near 1 cm^{-1}, has almost disappeared at submillimeter wavelengths. Values of the two components of $\hat{n}$ for a wide range of liquids have been reported at 29.712 cm^{-1} by Chamberlain and his colleagues.[83,99,100]

The absorption and refraction of several solids have been observed at

[93] G. W. Chantry and H. A. Gebbie, *Nature (London)* **208**, 378 (1965).

[94] G. W. Chantry, H. A. Gebbie, B. Lassier, and G. Wyllie, *Nature (London)* **214**, 163 (1967).

[95] N. E. Hill, *Proc. Phys. Soc.* **82**, 723 (1963).

[96] J. Chamberlain, *Chem. Phys. Lett.* **2**, 464 (1968).

[97] N. E. Hill, *Chem. Phys. Lett.* **2**, 5 (1968).

[98] J. Chamberlain, G. W. Chantry, H. A. Gebbie, N. W. B. Stone, T. B. Taylor, and G. Wyllie, *Nature (London)* **210**, 790 (1966).

[99] M. Davies, G. W. F. Pardoe, J. Chamberlain, and H. A. Gebbie, *Trans. Faraday Soc.* **64**, 847 (1968).

[100] J. Chamberlain, unpublished work quoted in "Submillimeter Spectroscopy" (by G. W. Chantry). Academic Press, New York and London, 1971.

TABLE I. Absorption and Refraction Data for Some Solids at 29.712 cm^{-1} [a]

Material	Refractive index	Exponential absorption coefficient (cm^{-1}) $I/I_0 = e^{-\alpha x}$
Polyethylene	1.512	0.20
Polypropylene	1.500	0.27
TPX (poly 4-methylpentene-1)	1.456	0.31
Polystyrene	—	~2.0
Polytetrafluorethylene	1.391	—
Polyethylene terephthalate	1.69	~50.0
Polymethyl methacrylate	1.593	~5.0
Crystalline quartz	2.115	~0.1
Vitreous silica	1.93	1.8–2.6

[a] Data from the work of the NPL group, principally J. Chamberlain.

29.712 cm^{-1} by the submillimeter spectroscopy group at NPL. Some of their results[101] are listed in Table I.

The principal value of this data is to provide the basic information for decisions on the use of a given material for optical components in far infrared instrumentation. Thus for windows, lenses, etc. the low-loss nonpolar polymers are the choice, whereas for beam dividers the very lossy but more highly reflecting polyethylene terephthalate is chosen. The laser observations can only provide marginally valuable information for attempts to understand the submillimeter absorption of solids because the observed bands are very broad and can be well characterized by conventional spectroscopic methods.

When the pure rotational lines of gases come to be considered, however, the opposite is true, for at low pressures these resonances are very sharp. In general, the line-center frequency will not be within the tuning range of the laser, and although this is not a serious limitation for refraction spectroscopy, where already Chamberlain and co-workers have measured[102] the refractive index of air at 337 μm and Strauch and

[101] G. W. Chantry and J. Chamberlain, Far infrared spectra of polymers, in "Materials Science of Polymers" (A. D. Jenkins, ed.). North Holland Publ., Amsterdam, 1972.

[102] J. Chamberlain, F. D. Findlay, and H. A. Gebbie, *Nature (London)* **206**, 886 (1965).

colleagues have carried out an interesting intracavity dispersion measurement[103] on D_2O, it does limit the scope of absorption spectroscopy. For this reason most of the interest has been in the measurement of the absorption coefficients of various gases as functions of pressure and relating these to the predictions of pressure-broadening theory.

If the predominant mechanism contributing to the line shape is pressure broadening, the line profile will be Lorentzian in the region ~ 1 cm^{-1} from the center frequency. Thus for the transition $(i \rightarrow j)$, the absorption coefficient at wave number $\bar{\nu}$ is given by

$$\alpha_{ij}(\bar{\nu}) = \frac{A_{ij}}{\pi} \cdot \frac{\Delta \bar{\nu}_{ij}}{(\bar{\nu} - \bar{\nu}_{ij})^2 + (\Delta \bar{\nu}_{ij})^2}, \qquad (2.4.18)$$

where A_{ij} is the integrated line strength, $\bar{\nu}_{ij}$ is the line center frequency, and $\Delta \bar{\nu}_{ij}$ is the width of the line at half height. In a gas mixture, the line-width parameter $\Delta \bar{\nu}_{ij}$ arises from both self and foreign-gas broadening, and in the microwave region it has been found that the assumption

$$\Delta \bar{\nu}_{ij} = P_a \sigma_{ija} + \sum_k P_k \sigma_{ijk} \qquad (2.4.19)$$

fits most of the observed phenomena. Dropping the suffixes,

$$\Delta \bar{\nu} = P_a \sigma_a + \sum_k P_k \sigma_k, \qquad (2.4.20)$$

where P_a is the pressure of the absorbing gas in torrs, P_k is the pressure of one of the broadening gases, σ_a is the line-width parameter for self-broadening in reciprocal centimeter reciprocal torrs, and the σ_k are line-width parameters for foreign gas broadening in the same units. This linear treatment can readily be generalized to the situation where there are several overlapping lines. It is reasonable to suppose that these equations will also apply in the higher-frequency submillimeter region and the first investigation[104] here was a study of the absorption of the HCN 890.76-GHz radiation by D_2O vapor. The $6_{15} \rightarrow 6_{24}$ line of D_2O lies very close to this frequency and the experiments, assuming pressure broadening parameters derived from other experiments on both D_2O and H_2O, were interpreted to evaluate the frequency separation between the laser

[103] R. G. Strauch, D. A. Stephenson, and V. E. Derr, *Infrared Phys.* **9**, 137 (1969).

[104] C. C. Bradley, W. J. Burroughs, H. A. Gebbie, and W. Slough, *Infrared Phys.* **7**, 127 (1969).

and the line center frequency. However, recent work by Benedict et al.[105] and by Duxbury and Jones[106] has shown that the frequency separation derived in this way was only in qualitative agreement with the true frequency separation ($\bar{\nu}_{\text{laser}} - \bar{\nu}_{15 \to 24} = 0.01$ cm^{-1}) and the conclusion must be that simple pressure-broadening theory cannot be reliably used even at only 0.01 cm^{-1} from the center frequency. Work on overlapping lines by Duxbury and Burroughs[107] has shown, though, that qualitative information about pressure broadening can be obtained, even when there are as many as 20 lines involved in the absorption process. Burroughs et al.[108] and Birch et al.[109] have studied atmospheric absorption by water vapor in the so-called "window" regions. Here the absorption is many times greater than predicted by simple collision broadening theories and this enhancement has been explained as due to very strong intermolecular interactions resulting in the formation of short-lived dimers and also to quadrupole induced absorption.

Diagnosis of plasmas by means of submillimeter lasers is assuming some importance now that relatively dense plasmas are being studied under laboratory conditions. The plasma is characterized by a quantity called the *plasma frequency* given by[89]

$$\nu_{\text{p}} = (n_{\text{e}} \cdot e^2/\pi m_{\text{e}})^{1/2} \qquad (2.4.21)$$

where n_{e} is the number of electrons per cubic centimeter, e is the electronic charge (4.774×10^{-10} esu), and m_{e} is the electronic mass (9.00×10^{-28} gm). If the electron density is $10^{16} \cdot$ cm^{-3}, ν_{p} is approximately 10^{12} Hz, i.e., 33 cm^{-1}. When the electron collision frequency ν_{e} is much less than ν_{p}, the length of plasma d in which a wave of frequency ν is reduced to $1/e$th of its original amplitude is given by[89]

$$d = \frac{2c}{\nu_{\text{e}}} \left(\frac{\nu}{\nu_{\text{p}}} \right)^2 \left[1 - \frac{\nu_{\text{p}}^2}{\nu^2} \right]^{1/2}. \qquad (2.4.22)$$

This quantity drops to zero for $\nu = \nu_{\text{p}}$ and is imaginary for $\nu < \nu_{\text{p}}$. The significance of d becoming imaginary is that the plasma reflects all

[105] W. S. Benedict, S. A. Clough, L. Frenkel, and T. E. Sullivan, *J. Chem. Phys.* **53**, 2565 (1970).

[106] G. Duxbury and R. G. Jones, *Mol. Phys.* **20**, 721 (1971).

[107] G. Duxbury and W. J. Burroughs, *J. Phys. B At. Mol. Phys.* **3**, 98 (1970).

[108] W. J. Burroughs, R. G. Jones, and H. A. Gebbie, *J. Quant. Spectrosc. Radiat. Transfer* **9**, 809 (1969).

[109] J. R. Birch, W. J. Burroughs, and R. J. Emery, *Infrared Phys.* **9**, 75 (1969).

electromagnetic waves of frequency less than ν_p—a familiar illustration being the reflection of radio waves by the plasma in the ionosphere. From this one can deduce that an electron density of 10^{16} cm^{-3} is the maximum that can be probed with 337-μm radiation. Within this restriction (and plasmas having $5.10^{15} > n_e > 5.10^{13}$ cm^{-3} have in fact been studied), it is possible to deduce not only the electron density but the collision frequency, and thus the electron temperature. The sensitivity of the method improves as ν approaches ν_p and therefore diagnoses with submillimeter radiation are superior (i.e., require much less length of plasma) than those made using He–Ne 3.39 μm laser radiation.

2.4.7.4. Electric Resonance Spectroscopy Using Submillimeter Lasers.
When a polar gas is subjected to an electric field, the pure rotational lines split up into a number of components whose separations depend on the field strength. If the unperturbed line frequency is near to the laser frequency, then at a sufficient field strength one of the components will come into exact coincidence with the laser and an absorption peak will be observed. This is called *electric resonance spectroscopy* and the splitting arises from the Stark effect, which is due to the removal of the M quantum number degeneracy of each J level by the field. The presence of the field removes the isotropy of space as far as the rotating dipole is concerned. If a molecule possesses a permanent magnetic moment—as, for example, O_2—then a similar splitting, due to the Zeeman effect, arises in a magnetic field and we have *paramagnetic resonance spectroscopy*. This phenomenon is also observed with semiconductors, where the continuum of states in the valence and conduction bands is destroyed by the presence of a magnetic field and the bands break up into a series of discrete levels. When the splitting of these levels is such that transitions between them coincide in frequency with the laser, we again have an absorption maximum.

None of these phenomena is peculiar to the submillimeter region and all are well known at lower frequencies. It is possible in the microwave region to tune the klystron sources by as much as 10% of the center frequency, or to use backward wave oscillators that have an even wider range. It is therefore usually possible to scan the entire region of interest, and for this reason the electric resonance experiments can be carried out with a fixed (dc) field and a variable frequency source. The sensitivity can be improved by employing modulation of the electric field (i.e., Stark modulation) and the signal-to-noise ratio can be improved by using synchronous detection. In the submillimeter version the source frequencies are fixed and so the electric field must be swept. This limits the amount of the spectrum that can be observed, for at a sufficiently high electric field the vapor will break down and arc discharges occur.

Paramagnetic resonance experiments are usually carried out, at microwave frequencies, by varying the magnetic field at a fixed klystron frequency in order that superheterodyne detection techniques may be employed. This arrangement applies to the submillimeter experiments except that, so far, superheterodyne methods have not been used.

If a molecular absorption line lies close to a laser transition, the application of an electric field will shift the Stark components of the absorption line either toward or away from the laser transition. If the components are shifted toward the laser transition, an increase and subsequent decrease in absorption will be observed as the component moves through resonance with the laser frequency. If spectra are obtained at low pressure, i.e., 0.1 Torr, there is likely to be very little absorption at the laser frequency due to the unperturbed absorption line because the line width at this pressure is small. In this case the only sort of absorption change that can occur is a resonant increase of the type described above. However, if there is appreciable absorption due to the unperturbed line, i.e., at pressures greater than 1 Torr, then the absorption can either increase or decrease on application of the electric field, depending on whether the envelope of the Stark components moves toward or away from the laser frequency. If the net absorption of radiation by the molecule in the unperturbed state is due to several molecular transitions close to the laser transition, then the Stark components of the various lines may shift either toward or away from the laser line. In this case the resulting change in absorption will be the resultant of the increases and decreases due to the contributing lines, and the result may be difficult to interpret.

A typical experimental arrangement used to study[106] the Stark effect in D_2O is shown schematically in Fig. 13. The continuous-wave HCN laser had a 4 m long 10 cm diameter planoconcave cavity and was operated at 2.5 kV and 0.35 A. The cavity was tuned to give the 890-GHz (337-μm) emission line of HCN. A polyethylene terephthalate (Melinex) beam splitter was incorporated at 45° into the laser cavity. The laser output when taken via this beam splitter was plane-polarized with the electric vector in the horizontal direction as indicated in Fig. 13. The radiation was directed through a 40 cm long absorption cell in which were placed two plane-parallel Stark electrodes 2.5 cm wide and of 30-cm effective length. The tapered ends shown in Fig. 13 greatly assisted the introduction of radiation into the space between the electrodes.

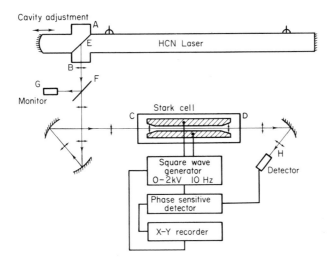

FIG. 13. Apparatus used by G. Duxbury and R. G. Jones [*Mol. Phys.* **20**, 721 (1971)] to study the Stark spectra of vapors in the far infrared.

In the orientation indicated in Fig. 13, the electric vector of the radiation is parallel to the applied Stark field in which case only the $\Delta M = 0$ transitions can be observed. In order to observe the $\Delta M = \pm 1$ transitions, the electrodes have to be rotated by $90°$. The experiments were carried out with two electrode spacings. A spacing of 300 μm allowed both directions of polarization to be transmitted through the cell but limited the number of Stark components observable due to the electrical breakdown of the gas. A spacing of 185 μm increased the breakdown field but cut off the direction of polarization perpendicular to the Stark field due to waveguiding effects. The spacers used were fused silica at 185 μm and Melinex at 300 μm.

The Stark components were observed by the Stark modulation method employing a zero-based square-wave modulation of the applied electric field. The modulation signal was detected by a Golay cell and a phase sensitive detector system and displayed as a function of the applied electric field on an X–Y recorder. The sense of the signal, i.e., whether it was due to increasing or decreasing absorption, was unambiguously determined by using a double-beam method similar to that described in Section 2.4.7.2. Stark splittings of the $J = 2 \rightarrow J = 3$ and $J = 5 \rightarrow J = 6$ rotation–inversion transitions of NH_3 have been detected by Uehara et al.[110]

[110] K. Uehara, T. Shimizu, and K. Shimoda, *IEEE J. Quantum Electron.* **QE 4**, 728 (1968).

The unperturbed transitions consist of widely (~ 40 GHz) separated doublets each component of which is multiple because of the removal of the K degeneracy by the centrifugal distortion term $2D_{JK}K^2(J + 1)$ in the expression for the line frequencies of the pure rotation spectrum of a symmetric rotor. The D_2O emission line at 171.6 μm (1747 GHz) lies some 16.3 GHz below the lower frequency component of the $2 \rightarrow 3$ line in ammonia so that on the application of a field of the order of 40 kV cm⁻¹, the Stark splittings would cause absorption lines to move over the laser frequency. Uehara and his colleagues used relatively high (~ 20 Torr) pressures of ammonia, which limits their resolving power and, in addition, using pulsed D_2O lasers, they were unable to employ the double beam and Stark modulation techniques so their sensitivity was not particularly high. Nevertheless they observed a distinct maximum of absorption with a field of 42.5 kV cm⁻¹. The D_2O laser lines at 84.1 and 84.29 μm (3564 and 3557 GHz) lie in between the two components of the $5 \rightarrow 6$ transition and in this case the expected behavior follows the more complex pattern mentioned earlier. The transmission of the gas was found to decrease monotonically (but asymptotically less and less)

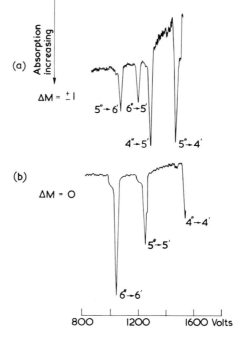

FIG. 14. Stark spectra of D_2O vapor at 890 GHz.

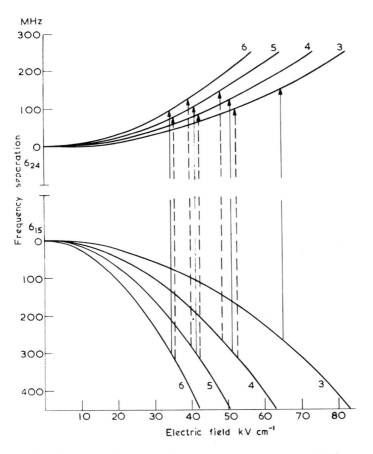

FIG. 15. Stark effect of the $6_{15} \rightarrow 6_{24}$ pure rotation line of D_2O.

with the field strength up to the highest (50 kV cm⁻¹) fields available to the experimenters.

More recently Duxbury and Jones[106] have obtained spectra of both D_2O and ND_3 using CW HCN and DCN lasers at 890 and 1540 GHz, respectively. Stark modulation was employed and consequently quite low pressures could be used. For D_2O these lay in the range 20–100 mTorr, and for ND_3 approximately 1 Torr. Typical spectra of the $6_{15} \rightarrow 6_{24}$ pure rotation transition in D_2O in both parallel and perpendicular polarization configurations are given in Fig. 14, and an energy level diagram is shown in Fig. 15. It is hoped by improving the uniformity of the electric field, with only a small square-wave modulation superimposed, that this technique will be applicable to precision measurements.

These will be of dipole moments and possibly polarizabilities of those light molecules which possess a rich spectrum in the submillimeter region. Perhaps it may also prove possible to use the Lamb-dip technique demonstrated at 10.6 μm by Brewer et al.[111] to resolve the hyperfine structure of far-infrared lines.

2.4.7.5. Paramagnetic Resonance Spectroscopy with Submillimeter Lasers. When a magnetic field is applied to a paramagnetic molecule in the gas phase, a splitting of the rotational energy levels occurs similar to that described in Section 2.4.7.4 for the case of an electric field. Again when the field is strong enough that one of the components comes into co-incidence with the laser frequency, we will observe a resonant absorption spectrum quite analogous to that described earlier for electric resonance. At low field strengths, the splitting of the lines is a linear function of H, but at higher field strengths, depending on the value of M, the splitting becomes a quadratic function of H. This is quite analogous to the Paschen–Back effect for the case of atoms.

Most of the work so far in this field has been carried out by Evenson and his collaborators. The first molecule studied[112] was oxygen, whose $N = 3$, $J = 4 \rightarrow N = 5$, $J = 5$ transition lies at 27.826 cm^{-1}, fairly close to the HCN laser line at 29.712 cm^{-1}. This transition owes its intensity to the magnetic dipole moment of the oxygen molecule and is consequently very weak. Gebbie et al.[113] required a path length of 100 m and an oxygen pressure of several atmospheres before they could observe it by conventional spectroscopic means. Observation with a laser source is, of course more comparable to a microwave spectroscopy experiment, where it is possible to actually observe the peak absorption of a line. Nevertheless a reasonable path length is required and Evenson et al. achieved this by using a small Fabry–Perot interferometer. This gave an effective absorbing path length of 2 m. The state $N = 3$, $J = 4$, $M = -4$ (i.e., $M = J$) shows a linear shift with magnetic field up to all fields currently available and at a field strength of 16.418 kG it shifts down in energy some 1.518 cm^{-1}. The upper state $N = 5$, $J = 5$, $M = -4$ shows also an initial negative shift but at higher field strengths a positive quadratic term dominates and at a field strength of 16.418 kG

[111] R. G. Brewer, M. Kelly, and A. Javan, *Phys. Rev. Lett.* **23**, 559 (1969).

[112] K. M. Evenson, H. P. Broida, J. S. Wells, R. J. Mahler, and M. Mizushima, *Phys. Rev. Lett.* **21**, 1038 (1968).

[113] H. A. Gebbie, W. J. Burroughs, and G. R. Bird, *Proc. Roy. Soc. (London)* **A310**, 579 (1969).

it shows an upward shift of 0.368 cm⁻¹. The transition

$$N = 3, \quad J = 4, \quad M = -4 \rightarrow N = 5, \quad J = 5, \quad M = -4$$

therefore coincides with the laser frequency at the field strength given above. The experimental arrangements were later improved by Wells and Evenson[114] by using an intracavity absorption cell and the sensitivity was much increased. With a variant of this system they were able to detect[115] the $^2\Pi_{1/2} J = \frac{1}{2} \rightarrow {}^2\Pi_{3/2} J = \frac{3}{2}$ line of OH using the 79-μm emission line of the water vapor laser. They have proposed that a relative Doppler shift of H_2O emitters and OH absorbers followed by spontaneous emission of the $^2\Pi_{3/2} J = \frac{3}{2} \rightarrow {}^2\Pi_{1/2} J = \frac{1}{2}$ line will create a population inversion of the hyperfine levels of the $^2\Pi_{3/2} J = \frac{1}{2}$ state. Stimulated emission between these hyperfine levels ($F = 2 \rightarrow 2$ and $1 \rightarrow 1$) will explain the galactic maser emission near a wavelength of 18 cm. The success of this experiment indicates that submillimeter EPR techniques may prove powerful in detecting free radicals that have so far not been observed by conventional EPR methods. In particular, nonlinear polyatomic free radicals that exhibit only magnetic dipole transitions in the microwave region could possibly be detected by searching for resonances with the Zeeman components of pure rotational transitions. Some of these would be of the allowed electric dipole type and therefore very much more intense. Also the absorption strength of lines rises initially as the cube of the frequency, and this effect will further enhance the intensity of the submillimeter lines compared with those in the microwave region. Already absorption in the stable NO_2 and NO radicals has been detected by Evenson and his colleagues.[116] The NH_2 radical, which is of astrophysical interest, is calculated to have one of the spin components of its $3_{03} \rightarrow 3_{12}$ transition within 2 GHz of the 964-GHz line of the HCN laser, from the energy levels given by Duxbury.[117] It is an excellent candidate for the type of search outlined above.

Recently several groups of workers have reported the use of submillimeter lasers as sources for EPR spectroscopy of rare earth compounds. A detailed discussion of EPR spectra of solids at microwave frequencies was given in Chapter 4.2 of Volume 3 (1st edition) of this series and so we will limit our account here to the novel features of the

[114] J. S. Wells and K. M. Evenson, *Rev. Sci. Instrum.* **41**, 226 (1970).

[115] K. M. Evenson, J. S. Wells, and H. E. Radford, *Phys. Rev. Lett.* **25**, 199 (1970).

[116] M. Mizushima, K. M. Evenson and J. S. Wells, *Phys. Rev.* **A5**, 2276.

[117] G. Duxbury, *J. Mol. Spectrosc.* **25**, 1 (1968).

newer work. The condition for resonance in an EPR experiment can be written

$$\nu = g\beta H, \qquad (2.4.23)$$

where ν is the resonant frequency, g is the gyromagnetic ratio of the electron in question, β is the Bohr magneton, and H the field strength. In a conventional EPR experiment—as, for example, the study of free radicals—g is approximately 2, β is 1.3997 MHz · G^{-1}, and therefore for ν to be 10 GHz, H must be of the order of 4 kG. The sensitivity and resolution are both improved by working at as high a field as possible, but to use submillimeter frequencies of the order 900 GHz, the field strengths required will be approximately 320 kG. Fields of this strength are now becoming available with superconducting magnets, and for the rare earth ions β is frequently much larger than 2, so the experiments are possible and the two improvements mentioned above are available. Prinz and Wagner[118] studied a single crystal of dysprosium phosphate at 4.2 K using the 171.76-μm line from a pulsed D_2O laser. For this material resonance was achieved at a field of 67.6 kG, from which $g_{\parallel}$ was deduced to be 19.55 $\pm$ 0.16. The resonance does not immediately satisfy Eq. (2.4.23) since there is a "built in" or saturation dipolar magnetic field of 4.1 kG which has to be added to the applied field and additionally there is a zero field splitting of 7.2 cm^{-1} which must be added to the laser frequency. Pulsed lasers are undesirable for this type of experiment since modulation of the magnetic field to improve the sensitivity is not possible. Kotthaus and Dransfeld[119] used a CW 337-μm HCN laser to repeat the measurements of Boettcher et al.[120] on holmium methyl sulphate. A highly homogeneous superconducting magnet was used giving a field of 40 kG and they were able, therefore, to use field modulation, synchronous detection, and first-derivative presentation to greatly improve the signal-to-noise characteristic. Eight lines were observed stretching between 39 and 42 kG and the line width of the first (hyperfine structure $M_I = +7/2$) was studied as a function of temperature to deduce the spin–lattice relaxation mechanism. The g value for the Ho^{3+} ion is 15.5.

Studies of cyclotron resonance in semiconductors using submillimeter lasers has yielded a great deal of information about the band structure of

[118] G. A. Prinz and R. J. Wagner, *Phys. Lett.* **30A**, 520 (1969).

[119] J. P. Kotthaus and K. Dransfeld, *Phys. Lett.* **30A**, 34 (1969).

[120] J. Boettcher, K. Dransfeld, and K. F. Renk, *Phys. Lett.* **26A**, 146 (1968).

these materials. Most of the experiments have been carried out at the Francis Bitter Magnet Laboratory in Cambridge, Massachusetts, by the group led by Lax.[121] Just as for EPR, the observations may in principle be made at microwave frequencies, but if the spectrum is to be clearly observed $2\pi\nu_c\tau$ must be greater than unity, where τ is the collision time of the spiraling electrons with the host lattice. For most semiconductors, τ is so short that meaningful observations can only be made at sub-millimeter frequencies. Promising early results were obtained by Button *et al.*[122] using a pulsed HCN laser, who observed cyclotron resonance in p type germanium at fields between 20 and 100 kG for a cyclotron resonance frequency $\nu_c = 890$ GHz. More refined work[123] using CW lasers has clearly shown up the complicated quantum structure of the degenerate valence band of InSb and the warped Fermi surfaces for tellurium. A very interesting recent study[124] has been done on the semipolar material CdTe, where it has been shown that the increase of cyclotron mass with cyclotron frequency was too large to be explained by band nonparabolicity. Calculations on the Fröhlich Hamiltonian showed quantitative agreement with experimental results, demonstrating that the polaron character of the resonating carrier is very important. Further experiments have been carried out on piezoelectric polaron cyclotron resonance in CdS by Button *et al.*[125] There can be little doubt that this type of spectroscopy is one of the most powerful methods for elucidating the finer details of the electronic structure of semiconductors.

2.4.7.6. Tuned Laser Spectroscopy. As discussed earlier, the absorption lines of a gas at low pressure are very sharp. It follows, therefore, that unless there is a close coincidence, the absorption coefficient of the gas at the laser frequency will be low and it is possible to include a gas cell filled with the gas actually inside the laser cavity without introducing serious additional losses. The effect of such an inclusion is that the radiation effectively traverses the specimen very many times and very

[121] See for example, C. C. Bradley, K. J. Button, B. Lax, and L. G. Rubin, *IEEE J. Quantum Electron.* **QE 4**, 733 (1968).

[122] K. J. Button, H. A. Gebbie, and B. Lax, *IEEE J. Quantum Electron.* **QE 2**, 202 (1966).

[123] K. J. Button, B. Lax, and C. C. Bradley, *Phys. Rev. Lett.* **21**, 350 (1968); **23**, 14 (1969).

[124] J. Waldman, D. M. Larsen, P. E. Tannenwald, C. C. Bradley, D. R. Cohn, and B. Lax, *Phys. Rev. Lett.* **23**, 1033 (1969).

[125] K. J. Button, B. Lax, and D. R. Cohn, *Proc. Int. Semiconductor Conf.*, Boston, 1970.

long effective path lengths result. The windows, unfortunately, introduce transmission and reflection losses, since there are no truly transparent far-infrared materials, but the reflection losses at least can be eliminated by mounting the windows at the Brewster angle. Windows are usually made of polyethylene terephthalate, despite its large absorption coefficient, because this polymer is very tough and quite thin (3-μm) films will stand up to moderate pressure differentials. Polypropylene film is also a useful material.

If the frequency separation of the line center and the laser is within the very narrow (5-MHz) tuning range of the laser, then, with the gas cell in position, the power output curve of the laser as a function of mirror separation will not be made up of smooth peaks (see Fig. 7). However, each peak will instead display a subsidiary minimum or dip. This dip occurs when the displaced laser frequency coincides with the absorption maximum of the gas line. Such a dip was observed by Duxbury and Burroughs[107] for the gas 1:1 difluoroethylene $CF_2 = CH_2$ included in the cavity of an HCN laser working at 337 μm. Later work by Duxbury and Jones[126] and by Bradley and Knight[72] using an external absorption cell showed that the center of the absorption line was ~0.5 MHz from the HCN laser frequency produced by their lasers, that the absorption coefficient at peak was 0.8×10^{-3} cm^{-1} and that the absolute frequency of the line was 890.7596 ± 0.0002 GHz.

At 890 GHz and a pressure of ~100 mTorr, the molecular absorption line shape is no longer Lorentzian, since the pressure-broadened and the Doppler-broadened widths are comparable. The resultant line width is a convolution of the Lorentzian and the Gaussian Doppler-broadened line-shape functions, the so-called *Voight line shape*. At much lower pressures, ~5 mTorr, the pressure-broadened contribution becomes much less than the Doppler broadening and the line shape becomes almost Gaussian. Thus in the submillimeter region, in contrast to what prevails in the microwave region, it is not possible to reduce the line width below a certain observable limiting value by reducing the pressure. This implies that we have a Doppler-limited resolution which is of the order of 1.4 MHz at 890 GHz for an average weight molecule at room temperature. A possible way around this difficulty is to try to utilize Lamb-dip techniques that have been pioneered in the infrared and visible regions and have, more recently, been successfully applied in the microwave region [see Section 2.4.7.7].

[126] G. Duxbury and R. G. Jones, *Phys. Lett.* **30A**, 498 (1969).

In this type of experiment[127] a standing wave is set up in the absorption cell. Only at line center of a Doppler-broadened line do molecules interact with running waves in both directions and therefore experience double the interaction of molecules whose velocity components are such that they absorb away from the line center. This results in an increased saturation at line center compared with the regions away from it and an observed decreased attenuation of the beam traversing the absorption cell. This decreased absorption, which is observed as an apparent emission feature at the line center, occupies only a small fraction of the total line. The apparent width is proportional only to pressure and thus at 5 mTorr could be only 0.1 MHz compared with the overall Doppler width at half height of 1.4 MHz. Unfortunately, the total absorption coefficient at these pressures is often only about 1×10^{-3} cm^{-1} and the Lamb-dip signal would probably be only $\sim$5–10% of this and a very stable set up would be required. In order to aid the design of such an experiment, Bradley et al.[128] have measured the saturation of the $CF_2 = CH_2$ line at 891 GHz in order to try to evaluate the dipole moment matrix element.

2.4.7.7. The CO_2 Laser as a Spectroscopic Source.

Apart from its shorter wavelength and much richer spectrum, the CO_2 laser is, in principle, very much the same as its longer wavelength analogs, the H_2O and HCN lasers, and like them may be used as a spectroscopic source. However, the high CW, power levels that can be obtained from CO_2 lasers and the enormous peak powers available from the Q-switched versions make possible a range of experiments which have as yet no submillimeter counterparts. Thus the Lamb-dip technique, which in essence confers very high resolving power by limiting the observation to a narrow central region of the Doppler-broadened profile, has been applied to the Stark spectrum of NH_2D studied with a CO_2 laser.[111] Rather interestingly, analogous techniques have been used by Winton and Gordy in the millimeter wave band.[129] A high Q Fabry–Perot cavity fed by the harmonically multiplied output of a klystron was employed and absorption lines of OCS and CH_3F between 1 and 3 mm were observed showing sharp Lamb dips at the line center. The presence of this sharp feature permitted much higher accuracy of line center frequency measurement, and Winton and Gordy quote accuracies of one part in 10^8.

[127] W. E. Lamb, Phys. Rev. **134**, A1429 (1964).

[128] C. C. Bradley, R. G. Jones, J. Birch, and G. Duxbury, Nature (London) **226**, 941 (1970).

[129] R. S. Winton and W. Gordy, Phys. Lett. **32A**, 219 (1970).

The same high precision is available at the CO_2 laser frequency, but is there even more valuable since the Doppler width increases linearly with frequency. Brewer et al.[111] were able to detect Stark splittings as small as 1.5 MHz when the Doppler width was 82 MHz, and these observations were moreover for an excited state. From an experimental point of view, a Lamb dip should be observed in an absorption line whenever the Doppler broadening is dominant and when one can use a high-power, very stable, source. Clearly, since the technique has been demonstrated at infrared and millimeter wavelengths, submillimeter studies should be possible when the necessary improvements to the laser stability have been accomplished.

The CO_2 laser radiates in a frequency region where many molecules have vibration–rotation transitions and as a result, far more coincidences can be found than is the case for submillimeter lasers. The coincidences with some lines, as yet unassigned, within the ν_3 band of SF_6 provided[67] the first means for passively Q-switching the laser. Subsequently several other compounds such as vinyl chloride[130] and methyl fluoride[68] were found to be "bleachable" and able, therefore, to act as Q-switching agents. The close coincidences can be investigated in any of the ways mentioned earlier in this section, but the high power available from the CO_2 laser makes another and more exciting prospect—stimulated fluorescence—possible. This has been observed in CH_3F, CH_3OH, CH_2CHCl, and NH_3 (see Section 2.4.9).

The dipole moments of CH_3F in its ground state and with ν_3 excited by one quantum have been measured by Brewer[131] using an interesting nonlinear effect. For molecules having near coincidences between the laser and the absorption line within the Doppler line width, or where it is useful to resolve hyperfine structure, resolution enhancement can be achieved by the Lamb-dip technique, by RF–infrared double resonance or by optical–optical double resonance methods. The dipole moments of CH_3F in its ground state and with ν_3 excited by one quantum have been measured by Brewer[131] using an interesting nonlinear effect. If methyl fluoride is placed in a Stark cell and exposed to two intense fields from two separate CO_2 lasers, a resonance will be observed when the Stark splitting, in one of the states between which the transition is occurring, just equals the frequency separation of the lasers. The reason for this is that a molecule will only interact strongly and nonlinearly with

[130] J. T. Yardley, Appl. Phys. Lett. **12**, 120 (1968).
[131] R. G. Brewer, Phys. Rev. Lett. **25**, 1639 (1970).

two fields when it Doppler shifts them equally. The frequency separation of the CO_2 lasers can be obtained by piezoelectric tuning of the cavity length of one of them coupled as usual to a feedback control loop. Using this method, Brewer observed Stark splitting of the ($J = 12, K = 2$) Q-branch line of CH_3F with an effective line width of only 100 kHz. With this high resolution, the doublet structure of the spectrum corresponding to differing dipole moments in the ground and excited slates was observed. From this, values of μ(ground) $= 1.8549 \pm 0.0010$ and μ(excited) $= 1.9009 \pm 0.0010$ were derived.

The CO_2 laser, operating as it does by vibrational transfer from N_2 molecules and from relaxation by collision with helium atoms, shows no perturbation of its active levels and direct spectroscopy of the emission can tell us a great deal about the vibration–rotation parameters. There are two types of experiment possible: firstly direct measurement[132] of the line frequencies using the so-called *metal–metal diode*,[79] and secondly, and rather more easily, difference frequency measurements between various lines in the spectrum. Bridges and Chang[133] beat the lines with one another in a bulk GaAs crystal and observed difference frequencies in the 50–80-GHz region. From this data they were able to deduce the rotational constants of the 00^01, 10^00, and 02^00 levels which were 2 orders of magnitude more accurate than could be derived from conventional spectroscopy. As an example, they give $B_0(00^01)$ as $0.38714044 \pm 0.00000037$ cm^{-1} as compared with the previous best value of 0.387132 ± 0.000040 cm^{-1}.

There is currently (see Section 2.4.9) considerable interest in applying stimulated Raman scattering to the production of coherent infrared sources. One of the most successful of these efforts has been the development of the tunable spin-flip laser by Patel and his group.[65,134] In the presence of a magnetic field, the conduction electrons in InSb have their levels split in two due to the interaction of the spin with the field (see Section 2.4.7.5). Raman scattering of an incident photon of frequency ν_0 will lead to the emission of a photon whose frequency (Stokes scattering) is given by

$$\nu_s = \nu_0 - g\beta H, \qquad (2.4.24)$$

where H is the magnetic field, β is the Bohr magneton, and g is the Landé

---

[132] L. O. Hocker, D. R. Sokoloff, V. Daneu, A. Szoke, and A. Javan, *Appl. Phys. Lett.* **12**, 401 (1968).

[133] T. J. Bridges and T. Y. Chang, *Phys. Rev. Lett.* **22**, 811 (1969).

[134] C. K. N. Patel, E. D. Shaw, and R. J. Kerl, *Phys. Rev. Lett.* **25**, 8 (1970).

splitting factor. At sufficiently high incident field, the spontaneous Raman scattering will go over into the very much more intense and coherent stimulated Raman scattering. The InSb crystal will therefore emit radiation whose frequency can be tuned by altering the applied magnetic field. Patel *et al.*[134] have observed a tunability range of 10.9 to 13.0 µm and the output was of the order 1 W with a half-width of 0.05 cm^{-1} when a Q-switched CO_2 laser was used as the pump. With this device, they have investigated part of the ν_2 band of NH_3 and succeeded in clearly resolving some groups of lines that had resisted conventional[135] techniques, even though these had a claimed resolution of 0.1 cm^{-1}.

These are a few examples of what has already been achieved with the CO_2 laser as a spectroscopic source. There can be little doubt that further advances will be made and it now appears that we are moving into a new era of infrared spectroscopy where the instrumental and natural (e.g., Doppler effect) difficulties that held up progress in the past have been vanquished.

2.4.8. Tunable Devices for the Generation of Coherent Submillimeter Power

2.4.8.1. Introduction. The various types of submillimeter laser mentioned in this chapter, although satisfactorily powerful, lack the wide frequency tunability which is necessary if they are to become routine spectroscopic sources. As we have seen, the HCN and H_2O lasers can have their output frequency altered by a few megahertz when the cavity length is changed, but this frequency change is negligible in comparison with the center frequency. The newer, optically pumped, lasers described in the previous section have a wider tuning range because of their inherently higher gain but even here the actual frequency change is only of the order 10 MHz. What is required is a frequency swing of some gigahertz. Tuning of the output frequency of the laser by the application of an axial magnetic field works well in the near infrared with atomic lasers but the results with submillimeter lasers have been disappointing, and with the HCN laser unobservable.[27] High-speed modulators, depending on the Faraday effect in ferrites, can generate sidebands[136] on the HCN laser emission, but again the frequency separation is only a few megahertz. Mixing of submillimeter laser radiation with power from a klystron has been used to measure the laser frequency (see Section 2.4.6)

[135] H. M. Mould, W. C. Price, and G. R. Wilkinson, *Spectrochim. Acta* **13**, 313 (1959).
[136] J. R. Birch and R. G. Jones, *Infrared Phys.* **10**, 217 (1970).

but so far no reradiated power at the sum and difference frequencies has been detected.

At the moment, the only practical route to tunable sources of coherent submillimeter radiation lies in difference frequency generation in nonlinear media using two optical lasers as the primary sources. This is a grieviously inefficient process with conversion ratios of the order 10^{-8}, but since the higher frequency lasers can often be operated CW at power levels in excess of 100 W and in the pulsed mode with peak powers of the order of megawatts, it will be realized that some microwatts of submillimeter power should be available and this would be adequate for most spectroscopic applications.

2.4.8.2. Difference Frequency Generation in Nonlinear Media. It has been mentioned earlier (Section 2.4.7.7) that the beat notes between two adjacent lines of the CO_2 laser can be obtained by nonlinear mixing in a GaAs crystal. The problem here as in all nonlinear difference frequency generation is phase-matching. If two intense electromagnetic fields of amplitudes E_1 and E_2 and of vacuum wave numbers $\bar{\nu}_1$ and $\bar{\nu}_2$ are applied to a medium that has a nonlinear polarizability, a wave at the difference frequency $\nu_3 = \nu_1 - \nu_2$ will be generated. The spatial characteristics of this wave will be given by

$$E(z) \propto E_1 E_2 \cos 2\pi(n_1 \bar{\nu}_1 - n_2 \bar{\nu}_2)z, \qquad (2.4.25)$$

where n_1 and n_2 are the respective refractive indices. If the wave is to propagate and to grow in amplitude, it follows that the term in brackets must equal $n_3 \bar{\nu}_3$, i.e.,

$$n_3 \bar{\nu}_3 = n_1 \bar{\nu}_1 - n_2 \bar{\nu}_2, \qquad (2.4.26)$$

which is the condition for phase-matching. Sometimes this equation can be satisfied and sometimes not, in which case a coherence length l may be defined by

$$l^{-1} = 2[n_1 \bar{\nu}_1 - n_2 \bar{\nu}_2 - n_3 \bar{\nu}_3], \qquad (2.4.27)$$

which is a measure of the mismatch. In most experimental situations, provided l is approximately equal to the length of the nonlinear element, satisfactory results may be expected.

Nonlinear generation is usually performed in a birefringent crystal, since by suitable choice of angle between the optic axis and the direction of propagation, the right-hand side of Eq. (2.4.26) can be made equal to the left-hand side. Essentially the two lasers are arranged so that their beams pass collinearly through the crystal, but are polarized at right angles: one beam propagates in a fashion determined by the value of the extraordinary index n_e at that angle and the other in a fashion determined by the ordinary index n_0. The far infrared index n_3 is usually high (due either to *reststrahlen* dispersion in ionic crystals or else to conduction edge effects ion semiconductors), but the wavenumber $\bar{\nu}_3$ is very much less than $\bar{\nu}_1$ or $\bar{\nu}_2$ and it follows therefore that Eq. (2.4.26) can be satisfied for relatively small angular displacements from the optic axis, even if, as is usually the case, $(n_0 - n_e)$ is small. Far-infrared generation by nonlinear mixing was first reported by Zernike and Berman,[137] who phase-matched the output of a pulsed neodymium laser in quartz. They obtained radiation near 100 cm^{-1} but gave no detailed analysis. This laser is known to emit a range of wavelengths between 1.059 and 1.073 μm so the observed difference frequency is reasonable. Quartz is moderately transparent in the far infrared, whereas many other crystals (such as ammonium dihydrogen phosphate ADP) that are valuable for sum frequency generation are virtually opaque due to reststrahlen absorption. This severely restricts the range of difference frequencies that can be generated and may explain the delay between the successful demonstration of sum frequency generation and the eventual production of difference frequency radiation.

Semiconductors are very attractive materials for the generation of difference frequencies, since they are fairly transparent in the far infrared, if pure, and at these wavelengths they have very large nonlinear coefficients. As examples, the coefficient of optical rectification χ^0 (in units of 10^{-8} esu), which is 0.64 for quartz, is 45 for CdS and 365 for ZnTe. Unfortunately the best material, ZnTe, is cubic so the conventional form of phase-matching is not available, but happily the dispersion in the visible region is such that a fairly large coherence length results. Rewriting Eq. (2.4.27) in the form

$$l^{-1} = 2\bar{\nu}_3[n - (dn/d\bar{\nu}) \cdot \bar{\nu}_L - n_3] \qquad (2.4.28)$$

we have n, the average refractive index at the laser frequency, equals 2.92, $dn/d\bar{\nu}$ equals -0.41×10^{-4} cm, $\bar{\nu}_L = 14{,}400$, and $n_3 = 3.2$, so that for difference frequency generation involving the beating of the radiation from ruby lasers, l will be approximately 1 mm. Because of this Yajima

---

[137] F. Zernike and P. R. Berman, *Phys. Rev. Lett.* **15**, 999 (1965).

and Inoue[138] were able to produce 29 cm^{-1} radiation at power levels (pulsed) of the order 10^{-3} W by beating the R_1 and R_2 lines from a single simultaneously Q-switched ruby laser.

Lithium niobate has a high nonlinear susceptibility and is birefringent. Faries et al.[139] were able to beat the radiation from two separate ruby lasers and obtain the difference frequency. For this material $n_e = 2.189$, $n_o = 2.273$, and $n_3 = 6.55$, so that phase-matching angles of the order of 10° are involved. The output frequency of a ruby laser can be altered by changing the temperature of the ruby rod, and it follows that by varying the temperature of one laser with respect to the other the difference frequency can be tuned. Faries and his colleagues were able to tune the output between 1.2 and 8.1 cm^{-1} when, with one laser at room temperature, the other was progressively cooled down to −40°C. The residual difference frequency, even when the two lasers were at the same temperature, arises because the lasers give out at least two modes and these modes are separated in wave number by about 2 cm^{-1}. Subsequently a similar variation of output frequency with laser cooling has been observed for the R_1/R_2 beating near 29 cm^{-1}, and with the use of liquid nitrogen as a coolant it has proved possible[140] to close the gap and obtain continuous tunability between 1.2 and 50 cm^{-1}.

The difference frequency generation from the CO_2 laser involves another method of phase-matching. When an electromagnetic wave is travelling in a waveguide, the phase velocity is a function of frequency and also of the refractive index of the medium filling the waveguide. This implies that a wave will "see" an effective refractive index that is determined by its frequency and the size of the guide. Clearly, therefore, with n_3 as a variable, the phase-matching equation can be satisfied. The coherence length is given by

$$l^{-1} = 2[n_1\bar{\nu}_1 - n_2\bar{\nu}_2 - [n_3{}^2 - (1/2\bar{\nu}_3a)^2]^{1/2}\bar{\nu}_3],\qquad(2.4.29)$$

where a is the width of the waveguide. Chang et al.[141] found that they could achieve perfect phase-matching near 54 GHz with $a = 2.77$ mm.

[138] T. Yajima and K. Inoue, Phys. Lett. 26A, 281 (1968); IEEE J. Quantum Electron. QE 5, 140 (1969).

[139] D. W. Faries, K. A. Gehring, P. L. Richards, and Y. R. Shen, Phys. Rev. 180, 163 (1969).

[140] P. L. Richards, private communication (1969).

[141] T. Y. Chang, N. Van Tran, and C. K. N. Patel, Appl. Phys. Lett. 13, 357 (1968).

The appropriate optical parameters were $(n_1 n_2)^{1/2} = 3.24$, $n_1 \bar{\nu}_1 - n_2 \bar{\nu}_2 = 3.28 \bar{\nu}_3$ and $n_3 = 3.47$. The millimeter wave pulses had peak powers of the order 10 μW.

Radiation at a much higher frequency ($\sim$100 cm^{-1}) has been obtained by Van Tran and Patel,[142] who beat the radiation from the two branches of the CO_2 laser at 9.6 and 10.6 μm (see Section 2.4.3) in an n-type InSb crystal. The transverse optical (i.e., *reststrahlen*) band of this crystal occurs at 183 cm^{-1} and the longitudinal optical mode at 197 cm^{-1}, so there will be strong dispersion near 100 cm^{-1} where the difference frequencies lie, and for this reason the coherence length will be quite short if a crystal is to be used in free space and in zero field. Magneto-plasma effects provide a way of altering n_3 such that the phase-matching condition can be satisfied. Essentially if a suitable magnetic field is applied so that the cyclotron resonance frequency occurs at an appropriate value *below* ν_3 and if the number of free carriers is sufficient to give a correct plasma frequency, then the negative going dispersion in ε' from these processes can correct the positive going dispersion due to the lattice modes. Van Tran and Patel found that the phase-matching condition could indeed be obtained merely by choosing the correct value for the magnetic field and so it is possible to "tune" the system for any value of the difference frequency provided it lies somewhere in the 100-cm^{-1} region. The fields required were of the order of 10 kG and free carrier concentrations had to be less than 3.4×10^{15} cm^{-3}, for at this particular value phase-matching was found for zero field. The experimentalists obtained peak powers of the order of 10^{-6} W for input peak powers of the order of 10^3 W. These difference frequency generators are not tunable in the sense that their frequencies may be continuously varied, but from the mixing of the large number of lines coming out of the CO_2 laser it ought to be possible to obtain a series of emission lines in the far infrared spaced by roughly 2 cm^{-1}. A spectroscopic source of this kind which could be likened to a fine comb would be perfectly adequate for delineating the broad-band absorptions characteristic of glasses, liquids, and amorphous solids.

Stimulated Raman scattering may provide another means for generating difference frequencies in the far infrared. The normal Raman effect consists of the inelastic scattering of photons by molecules, as a result of which the photon comes off, shifted either upward (anti-Stokes) or downward (Stokes) in frequency by an amount ν_i, where $h\nu_i$ is some

[142] N. Van Tran and C. K. N. Patel, *Phys. Rev. Lett.* **22**, 463 (1969).

characteristic energy difference of the scattering species. The effect is extremely weak with usually less than 10^{-6} of the incident power going into the Raman scattering. Stimulated scattering is found when a coherent beam of radiation from a Q-switched laser is focused into a suitable medium. Like all nonlinear effects there is a threshold, but if the focused intensity is greater than this threshold, stimulated Raman scattering, either Stokes or anti-Stokes, may be observed with power levels up to 10% of the incident power. Clearly, by beating the laser radiation with the stimulated Raman radiation, it should be possible to obtain the difference frequency ν_i. This has been done for a number of molecules that have vibration frequencies in the mid-infrared, but the most exciting advance as far as submillimeter generation is concerned lies in the observation of stimulated Raman scattering by the acoustic and optical phonons of crystals. Ionic crystals that have intense transverse optical absorptions bands in the far infrared and large nonlinear susceptibilities exhibit the phenomenon of "polariton" scattering. A polariton is a mixed quantized entity which for low wave vector behaves like a photon and which for high wave vector acts like a transverse optic phonon. Raman scattering by polaritons has been observed from quartz,[143] gallium phosphide,[144] and especially lithium niobate.[145] The dispersion curves of polaritons start off with a slope equal to the velocity of light in the medium, but the slope rapidly decreases as $\bar{k}$ increases. It follows that to conserve momentum, the scattered Raman frequency will be a sensitive function of angle and the true polariton scattering will be confined to a region extending no more than a few degrees from the laser beam. Yarborough and his colleagues[146] found that stimulated emission from the 248 cm^{-1} polariton mode of LiNbO$_3$ could be obtained with two polished surfaces of the crystal acting as the resonator. The Raman displacement varied rapidly from 42 to 200 cm^{-1} as the angle between the beam from a Q-switched ruby laser and the resonator axis was altered. A subsidiary face was cut at 60° on the crystal, and from this emerged the difference frequency radiation. Very high peak powers ($\sim$5 W) have been quoted for this so-called "idler" radiation. The device is most

[143] J. F. Scott, L. E. Cheesman, and S. P. S. Porto, *Phys. Rev.* **162**, 834 (1967).

[144] C. H. Henry and J. J. Hopfield, *Phys. Rev. Lett.* **15**, 964 (1965).

[145] H. E. Puthoff, R. H. Pantell, B. G. Huth, and M. A. Charon, *J. Appl. Phys.* **39**, 2144 (1968); R. Claus, H. W. Schrotter, H. H. Hacker, and S. Haussuhl, *Z. Naturforsch.* **24a**, 1733 (1969).

[146] J. M. Yarborough, S. S. Sussman, H. E. Puthoff, R. H. Pantell, and B. C. Johnson, *Appl. Phys. Lett.* **15**, 102 (1969).

attractive, for a large region of the far infrared can be swept with high-power coherent radiation merely by altering the angle of a crystal.

2.4.9. Miscellaneous Submillimeter Lasers

2.4.9.1. Other Vibration–Rotation Lasers. Stimulated emission from H_2O and HCN and their isotopic variants has proved so theoretically interesting and so practically valuable that vigorous searches have been instituted in order to find more species which can be induced to radiate. However, when the stringent requirements for resonance perturbation are borne in mind, it will be realized that lasing systems must be rare and, to date, only four more vibration–rotation lasers have been discovered: these involve electrical discharges through H_2S, OCS, SO_2, and NH_3. The emission[147] from H_2S probably arises from a similar type of resonance perturbation to that which is found for H_2O, but since H_2S is not very stable in an electric discharge, the H_2S laser is unlikely to come into such widespread use as the H_2O laser. Very low (~ 2 Hz) repetition frequencies and very high pumping speeds (500 liters/min) coupled with relatively low discharge voltages (3.6 kV) are essential if this device is to radiate. Hassler and Coleman[147] in 1969 observed 24 emission lines from H_2S in the region 33–225 μm. Of these, 21 have been assigned to transitions in the $\nu_1 \rightarrow 2\nu_2$ system but the details have not been published. A considerable difficulty facing anyone attempting an assignment is that the ν_3 band has not so far been observed in absorption, so the part of the scheme involving ν_3 cannot at present be verified. Hassler and Coleman[147] have also obtained emission from OCS at 123 and 132 μm. A buffer gas (He, N_2, or O_2) was necessary and again very low repetition frequencies were found essential. No assignment of the OCS lines has yet been made.

The SO_2 laser was independently discovered by Dyubko *et al.*[148] at Kharkov, by Hassler and Coleman[147] in Illinois, and by Hard[149] in Massachusetts. Four lines have been found between 140 and 215 μm; the shortest wavelength line (140.85 μm in air) can be obtained in both pulsed and CW operation as can the line at 192.67 μm, while the remaining two at 151.16 and 215.27 μm have so far been observed only from pulsed lasers. Addition of helium to the gas stream greatly stabilizes the laser and increases the power output. The pulsed laser ran best with

[147] J. C. Hassler and P. D. Coleman, *Appl. Phys. Lett.* **14**, 135 (1969).
[148] S. F. Dyubko, V. A. Svich, and R. A. Valitov, *JETP Lett.* **7**, 320 (1968).
[149] T. M. Hard, *Appl. Phys. Lett.* **14**, 130 (1969).

a gas mixture of 0.4 Torr SO_2 plus 0.4 Torr He, while for CW operation 0.6 Torr SO_2 plus 0.3 Torr He was found optimal. After running for a while, a deposit of sulfur forms inside the laser, but this can be readily removed by flowing air through the apparatus while maintaining the discharge. The SO_2 laser can apparently give output powers within an order of magnitude of that given by the H_2O laser at 118.6 μm. For this reason it is likely to come into much more common use in the near future. No assignment of the four lines has yet been attempted because the infrared data is too scanty, but a resonance of v_1 and v_3 with $2v_2$ is known to occur and may be involved in the inversion mechanism.

Laser oscillation at wavelengths between 21 and 32 μm has been reported from a pulsed electrical discharge through ammonia by Mathias *et al.* and several further lines of the system have been observed by Akitt and Wittig.[150] Some of these lines have been shown by Lide to be members of the P branch of the transition $v_2 = 3^s \rightarrow v_2 = 2^a$. The v_2 levels of ammonia are split into a symmetric and an antisymmetric level by the inversion tunneling and Lide points out that the level $v_2 = 3^s$ has a total energy of 2384 cm^{-1}, which is close to the fundamental frequency of nitrogen. The laser mechanism may therefore involve the initial production of nitrogen by the electrical decomposition of ammonia followed by selective collisional excitation of ammonia molecules. However, it must be realized that the majority of the observed lines are so far unassigned and a confident account of the probable mechanism would be premature. The pulsed ammonia laser is technically interesting since it overlaps the shortest wavelength water laser lines and extends toward the 10-μm region where the CO_2 and N_2O lasers radiate. Emission from NH_3 at very much longer wavelengths has been observed using stimulated fluorescence techniques and this will be described in the next section.

2.4.9.2. Optically Pumped Lasers.

As has been remarked earlier, many molecules have absorption bands in the 10-μm region and some of these have been used to passively Q-switch the CO_2 and N_2O lasers. For this to be possible it is necessary that the lines of the absorption band be in close coincidence with those of the laser and also that the relaxation times are such that saturation occurs. When a vibration–rotation transition is saturated, there will exist a rotational population inversion in one or both of the vibrational states involved, provided the relaxation times are

---

[150] L. E. S. Mathias, A. Crocker, and M. S. Wills, *Phys. Lett.* **14**, 33 (1965); D. P. Akitt and C. F. Wittig, *J. Appl. Phys.* **40**, 902 (1969).

favorable. If the molecule has a permanent electrical dipole moment, radiative rotational transitions can occur and laser action in the far infrared is possible.

Chang and his colleagues[151,152] have observed stimulated emission from three gases, methanol, vinyl chloride, and methyl fluoride, pumped by the CO_2 laser. The initial observations were of six frequencies in the 600-GHz (20-cm^{-1}) region obtained from methyl fluoride CH_3F using a Q-switched CO_2 laser as the source. The strong $P(20)$ line of the CO_2 laser at 9.55 μm has a very close coincidence with the $Q(12)$ line of the

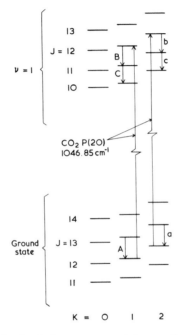

FIG. 16. Stimulated fluorescence and laser action in CH_3F.

ν_3 (CF stretch) band of CH_3F. An energy level diagram is shown in Fig. 16. The coincidence is virtually perfect for just two lines of the K fine structure of $Q(12)$, namely those with $K = 1$ and $K = 2$. The pumping action of the laser overpopulates $J(12)K(1)$ and $J(12)K(2)$ of the upper-state leading to stimulated emission B, b and cascade emission C, c. It also depopulates $J(12)K(1)$ and $J(12)K(2)$ of the lower state leading

151 T. Y. Chang and T. J. Bridges, *Opt. Commun.* 1, 423 (1970).

152 T. Y. Chang, T. J. Bridges, and E. G. Burkhardt, *Appl. Phys. Lett.* 17, 249 (1970).

to stimulated emission A, a. The wavelengths and wave numbers of these transitions are given in the Appendix. Later, Chang *et al.*[152] observed continuous wave emission of the B, b transitions.

The experimental arrangements used by these authors featured an ~80-cm Fabry–Perot cavity that contained the gas. The cavity is plano-concave and CO_2 pump radiation is coupled in through a 1.2-mm hole in the center of the plane mirror. The concave mirror has a radius of curvature of 190 cm and is mounted on a motor-driven micrometer stage to facilitate cavity length tuning. The CO_2 radiation after entering the cavity expands into a diverging beam and becomes trapped for a large number of round trips. The submillimeter output is taken from the same coupling hole via a cylindrical horn to reduce the divergence. The beam then passes via a 45° mirror to the detector, which is either a Rollin[153] InSb detector or else a point contact diode. The absolute time frequencies of the methyl fluoride emission lines were measured in the usual way by mixing the submillimeter signal with harmonics of a microwave signal in the point contact diode. The rotational parameters of the ground state of CH_3F are known from the work of Winton and Gordy[129] (see Section 2.4.7.7) and by comparing the time frequencies of the emission lines with the calculated values, the two ground state transitions can be identified. From the frequencies of the other four lines, more accurate rotation and distortion constants for the ν_3 state of CH_3F may be deduced. The polarization of the submillimeter laser output is perpendicular to that of the pumping radiation, which agrees with theory since the absorption is a $\Delta J = 0$ transition and the emission is a $\Delta J = -1$ process. With the Q-switched pump, the bandwidth of the radiation varies from 60 MHz at a gas pressure of 50 mTorr, up to at least 160 MHz at higher pressure, where it exceeds the free spectral range of the cavity. The active medium in this type of submillimeter laser is un-ionized and it is possible therefore to calculate the gain as a function of the molecular parameters. For CH_3F this gives the gain in the small signal limit as 27 dB m^{-1}, which is much larger than that of the HCN laser at 337 μm, where the gain is only 2 dB m^{-1}. The superiority of optically pumped to electrically excited lasers must in part be due to the absence of any disruptive discharge in the former.

The observed transitions for optically pumped methanol CH_3OH and vinyl chloride are also given in the Appendix. Most of these are not yet assigned due to the lack of reliable high-resolution infrared spectroscopic

[153] M. A. Kinch and B. V. Rollin, *Brit. J. Appl. Phys.* **14**, 672 (1963).

data for these two molecules. The 570.5-μm line of CH_3OH has, however, been identified as the $J(11) \rightarrow J(10)$ transition pumped via an $R(10)$ absorption line. The output of the cavity is always strongly linearly polarized parallel or perpendicular to the pump polarization. Now, for a symmetric top, perpendicular polarization implies a Q branch transition, whereas parallel polarization imples an R or P branch transition. It is possible that a similar classification may be useful in an approximate manner for this asymmetric rotor.

CW laser action at 81.5 and 263.4 μm has been observed in ammonia gas optically pumped by the radiation from an N_2O laser.[154] The lines have a similar origin to those at higher frequency observed using electrical excitation. The 81.5-μm line has been assigned to the transition $J(8)K(7)^s \rightarrow J(7)K(7)^a$ in the $\nu_2 = 1$ state of NH_3 and the 263.4-μm line has been assigned to the inversion transition $J(7)K(7)^a \rightarrow J(7)K(7)^s$. This latter is a cascade transition, but can oscillate independently. The independent oscillation is attributable to stepwise relaxation produced by resonant collisions among the NH_3 molecules. A similar effect occurs in the CH_3F laser.

Emission in the 13–23-μm region from BCl_3 has been observed by Karlov and his colleagues.[155] They added the BCl_3 directly to the discharge tube of the laser and because of this, it is not possible to say definitely whether this was an example of optical pumping or not. It is quite possible that some other inversion mechanism may be involved. The observed lines are vibration–rotation transitions in $^{10}BCl_3$ and $^{11}BCl_3$.

2.4.9.3. Noble Gas Lasers. Submillimeter emission from noble gas lasers has been observed as far out as 133 μm, principally by Patel and his colleagues.[156] Neon is the most prolific emitter and the observed lines come from the higher analogs of the well-known $3s \rightarrow 2p$, etc. transitions. The lines are usually weak ($\sim 10^{-8}$ W) and are unlikely for that reason to serve a spectroscopic purpose, but there are many of them and they are easily produced, so they may serve as secondary frequency or wavelength standards. Emission at 95.8 and 216 μm from pure helium has been obtained (CW) by Levine and Javan[157] and (pulsed) by Crocker

[154] T. Y. Chang, T. J. Bridges, and E. G. Burkhardt, *Appl. Phys. Lett.* **17**, 357 (1970).

[155] N. V. Karlov, Yu. B. Konov, Yu. N. Petrov, A. M. Prokhorov, and O. M. Stel'-makh, *JETP Lett.* **8**, 12 (1968).

[156] C. K. N. Patel, W. L. Faust, R. A. McFarlane, and C. G. B. Garrett, *Proc. IEEE* **52**, 713 (1964).

[157] J. S. Levine and A. Javan, *Appl. Phys. Lett.* **14**, 348 (1969).

et al.[158] The transitions can readily be assigned to $3^1P \rightarrow 3^1D$ and $4^1P \rightarrow 4^1D$, respectively. The laser has to be exceedingly clean internally for these lines to be observed and additionally very high current densities are necessary but when these requirements are satisfied the emission can be quite intense ($\sim$1.5 mW). No sealing off difficulties should arise since the system is atomic and there seems no reason why pure helium sub-millimeter lasers should not become routine laboratory sources. Absolute frequency measurements of these two lines should prove considerably interesting since the frequencies can in principle be calculated from first principles.

2.4.9.4. Chemical Lasers. Emission is observed from pulsed discharges through the boron halides in the conventional laser arrangement. Very many lines are obtained between 11 and 41 μm but unlike the special case outlined in Section 2.4.9.2, there is no doubt that the BX_3 molecules are *not* the emitters. All the observed frequencies fit known transitions in the hydrogen halides. With BF_3 as the starting gas, the strongest line at 819.062 cm^{-1} fits the pure rotational transition

$$J = 22 \rightarrow J = 21$$

in the vibrational ground state of HF. The hydrogen is presumably abstracted from impurities on the tube walls or else from the tubes in the gas handling system. Addition of traces of water vapor greatly enhances the emission. Deutsch in 1967 observed[159] very many pure rotational lines of HF in various vibrationally excited states when the reaction

$$CF_4 + H_2 \rightarrow CF_3H + HF$$

was taking place in a laser cavity under the influence of an electric discharge (pulsed). Gas pressures (measured at the exit port) were between 0.75 and 5.0 Torr. There seems to be no reason why very many more chemical lasers should not be constructed operating on pure rotational transitions. High powers are in principle possible since one is not restricted to the low concentrations that are usually dictated by discharge conditions. Low-pressure flames in a laser cavity may yield some very interesting emission lines.

[158] L. E. S. Mathias, A. Crocker, and M. S. Wills, *IEEE J. Quantum Electron.* **QE 3**, 170 (1967).

[159] T. F. Deutsch, *Appl. Phys. Lett.* **11**, 18 (1967).

Appendix

TABLE A.I. Observed Laser Lines in $H_2{}^{16}O$

$\bar{\nu}$ (cm^{-1})	λ (µm)	Upper level		Lower level		Remarks[a]	Peak pulsed output power (W)	Reference
45.407	220.230	100	5_{23}	020	5_{50}	P, CW	0.002	b
83.28	120.08	001	6_{42}	001	6_{33}	P		c
84.3234	118.591	001	6_{42}	020	6_{61}	P, CW	0.001	d
86.72	115.32	020	8_{35}	020	8_{26}	P, CW	0.007	
110.42	90.565					P		
111.18	89.947					P		
111.39	89.772	100	9_{54}	020	9_{63}	P	0.006	
114.32	87.469	100	8_{53}	020	8_{62}	P		
114.52	87.323					P		
115.64	86.478	100	9_{54}	100	9_{45}	P		
115.87	86.301					P		
116.87	85.564	100	7_{52}	020	7_{61}	P		
126.44	79.087	020	8_{44}	020	8_{35}	P, CW	0.006	
127.48067	78.44326	100	8_{08}	020	8_{35}	P, CW	0.007	
136.24	73.401	100	8_{17}	100	8_{08}	P, CW	0.002	
137.26	72.856	100	8_{17}	020	8_{44}	P		
146.32	68.344	020	5_{41}	020	5_{32}	P		
148.88	67.169	100	6_{25}	020	5_{50}	P	0.01	
		020	4_{41}	020	4_{32}	P		
149.45	66.903					P		
149.52	66.880					P		
173.01	57.799	020	9_{54}	020	9_{45}	P		
173.43	57.659	100	9_{19}	020	8_{44}	P	0.02	
181.53	55.088	020	5_{50}	020	5_{41}	P, CW	0.06	
181.82	55.000	020	6_{52}	020	6_{43}	P		
182.31	54.853					P		b
185.49	53.910					P	0.008	c
203.83	49.06	100	7_{43}	020	7_{52}	P	0.001	
205.44	48.676					P	0.07	
207.5	48.19	100	9_{45}	020	9_{54}	P, G	0.028	
209.70	47.687	020	6_{61}	020	6_{52}	P, CW	0.04	
210.67	47.468	001	6_{33}	020	6_{52}	P, CW	0.06	
210.67	47.39					G,	0.024	
211.67	47.244	020	9_{63}	020	9_{54}	P, CW	0.08	
217.82	45.91					P, G	0.006	
219.70	45.517					P	0.007	

TABLE A.I (*continued*)

$\bar{\nu}$ (cm^{-1})	λ (μm)	Identification Upper level	Lower level	Remarks[a]	Peak pulsed output power (W)	Reference
235.24	42.51			P, G	0.006	
246.08	40.638	020 4_{41}	020 3_{30}	P	0.01	
247.2	40.45	100 $13_{1,13}$	020 $12_{2,10}$	P		
251.92	39.695	001 7_{44}	020 6_{61}	P	0.10	
262.56	38.086			P		
264.21	37.848	100 $12_{0,12}$	020 11_{39}	P	0.003	
273.18	36.606	010 $13_{11,2}$	010 $13_{10,3}$	P	0.009	
279.07	35.833	100 $12_{1,12}$	020 11_{29}	P	0.10	
285.58	35.017	100 7_{34}	100 6_{25}	P		
		020 $12_{3,10}$	020 11_{29}	P		
289.02	34.60			P, G	0.35	
302.76	33.029	100 5_{14}	020 4_{41}	P, CW	7.0	
303.73	32.924	020 5_{50}	020 4_{41}	P	0.41	b
351.48	28.451	100 6_{42}	020 5_{51}	P		c
352.66	28.356	020 8_{44}	020 7_{35}	P	0.01	
353.73	28.270	100 8_{08}	020 7_{35}	P	0.60	
356.46	28.054	020 6_{61}	020 5_{50}	P		
357.51638	27.97075	001 6_{33}	020 5_{50}	P, CW, S	3.0	
375.09	26.660	100 7_{43}	020 6_{52}	P	0.50	
400.54	24.966	100 8_{44}	020 7_{53}	P		
427.99	23.365	100 9_{45}	020 8_{54}	P	0.10	
432.3	23.13	020 9_{63}	020 8_{54}	P	0.04	
590.60	16.932	010 $13_{11,2}$	010 $12_{10,3}$	P	0.02	
836.1	11.96			P		
845.3	11.83			P		
1297.19	7.7090	020 6_{61}	010 7_{70}	P		
1316.38	7.5966	020 5_{50}	010 6_{61}	P		
1340.70	7.4588	020 4_{41}	010 5_{50}	P		
2096.4	4.77			P		
4387.60	2.2792			P		

[a] Key: P, pulsed; CW, continuous wave; G, grating resonator; S, strongest emission line in most lasers.

[b] B. Hartmann and B. Kleman, *Appl. Phys. Lett.* **12**, 168 (1968); W. S. Benedict, *ibid.* **12**, 170 (1968); M. A. Pollack and W. J. Tomlinson, *ibid.* **12**, 173 (1968).

[c] W. S. Benedict, M. A. Pollack, and W. J. Tomlinson, *IEEE J. Quantum Electron.* **QE 5**, 108 (1969).

[d] V. Daneu, L. O. Hocker, A. Javan, D. Ramachandra Rao, A. Szoke, and F. Zernike, *Phys. Lett.* **29A**, 319 (1969).

TABLE A.II. Observed Laser Lines in $H_2^{18}O$ [b]

$\bar{\nu}$ (cm^{-1})	λ (µm)	Identification Upper level	Identification Lower level	Remarks[a]	Correct relative output power
178.16	56.129	020 5_{50}	020 5_{41}	P	0.04
202.31	49.430	100 8_{45}	020 8_{54}	P	0.08
205.07	48.765	100 7_{44}	020 8_{53}	P	0.05
205.74	48.604	020 6_{61}	020 6_{52}	P	0.24
206.76	48.366	100 6_{43}	020 6_{52}	P	0.28
282.62	35.383	100 $12_{1,12}$	020 11_{29}	P	0.08
300.23	33.308	020 5_{50}	020 4_{41}	P	0.08
353.42	28.295	100 6_{43}	020 5_{50}	P	1.00
376.01	26.595	100 7_{44}	020 6_{51}	P	0.38
397.42	25.162	100 8_{45}	020 7_{52}	P	0.38

[a] Key: P, pulsed.
[b] W. S. Benedict, M. A. Pollack, and W. J. Tomlinson, *IEEE J. Quantum Electron.* **QE 5**, 108 (1969).

TABLE A.III. Observed Laser Lines in D_2O

$\bar{\nu}$ (cm^{-1})	λ (µm)	Identification Upper level	Identification Lower level	Remarks[a]	Peak pulsed output power (W)	Reference
58.25	171.67	100 $11_{0,11}$	020 11_{38}	P, CW		[b]
58.80	170.08	020 11_{47}	020 11_{38}	P		[c]
89.49	111.74	100 13_{68}	020 13_{77}	P		
90.51	110.49	100 12_{66}	020 12_{75}	P		
91.84	108.88	100 11_{65}	020 11_{74}	P		
92.67	107.91	100 13_{68}	100 13_{59}	P		
92.823	107.731	100 11_{66}	020 11_{75}	P, CW	0.01	
96.78	103.33			P		
101.01	99.00			P		
118.6535	84.2791	100 $12_{1,12}$	020 11_{47}	P, CW	0.05	[d]
118.89	84.111	020 11_{66}	020 11_{57}	P	0.02	

TABLE A.III (*continued*)

$\bar{\nu}$ (cm^{-1})	λ (μm)	Identification Upper level	Lower level	Remarks[a]	Peak pulsed output power (W)	Reference
119.43	83.730	020 10_{65}	020 10_{56}	P		
127.94	78.16			P		
131.05	76.305			P	0.009	
134.18	74.526	100 13_{59}	020 13_{68}	P	0.04	
136.35	74.341	100 12_{58}	020 12_{67}	P	0.04	
137.44	72.757	020 11_{75}	020 11_{66}	P	0.08	
138.07	72.427	020 10_{73}	020 10_{61}	P	0.02	
		020 10_{74}	020 10_{65}	P		
138.99	71.944	100 11_{57}	020 11_{66}	P, CW	0.008	
163.45	61.182			P		
175.96	56.830	100 $16_{0,16}$	020 $15_{1,15}$	P	0.10	
.182.72	54.73			P		
197.20	50.71			P		
204.92	48.80			P		
239.28	41.79	100 12_{66}	020 11_{75}	P		
243.94	40.994	020 10_{64}	020 9_{55}	P	0.02	b
252.97	39.53	100 9_{55}	020 8_{62}	P		c
264.13	37.860	100 10_{56}	020 9_{63}	P		
264.63	37.788	100 10_{55}	020 9_{64}	P		
273.78	36.526	020 11_{75}	020 10_{64}	P	0.04	
275.30	36.324	100 11_{57}	020 10_{64}	P	0.40	
277.04	36.096	100 11_{56}	020 10_{65}	P		
285.05	35.081	100 12_{58}	020 11_{65}	P	0.30	
295.02	33.896	100 13_{59}	020 12_{66}	P	0.30	

[a] Key: P, pulsed; CW, continuous wave.

[b] B. Hartmann and B. Kleman, *Appl. Phys. Lett.* **12**, 168 (1968): W. S. Benedict, *ibid.* **12**, 170 (1968); M. A. Pollack and W. J. Tomlinson, *ibid.* **12**, 173 (1968).

[c] W. S. Benedict, M. A. Pollack, and W. J. Tomlinson, *IEEE J. Quantum Electron.* **QE 5**, 108 (1969).

[d] V. Daneu, L. O. Hocker, A. Javan, D. Ramachandra Rao, A. Szoke, and F. Zernike, *Phys. Lett.* **29A**, 319 (1969).

TABLE A.IV. CW and Pulsed Laser Transitions of HCN

λ_{vac} (μm)	$\bar{\nu}_{vac}$ (cm^{-1})	Identification		Remarks[a]	Peak pulsed output power (W)	Ref.
		Upper state	Lower state			
372.5283	26.8436	$04^00, J=9$	$04^00, J=8$	P, CW	0.6	b
336.5578	29.7126	$11^10, J=10(+)$	$04^00, J=9$	P, CW(VH)	0.6(10)	b,c,d
335.1831	29.8344	$04^00, J=10$	$04^00, J=9$	CW		b
310.8870	32.1660	$11^10, J=11(-)$	$04^00, J=10$	P, CW	1	b,d
309.7140	32.2879	$11^10, J=11(-)$	$11^10, J=10(+)$	P, CW	0.4	b
284.	35.2	$11^10, J=12(+)$	$11^10, J=11(-)$	CW		b
222.949	44.853			P	0.08	c
211.001	47.393			P, CW	0.2	c
201.059	49.737			P	0.05	c
134.932	74.111	$12^00, J=25$	$05^10, J=24(+)$	P	0.8	c
130.838	76.430	$12^00, J=26$	$05^10, J=25(-)$	P	4	c
128.629	77.743	$12^00, J=26(-)$	$05^10, J=25(+)$	P, CW	9	c
126.164	79.262	$12^00, J=27(+)$	$05^10, J=26(-)$	P	3	c
116.132	86.109			P	0.5	c
112.066	89.233			P	0.2	c
101.257	98.759			P	0.2	c
98.693	101.325			P	0.8	c
96.401	103.733			P	0.2	c
81.544	122.618			P	0.1	c
77.001	129.868			P	0.003	c
76.093	131.418			P	0.005	c
73.101	136.796			P	0.008	c
71.899	139.084			P	0.3	c

[a] Key: P, pulsed; CW, continuous wave; VH, very high.

b L. O. Hocker and A. Javan, *Phys. Lett.* **25A**, 489 (1967).

c L. E. S. Mathias, A. Crocker, and M. S. Wills, *IEEE J. Quantum Electron.* **QE 4**, 205 (1968).

d L. O. Hocker, A. Javan, D. Ramachandra Rao, L. Frenkel, and T. Sullivan, *Appl. Phys. Lett.* **10**, 147 (1967).

TABLE A.V. CW and Pulsed Laser Transitions of DCN

λ_{vac} (μm)	$\bar{\nu}_{vac}$ (cm^{-1})	Identification		Remarks[a]	Autput power[a]	Ref.
		Upper state	Lower state			
204.3872	48.9267	$09^10, J = 20(+)$	$09^10, J = 19(-)$	P, CW	M	b
194.7644	51.3441	$09^10, J = 21(-)$	$09^10, J = 20(+)$	CW	M	c
194.7027	51.3604	$22^00, J = 21$	$09^10, J = 20(+)$	P, CW	H	
190.0080	52.6294	$22^00, J = 22$	$22^00, J = 21$	CW	M	
189.9490	52.646	$22^00, J = 22$	$09^10, J = 21(-)$	P, CW	H	
181.789	55.009	$22^00, J = 23$	$22^00, J = 22$	P	L	

[a] Key: P, pulsed; CW, continuous wave; H, high; M, medium; L, low.
[b] L. E. S. Mathias, A. Crocker, and M. S. Wills, *IEEE J. Quantum Electron.* **QE 4**, 205 (1968).
[c] R. G. Jones, C. C. Bradley, J. Chamberlain, H. Gebbie, N. W. B. Stone, and H. Sixsmith, *Appl. Opt.* **8**, 701 (1969).

TABLE A.VI. Pulsed Laser Transitions of HC^{15}N [b]

λ_{vac} (μm)	$\bar{\nu}_{vac}$ (cm^{-1})	Remarks[a]	Output power[a]
165.150	48.927	P	L
138.767	51.359	P	L
113.311	52.359	P	L
110.240	55.009	P	L

[a] Key: P, pulsed; L, low.
[b] L. E. S. Mathias, A. Crocker, and M. S. Wills, *IEEE J. Quantum Electron.* **QE 4**, 205 (1968).

TABLE A.VII. CW and Pulsed Laser Transitions of SO$_2$

λ_{air} (μm)	$\bar{\nu}_{vac}$ (cm^{-1})	Remarks[a]	Output power[a]	Reference
140.85	70.98	P, CW	H	b
151.16	66.14	P	L	c
192.67	51.89	P, CW	M	d
215.27	46.44	P	L	

[a] Key: P, pulsed; CW, continuous wave; H, high; M, medium; L, low.
[b] J. C. Hassler and P. D. Coleman, *Appl. Phys. Lett.* **14**, 135 (1969).
[c] S. F. Dyubko, V. A. Svich, and R. A. Valitov, *JETP Lett.* **7**, 320 (1968).
[d] T. M. Hard, *Appl. Phys. Lett.* **14**, 130 (1969).

TABLE A.VIII. Pulsed Laser Transitions of H_2S [b]

λ_{air} (μm)	$\bar{\nu}_{air}$ (cm^{-1})	Output power[a]
225.3	44.39	H
192.9	51.84	M
162.4	61.58	H
140.6	71.12	M
135.5	73.80	L
130.8	76.45	L
129.1	77.46	L
126.2	79.24	L
116.8	85.62	L
108.8	91.91	L
103.3	96.81	H
96.38	103.8	L
92.00	108.7	L
87.47	114.3	H
83.43	119.9	VL
80.50	124.2	H
73.52	136.0	VL
61.50	162.6	H
60.29	165.9	H
56.84	175.9	M
52.40	190.8	M
49.62	201.5	L
33.64	297.3	H
33.47	298.8	L

[a] Key: H, high; M, medium; L, low; VL, very low.
[b] J. C. Hassler and P. D. Coleman, *Appl. Phys. Lett.* **14**, 135 (1962).

TABLE A.IX. Pulsed Laser Transitions of OCS [b]

λ_{air} (μm)	$\bar{\nu}_{air}$ (cm^{-1})	Output power[a]
132	75.8	L
123	81.3	L

[a] Key: L, low.
[b] J. C. Hassler and P. D. Coleman, *Appl. Phys. Lett.* **14**, 135 (1969).

TABLE A.X. Pulsed Laser Transitions in NH_3 [b]

λ_{vac} (μm)	ν_{vac} (cm^{-1})	Identification	Remarks[a]	Relative output power[a]
32.13	311.2		(1)–(3)	H
31.951	313.3		(1)–(3)	H
30.69	317.8		(1)	L
26.282	380.49	$3^s \rightarrow 2^a$, ν_2 $P(6)$ $K = 0, J = 5-6$	(1), (3)	L
25.12	398.1		(1)–(3)	H
24.918	401.32	$P(5)$ $K = 1, J = 4-5$	(1)–(3)	H
23.86	419.1		(1)–(3)	H
23.675	422.39	$P(4)$ $K = 0, J = 3-4$	(1)–(3)	H
22.71	440.3		(1)–(3)	H
22.563	443.20	$P(3)$ $K = 1, J = 2-3$	(1)–(3)	H
22.542	443.62	$P(3)$ $K = 2, J = 2-3$		
21.471	465.74	$(P(2)$ $K = 1, J = 1-2)$	(1)–(3)	M
18.21	549.1		(1)–(3)	M
15.47	646.4		(2), (3)	H
15.41	648.9		(1)	L
15.08	663.1		(1), (3)	L
15.04	664.9		(1)	L
14.78	676.6		(1), (3)	H

[a] Key: (1) nonfrequency selective; (2) frequency selective, 22-μm grating; (3) frequency selective, 30-μm grating; H, high power; M, medium power; L, low power.

[b] L. E. S. Mathias, A. Crocker, and M. S. Wills, *Phys. Lett.* **14**, 33 (1965); D. P. Akitt and C. F. Wittig, *J. Appl. Phys.* **40**, 902 (1969).

TABLE A.XI. CW Laser Transitions in Helium

λ_{vac} (µm)	$\bar{\nu}_{vac}$ (cm^{-1})	Assignment	Remarks[a]	Relative intensity	Reference
216.3	46.23	$4^1P - 4^1D$	CW	M	[b]
95.8	104.4	$3^1P - 3^1D$	P, CW	H	[b,c]

[a] Key: CW, continuous wave; P, pulsed; M, medium; H, high.
[b] J. S. Levine and A. Javan, *Appl. Phys. Lett.* **14**, 348 (1969).
[c] L. E. S. Mathias, A. Crocker, and M. S. Wills, *IEEE J. Quantum Electron.* **QE 3**, 170 (1967).

TABLE A.XII. CW Laser Transitions in Neon [b]

λ_{vac} (µm)	$\bar{\nu}_{vac}$ (cm^{-1})	Assignment	Remarks[a]	Relative intensity[a]
132.8	75.30	None	CW	VL
126.1	79.30	None	CW	VL
124.4	80.39	$8p_{7,6} - 8d_1''$ or $8p_{7,6} - 8d_1'$	CW	VL
106.02	94.32	$9p_3 - 9d_2$	CW	VL
93.02	107.5	None	CW	VL
89.93	111.2	$7p_9 - 7d_4$	CW	VL
88.46	113.0	$7p_7 - 7d_1$	CW	VL
86.9	115.0	$7p_4 - 7s_1''''$	CW	VL
85.01	117.6	$7p_6 - 7d_1$		
72.15	138.6	$7p_1 - 7s_1'$	CW	VL
55.68	179.6	$6p_4 - 6s_1''''$	CW(He–Ne)	VL
53.47	187.0	$6p_6 - 6d_1$		
52.39	190.9	$6p_2 - 6s_1$	CW	VL
50.76	197.0	$6p_6 - 6d_3$	CW	VL

[a] Key: P, pulsed; CW, continuous wave; H, high power; M, medium power; L, low power; VL, very low power.
[b] C. K. N. Patel, W. L. Faust, R. A. McFarlane, and C. G. B. Garrett, *Proc. IEEE* **52**, 713 (1964).

TABLE A.XIII. Optically Pumped CW and Pulsed Laser Transitions of CH_3F Using a CO_2 Laser

λ_{vac} (μm)	$\bar{\nu}_{vac}$ (cm^{-1})	Identification	Remarks[a]	Relative intensity[a]	λ pump (μm)	Reference
496.071	20.1584	ν_3, $K = 1$, $J = 12 - 11$	P, CW	H	9.5524	[b]
496.105	20.1570	ν_3, $K = 2$, $J = 12 - 11$	P, CW	H		[c]
541.113	18.4804	ν_3, $K = 1$, $J = 11 - 10$	P			
541.147	18.4793	ν_3, $K = 2$, $J = 11 - 10$	P			
451.903	22.1286	G.S. $K = 1$, $J = 13 - 12$	P			
451.924	22.1276	G.S. $K = 2$, $J = 12 - 12$	P			

[a] Key: P, pulsed; CW, continuous wave; H, high; G.S., ground state.
[b] T. Y. Chang and T. J. Bridges, *Opt. Commun.* **1**, 423 (1970).
[c] T. Y. Chang, T. J. Bridges, and E. G. Burkhardt, *Appl. Phys. Lett.* **17**, 249 (1970).

TABLE A.XIV. Optically Pumped CW Laser Transitions of $H_2C:CHCl$ Using a CO_2 Laser [b]

λ_{vac} (μm)	$\bar{\nu}_{vac}$ (cm^{-1})	Identification	Remarks[a]	Relative intensity[a]	λ Pump
634.4	15.76	CH$_2$ rock, pure rotational	CW	M	9.5524
507.7	19.70	CH wag, pure rotational	CW	M	10.6114
386.0	25.91	CH wag, pure rotational	CW	L	10.6114

[a] Key: CW, continuous wave; M, medium; L, low.
[b] T. Y. Chang, J. Bridges, and E. G. Burkhardt, *Appl. Phys. Lett.* **17**, 249 (1970).

TABLE A.XV. Optically Pumped CW Laser Transitions of CH_3OH Using a CO_2 Laser [b]

λ_{vac} (μm)	$\bar{\nu}_{vac}$ (cm^{-1})	Identification	Remarks[a]	Relative intensity[a]	λ pump
570.5	17.53	$\nu_5(R10), J = 11 - 10$	CW	H	9.5198
369.1	27.09	ν_5	CW	H	9.5198
223.5	44.74	ν_5	CW	M	9.5198
164.3	60.86	ν_5	CW	M	9.5198
264.6	37.79	ν_5	CW	M	9.6760
263.7	37.92		CW	M	
254.1	39.35		CW	M	
253.6	39.43		CW	L	
237.6	42.09		CW	L	
699.5	14.30		CW	VL	
292.5	34.19		CW	VL	
190.8	52.41		CW	VL	
185.5	53.91		CW	VL	
70.6	142.		CW	L	
202.4	49.41	ν_5	CW	M	9.6948
170.6	58.62		CW	H	
417.8	23.93		CW	M	
392.3	25.49		CW	H	
118.8	84.18		CW	H	
292.2	34.22	ν_5	CW	VL	9.7140
278.8	35.87		CW	VL	
198.8	50.30		CW	L	
193.2	51.76		CW	VL	

[a] Key: CW, continuous wave; H, high; M, medium; L, low; VL, very low.
[b] T. Y. Chang, T. J. Bridges, and E. G. Burkhardt, *Appl. Phys. Lett.* **17**, 249 (1970).

TABLE A.XVI. Optically Pumped CW Laser Transitions of NH_3 Using an N_2O Laser [b]

λ (μm)	$\bar{\nu}$ (cm^{-1})	Identification	Remarks[a]	Relative intensity[a]	λ pump
81.5	123.0	ν_2, $K = 7$, $8^s \rightarrow 7^a$	CW	H	10.78
263.4	37.97	ν_2, $K = 7$, $7^a \rightarrow 7^s$	CW	M	

[a] Key: CW, continuous wave; H, high; M, medium.
[b] T. Y. Chang, T. J. Bridges, and E. G. Burkhardt, *Appl. Phys. Lett.* **17**, 357 (1970).

Addendum

Duxbury and Jones[160] have extended their observations of electric resonance spectroscopy and have studied the 1.5397-THz line (J = 5 ← 4) of ND_3 using a DCN (195 μm) laser as the source, and Brittain et al.[161] have successfully applied the small square-wave modulation technique to the study of the 890-GHz D_2O transition. The forecast that submillimeter EPR techniques might prove very powerful in detecting some of the free radicals that had evaded conventional techniques has turned out correct, for Evenson and his colleagues[162] have successfully observed the CH radical via its pure rotation spectrum. Curl et al.[163] have studied the laser magnetic resonance spectrum of NO_2 using the 337-μm and the 311-μm lines of the HCN laser.

Electric resonance (or Stark-shift spectroscopy) of gases using the CO_2 laser as a source has continued to yield interesting results. Shimizu[164,165] has carried out extensive investigations of the ν_2 bands of $^{14}NH_3$ and $^{15}NH_3$. The ν_3 bands of $^{12}CH_3F$ and $^{13}CH_3F$ have been investigated in detail by Duxbury et al.[166] These latter have also shown that the overall sensitivity of spectrometers, operating in this manner, is at least 10^4

[160] G. Duxbury and R. G. Jones, *Chem. Phys. Lett.* **8**, 439 (1971).
[161] A. H. Brittain, A. P. Cox, G. Duxbury, T. G. Hersey, and R. G. Jones, *Mol. Phys.* **24**, 843 (1972).
[162] K. M. Evenson, H. E. Radford, and M. M. Moran, *Appl. Phys. Lett.* **18**, 426 (1971).
[163] R. F. Curl, K. M. Evenson, and J. S. Wells, *J. Chem. Phys.* **56**, 5143 (1972).
[164] F. Shimizu, *J. Chem. Phys.* **52**, 3572 (1970).
[165] F. Shimizu, *J. Chem. Phys.* **53**, 1149 (1970).
[166] G. Duxbury, S. M. Freund, T. Oka, and J. T. Tiedje (to be published).

times higher than that of a conventional infrared spectrometer. This was demonstrated by the ready observation of the hot band lines $(2\nu_3 \leftarrow \nu_3)$ on an oscilloscope with a brief time constant (e.g., 10 msec). The calculated[167] ultimate sensitivity of this type of spectrometer is so high that the minimum detectable number of absorbing molecules may be as low as 10^2 to 10^6.

Double-resonance and two-photon absorption techniques are rapidly making an impression in the infrared region. The numerous near coincidences of vibration–rotation lines with emission lines of the CO_2 laser, coupled with the inherent power of this laser, have made a series of fascinating experiments possible. Infrared–microwave double resonance has been used by Shimizu and Oka[168] to study $^{14}NH_3$ and $^{15}NH_3$. The technique has been extended to cover radio-frequency double resonance by Shimizu,[169] who has obtained rf resonances between the Stark levels of the $P(3, 2)$ line of the ν_2 band of PH_3. An N_2O laser was used here and the resonance was monitored by observing the change in transmission of the sample with respect to the laser radiation as the radiofrequency source was tuned. Since the line width of the resonance is governed by the radio frequency used, very high resolution is possible. The precision of the dipole moment measurements carried out in this way is comparable with that of those determined using the optical-optical method described earlier.

The tunability of the CO_2 laser as a spectroscopic source has been enhanced by the use of the two-photon technique pioneered by Oka and his colleagues.[170] In these experiments radiation from a tunable microwave source is "added" by means of the nonlinearity of the molecular absorption to a fixed-frequency laser emission. In this way absorption lines lying several GHz from the laser line center may be studied. Further work by Freund and Oka[171] has shown that it is possible to obtain an inverse Lamb dip on the two-photon transition, thus enhancing the resolution as explained previously. The latest advance in this field has come in the work of Oka et al.[172] who used two two-photon pumping sources and were able to observe infrared–infrared double resonance. An alternative approach to the problem of making the CO_2 laser tunable

[167] K. Shimoda, Int. J. Appl. Phys. (to be published).

[168] T. Shimizu and T. Oka, J. Chem. Phys. 53, 2536 (1970).

[169] F. Shimizu, Chem. Phys. Lett. 18, 382 (1973).

[170] T. Oka and T. Shimizu, Appl. Phys. Lett. 19, 88 (1971).

[171] S. M. Freund and T. Oka, Appl. Phys. Lett. 21, 60 (1972).

[172] T. Oka, S. M. Freund, J. W. C. Johns, and A. R. McKellar (to be published).

over an appreciable range has been developed by Corcoran *et al.*[173] who have constructed a spectrometer in which the probing radiation is provided by mixing CO_2 laser radiation and the radiation from a millimeter-wave klystron in a gallium-arsenide-loaded waveguide.

An assignment of the lines observed from the SO_2 laser has now been published by Hubner *et al.*[174] As expected, the mechanism proposed involves irregular perturbation between ν_1 and $2\nu_2$. Ten lines are now known from this laser, one of them at 206 μm having been forecast on the basis of the assignment. Its subsequent discovery lends strong support to the proposed assignment.

The number of optically pumped lasers has been extended both by the application of the techniques described earlier and by the introduction of the transversely excited atmospheric pressure (TEA) CO_2 laser. This laser gives out a regular series of pulses whose peak heights can be in the megawatt region. Chang and McGee[175] have observed 21 new laser lines, with wavelengths ranging from 192.8 μm to 1.8144 mm, emitted from $^{12}CH_3F$ and $^{13}CH_3F$, acetonitrile CH_3CN, and methyl acetylene $CH_3C\equiv CH$ pumped by a CO_2 laser. Dyubko *et al.*[176] have observed 18 new lines from 1:1 difluorethylene CF_2CH_2, vinyl cyanide CH_2CHCN, and methylamine CH_3NH_2 again with CO_2 laser pumping. Wagner and Zelano[177] have obtained laser action in formic acid $HCOOH$, methyl chloride CH_3Cl, methyl bromide CH_3Br, and ethyl fluoride CH_3CH_2F using an electrically pulsed CO_2 laser. For these emissions, photon efficiencies of the order 1% have been claimed.

The first use[178] of the TEA laser in this connection was as a more powerful pump for the 496-μm line of CH_3F. The original work[151,152] gave a power output of 0.1 W for this line and the power output was found to increase to 10 W under TEA excitation.[178] However the output power of the TEA laser used was circa 1 MW as compared with the 1.5 kW of Chang and Bridges. The pumping efficiency of the more powerful laser is therefore less than that of the less powerful. So far no complete

[173] V. J. Corcoran, J. M. Martin, and W. T. Smith, Montreal Conf. Quantum Electronics, Paper M1, 1972.

[174] G. Hubner, J. C. Hassler, P. D. Coleman, and G. Steenbeckeliers, *Appl. Phys. Lett.* **18**, 511 (1971).

[175] T. Y. Chang and J. D. McGee, *Appl. Phys. Lett.* **19**, 103 (1971).

[176] S. F. Dyubko, V. A. Svich, and L. D. Fesenko, *J.E.T.P. Lett.* **16**, 592 (1972).

[177] R. J. Wagner and A. J. Zelano, Montreal Conf. Quantum Electronics (1972).

[178] F. Brown, E. Silver, C. E. Chase, K. J. Button, and B. Lax, *IEEE J. Quantum Electron.* **QE 8**, 499 (1972).

explanation has been advanced for this. The 570 μm line of methanol can also be enhanced in power by TEA pumping yielding an output of circa 20 mW. The high electric fields produced by the TEA lasers allow off-resonance pumping of absorption lines by the laser radiation. This has been demonstrated by Fetterman et al.[179] for the examples of ammonia, methyl chloride, 1:1 difluorethane, CH_3CHCl_2, 1:2:3 trifluoroethane CH_3CF_3, methanol, and methyl fluoride. The absorption line centers of the transitions pumped were up to 1 GHz away from the center of the pumping TEA laser transitions.

The attempts to determine a more accurate value for the velocity of propagation of electromagnetic radiation have recently succeeded. Evenson and his colleagues[180] at the National Bureau of Standards have measured the frequency of the 3.39 μm line of the He–Ne laser stabilized to the saturated molecular absorption in methane. Their value is 88.376181627(50) THz. The wavelength of this line relative to the ^{86}Kr standard has been determined by Burger and Hall[181] also of NBS. These workers encountered an unexpected difficulty in that the fundamental standard line of ^{86}Kr was found to be asymmetric. One can therefore define two standard wavelengths—that of the peak of the line and that of the center-of-gravity of the line. In terms of these two, Burger and Hall report $\lambda_{max} = 3.392231404$ μm and $\lambda_{cg} = 3.392231376$ μm. Combining the latter with the frequency quoted by Evenson et al. gives $c_{cg} = 2.99792456_2 \times 10^5$ km sec^{-1}.

[179] H. R. Fetterman, H. R. Schlossberg, and J. Waldman, Opt. Commun. 6, 156 (1972).

[180] K. M. Evenson, J. S. Wells, F. R. Petersen, B. L. Danielson, and G. W. Day, Appl. Phys. Lett. 22, 192 (1973).

[181] R. L. Burger and J. L. Hall, Appl. Phys. Lett. 22, 196 (1973).

3. LIGHT SCATTERING[*][†]

3.1. Introduction

This chapter, of course, cannot be a complete treatment of light scattering even if it is limited to materials of a noncrystalline nature. The subject of light scattering is such an extensive field that in order to have a comprehensive and complete treatment of the subject a number of large volumes would be necessary. We shall attempt to present some of the historical high spots of the subject and bring some of the newer discoveries made in the laser era to the attention of the reader. Our approach to the subject will be largely a description of experimental observations and methods. We should point out that many books and extensive articles have been written on the various aspects of the theory of light scattering. Among these are the following which the authors have found particularly useful: (1) Rayleigh Streuung and Raman Effect, by Placzek, "Handbuch der Radiologie" (E. Marx, ed.). Vol. 6. Akademische Verlagsges, 1934. (2) "Scattering of Light and the Raman Effect," by S. Bhagavantam. Chem. Publ., Brooklyn, New York, 1942. (3) "Infrared and Raman Spectra," by G. Herzberg. Van Nostrand-Reinhold, Princeton, New Jersey, 1951. (4) "Non Linear Optics," by N. Bloembergen. Benjamin, New York, 1965. (5) "Molecular Scattering of Light," by I. L. Fabelinskii (translated from the Russian by Robert T. Beyer). Plenum Press, New York, 1968.

3.2. Spontaneous Rayleigh and Brillouin Scattering

3.2.1. General Rayleigh Scattering

Tyndall[1] prepared artificial atmospheres containing small particles and was able to show in the laboratory that these small particles were

---

[1] J. Tyndall, *Phil. Mag.* **37**, 384 (1869).

[†] See also Volume I, Chapter 7.7 and Volume 9A, Part 3.

[*] Part 3 is by D. H. Rank and T. A. Wiggins.

responsible for the scattering of light in these atmospheres. This scattering by particles is still known as the *Tyndall effect*. Tyndall, however, was incorrect in his assumption that the blue color of the sky was caused by scattering from small foreign particles. Lord Rayleigh and later Schuster[2] showed that the measured opacity of the dust-free atmosphere was in agreement with what would be expected if the light scattering was caused by molecules alone.

Lord Rayleigh[3] was able to show theoretically that particles which were small compared to the wavelength of light produced scattering in which the scattered intensity was inversely proportional to the fourth power of the wavelength.

Very extensive work was performed on the depolarization of the light scattered by dust-free air by Cabannes[4] for the first time in 1915. Several years later Lord Rayleigh[5] made systematic investigations on this subject using several other dust-free gases besides air. The amount of depolarization of light scattered from a gas is very small and extreme care must be taken to avoid systematic experimental errors. In the prelaser era in order to obtain sufficient intensity it was necessary to use a condensing lens to concentrate the light incident on the gas. The use of such a condensing lens gave rise to convergence error, for which it was necessary to correct, in order to obtain reliable depolarization measurements.

The experimental arrangements used by the early investigators in the study of light scattering by gases are discussed in detail by Bhagavantam[6] in Chapter IV of his book, "The Scattering of Light and the Raman Effect." Various types of optical crosses are described for use with both low- and high-pressure gases. It cannot be too strongly emphasized that the avoidance of parasitic light is absolutely essential for the success of experiments dealing with light scattering of gases. Theoretically, if the absolute scattering power of a gas is determined, it is possible to calculate Avogadro's number in terms of the scattering power and other experimentally determined parameters of the gas. Cabannes[7] has performed these experiments for argon gas and obtained 6.90×10^{23} for

[2] A. Schuster, "Optics." Arnold, London, 1909.

[3] Lord Rayleigh, *Phil. Mag.* **41**, 107, 274 (1871).

[4] J. Cabannes, *C. R. Acad. Sci. Paris* **160**, 62 (1915).

[5] Lord Rayleigh, *Proc. Roy. Soc.* **94**, 453 (1918).

[6] S. Bhagavantam, "The Scattering of Light and the Raman Effect." Chem. Publ., Brooklyn, New York, 1942.

[7] J. Cabannes, *Ann. Phys.* **15** [9], 5 (1921).

Avogadro's number. Daure[8] has made a similar determination for ethyl chloride and obtained a value of 6.50×10^{23} for Avogadro's number. Considering the difficulties of these experiments, the results obtained give remarkable agreement with the determination of this important constant by other methods.

If we turn to scattering in the liquid state, we find that the absolute intensity of the scattered light is not proportional to the density of the material as we proceed from gas to liquid. Lord Rayleigh[9] found that molecule for molecule the scattering of ether vapor was seven times as strong in the gas as in the liquid phase. Since the work of Lord Rayleigh many investigators have subsequently verified his results with other liquids. Thus far we have been concerned only with the intensity and polarization of 90° Rayleigh scattering in gases and liquids. We shall now turn our attention to the fine structure of Rayleigh scattering, which is known as the *Brillouin effect*.

3.2.2. Brillouin Effect

Einstein and Smoluchowski have shown that light is scattered in a medium, essentially from the thermal density fluctuations. In order to explain the behavior of the specific heats of solids Debye was led to postulate naturally occurring sonic waves propagated throughout the medium. These sonic disturbances are essentially thermal noise and are an inevitable consequence of the Einstein–Smoluchowski density fluctuations in an elastic medium. In a brilliant flash of intuition Brillouin[10,11] predicted the logical consequences of the above situation with regard to a fine structure that should be observable in liquids, gases, and solids in the spectrum of the Rayleigh scattered light.

The Brillouin effect can be considered in a very simple manner. The statistical density fluctuations in a liquid, solid, or gas give rise to sonic disturbances (Debye waves or phonons). These waves are propagated in all possible directions and only become observable by means of light scattering when the Bragg condition (coherence condition) applies. The pertinent equation is $\lambda_l = 2\lambda_s \sin \Phi/2$, where λ_l is the wavelength of light, λ_s is the wavelength of the sonic disturbance, and Φ is the angle of scattering. These sonic wave trains propagated in all directions can be

---

[8] P. Daure, *C. R. Acad. Sci. Paris* **180**, 2032 (1925).

[9] Lord Rayleigh, *Proc. Roy. Soc.* **95**, 155 (1918).

[10] L. Brillouin, *C. R. Acad. Sci. Paris* **158**, 1331 (1914).

[11] L. Brillouin, *Ann. Phys.* **17** [9], 88 (1922).

considered as systems of moving mirrors. Light reflected from these moving mirrors will undergo a Doppler shift of magnitude $\Delta\nu = \pm(2\nu v/c)$ $\times \sin \Phi/2$, where $\Delta\nu$ is the frequency shift of the scattered light, ν is the frequency of the incident light, v is the velocity of sound, and c is the velocity of light in the medium in question. Furthermore it is easy to see that $\Delta\nu \equiv \nu_s$, where ν_s is the frequency of the sonic disturbance. If visible light is used, the sonic frequency chosen by the Bragg condition will lie in the gigahertz frequency region. The moving mirror picture also leads to the conclusion that the modified Brillouin components are completely polarized for $90°$ scattering.

Somewhat later and independently Mandelstam[12] arrived at results similar to those obtained by Brillouin derived from somewhat different considerations. In 1930, Gross[13] in Leningrad, in a brilliantly executed experiment, observed the Brillouin effect experimentally. Gross performed his experiment using a low-pressure Hg arc as the light source and a 30-step Michelson echelon as the high-resolution spectrograph. Gross was able to verify the angular dependence of the frequency shift and the other general results predicted by Brillouin's theory. Gross also noted that the central undisplaced component was apparently missing in the light scattered from crystal quartz. Landau and Placzek[14] proposed that the intensities of the modified and unmodified Brillouin components obey the relationship $I_c/2I_B = \gamma - 1$, where I_c and I_B are the intensities of the unmodified and modified components, respectively, and γ is the ratio of the specific heat at constant pressure to the specific heat at constant volume. The Landau–Placzek relationship is not always correct, but in the case of many substances it is an adequate representation of the results of experiment. Fabelinskii[15] (p. 364) has discussed the problem of the Landau–Placzek formula in considerable detail.

Experimental work on the Brillouin effect was very difficult before the invention of laser light sources. The low-pressure Hg arc was probably the best source available when all aspects are considered. The isotope effect and hyperfine structure cause considerable difficulty, since the frequency shifts that one wished to observe are small compared to the spread of the hyperfine pattern. Luckily, the far-flung satellites in the hyperfine pattern of the Hg lines are usually weak compared to the

[12] L. I. Mandelstam, *J. Russ. Phys. Chem. Soc.* **58**, 831 (1926).

[13] E. Gross, *Nature (London)* **126**, 201 (1930).

[14] L. D. Landau and G. Placzek, *L. Phys. Sowjetunion* **5**, 172 (1934).

[15] I. L. Fabelinskii, "Molecular Scattering of Light" (translated from the Russian by R. T. Beyer). Plenum Press, New York, 1968.

central component. Thus it is not possible to observe Brillouin components of intensity much smaller than the central unmodified component with Hg arc excitation when one is dealing with liquids where, in general, the Brillouin shift is approximately 0.2–0.3 cm^{-1}. However, because of the existence of the even isotopes of Hg, the central components of the Hg arc lines have an effective width of the order of 0.06 cm^{-1}. One of us[16] has summarized the work on the Brillouin effect that was performed in the period prior to the invention of the laser.

3.2.3. Experimental Techniques

3.2.3.1. Preparation of Samples (Liquids). The study of Brillouin spectra requires samples that are exceptionally free of dust or other small particles. The slightest amounts of such foreign material will give rise to sufficient parasitic light to vitiate the results seriously if meaningful intensity measurements are to be made. It does not take a very large amount of parasitic light to completely mask the Brillouin components.

For 90° scattering, one of the most desirable cells is an optical cross. This type of cell can be made completely of glass or quartz and thus is inherently a clean device. If desired, two of the arms of the cross can be provided with Brewster windows and the device can then be inserted in the resonant cavity of a laser light source. The cross is provided with a side tube for filling. This side tube is connected to a bulb that has nearly double the volume capacity of the main body of the cross. A typical one is shown in Fig. 1.

After the above-mentioned apparatus has been properly cleaned and dried, the bulb is filled to about three-quarters of its volume capacity with the desired liquid, which has been filtered through a Millipore filter. The apparatus is then evacuated and sealed off. The liquid is then vacuum distilled without ebullition from the bulb into the main body of the cross. In this manner no dirt nor dust particles are transferred from the bulb to the cross in the process of distillation. However, there will always be a certain amount of dust or dirt on the walls of the cross and connecting tubing. The liquid is then poured back into the bulb from the cross. This procedure tends to wash the dust or dirt from the walls of the cross and transfer said material to the bulb. If this procedure is repeated several times, one finally obtains a cross filled with liquid that

[16] D. H. Rank, Brillouin effect in liquids in the pre-laser era, *J. Acoust. Soc. Amer.* **49**, 937 (1971).

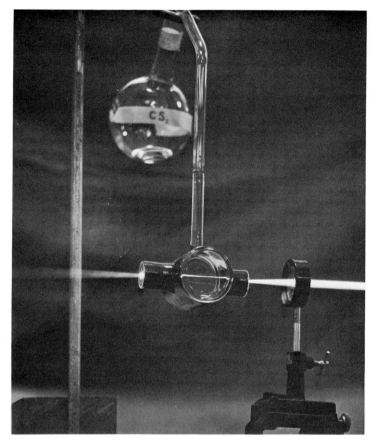

FIG. 1. Typical optical cross and auxiliary bulb.

is optically empty. No trace of dust or particles can be seen in the liquid and the scattering one obtains is truly scattering from the molecules of the liquid itself.

If one replaces the optical cross by a spherical bulb of good quality, it is easily possible to make measurements at scattering angles of 25° and 155°. It may be desirable to make measurements at angles different from 90°, particularly if investigation of the frequency dependence of the sound velocity is desired. A description of an accurate method of using such a device for scattering at angles other than 90° is given in a paper by Eastman[17] and his co-workers. The Bragg coherence condition will

[17] D. P. Eastman, A. Hollinger, J. Kenemuth, and D. H. Rank, *J. Chem. Phys.* **50**, 1567 (1969).

choose different sonic frequencies as the angle of scattering is changed. Of course, $\Delta\nu$, the Brillouin frequency shift (Doppler shift), varies as the sin $\Phi/2$ and is identically equal to ν_s, the frequency of the sound whose velocity is being measured.

3.2.3.2. Measurement of Brillouin Frequency Shifts.

Study of Brillouin frequency shifts in liquids or gases requires the use of instruments of high resolution. The light available in spontaneous scattering is weak, even if powerful lasers are used as the light source. Gross,[13] the discoverer of the Brillouin effect, used a Michelson echelon as the high-resolution instrument. Several of the early workers following Gross also used echelons. However, the echelon has some very undesirable characteristics and it is no longer used for this purpose. Modern diffraction gratings have ample resolution and all their spectral properties would seem to make them ideally suited for study of the Brillouin effect. However, the intensity of the spectra produced by high-resolution modern grating spectrographs is somewhat weaker than is desirable for many problems in Brillouin scattering. Benedek *et al.*[18] have used a 12-m focal length grating spectrograph in some studies of the spontaneous Brillouin effect. The advantages of the grating spectrograph for work on the Brillouin effect lie in the fact that this is the only high-resolution instrument which has a free spectral range very large compared to the extent of Brillouin patterns. For the study of stimulated Brillouin spectra and optical mixing (which will be treated in Section 3.7.1) the grating spectrograph is ideal since frequently large intensity of the scattered light is available.

Fabry–Perot etalons have been the most frequently used high-resolution instruments for the study of Brillouin scattering. These instruments have been used both with photographic and photoelectric recording of the spectra. When the Fabry–Perot fringes are photographed it is easy to measure the Brillouin shifts without obtaining the wavelengths of the Brillouin components. It is well known that the free spectral range in wave numbers (separation of neighboring orders) of a Fabry–Perot etalon is $\Delta\nu = 1/2t$, where t is the separation of the plates in centimeters. It is easy to derive the expression for the fractional order of interference ε_p for the pth ring by modifying the expression given by Williams[19]

$$\varepsilon_p = d_p{}^2/(d_p{}^2 - d_{p-1}^2).$$

[18] G. B. Benedek, J. B. Lastovka, K. Fritsch, and T. Greytak, *J. Opt. Soc. Amer.* **54**, 1284 (1964).

[19] W. E. Williams, "Applications of Interferometry." Dutton, New York, 1929.

Here d_p is the linear diameter of the pth ring. It can be seen from the above expression for ε_p that all the necessary information for obtaining Brillouin shifts is contained in the measurement of the ring diameters, and t is the separation of the etalon plates.

In the expression for ε_p it should be noted that $d_p{}^2 - d_{p-1}^2$, which we shall call the *denominator*, is the difference of the squares of the diameters of adjacent rings in the Fabry–Perot pattern. In a Brillouin pattern the denominators will all be the same for the central and the modified Brillouin components. (The value of the denominators in a Fabry–Perot pattern is a linear function of wavelength, and over the small wavelength range encompassed by a Brillouin pattern the change in value is completely negligible.) Thus one can see that upon measurement of the ring diameters one can use the mean value of all the denominators obtained from both the unmodified and modified members of the Brillouin triplet. The denominators are the least precisely determined quantities in the equation for ε_p. As long as one does not change the adjustment of the etalon or the focus of the lens forming the images of the fringes on the photographic plate, denominators obtained from several plates may be averaged, thus improving the precision of the ε_p determination. When the mean denominator is inserted in the expression for ε_p one obtains an ε_p value for each ring measured in the different orders and the mean of the ε_p value is used to determine the fractional order for each of the modified and the unmodified Brillouin components. Once the fractional orders are known, the Brillouin shifts can be immediately determined from $\Delta\nu$, the previously determined interorder separation.

Photoelectric detection, when possible, gives rise to some distinct advantages, particularly when intensity measurements of Brillouin components are desired. The equation for the condition responsible for the Fabry–Perot rings (constructive interference) is $m\lambda = 2\mu t \cos \theta_p$, where m is the order of interference, θ_p is the semi-angular diameter of the rings, λ is the wavelength, and μ is the index of refraction of the medium between the interferometer plates which have a separation t. Upon rearranging and differentiating we obtain the equation

$$d\nu = -(m/2t\mu^2 \cos \theta_p)\, d\mu,$$

where $d\nu$ is expressed in wave numbers. This equation shows that $d\nu$ is directly proportional to $d\mu$. To a high degree of approximation the change in the index of refraction of air is directly proportional to the pressure. Thus if one can change the air pressure between the interferometer plates linearly with time, one has a scanning device which is linear to wave numbers.

Pressure scanning of interferometer patterns has been used by many investigators who have studied the Brillouin effect. Among the first of these were Chiao and Stoicheff.[20] Rank and Shearer[21] have described a method involving a leak through a capillary where the flow of gas is supersonic. Under these conditions the change in pressure is directly proportional to time, practically independent of the back pressure, thus yielding a linear frequency scan. This system has been in use in our laboratory for many years and has proved to be eminently satisfactory. It is, of course, easy to see that the frequency calibration of such an interferometer system is automatic in that the pattern is repeated a number of times during a scan and the orders are $1/2t$ cm^{-1} apart. It is pertinent to remark that the optical system that projects the interference pattern on the detector must be provided with diaphragms so that the detector can see only a small fraction of the central ring (considerably less than 0.1 order in general). If the fractional order reaching the detector is too large, both the contrast and the resolving power of the interferometer are impaired.

If one attempts to measure the velocity of sound from the separation of the Brillouin components, care must be taken to make measurements at a given temperature or determine the temperature coefficient of the sound velocity so that measurements made of different substances may be compared. It is even more important to correct for the temperature dependence of the sound velocity when one attempts to measure the frequency dispersion of velocity by changing the scattering angle.

3.2.3.3. Fabry–Perot Interferometers.

The most recent treatment of the theory and methods of interpretation of the interference spectra produced by the Fabry–Perot interferometer has been given by Fabelinskii[15] in Appendix II of his book. In Section 3.2.3.2 we have described the method that we consider to be the ideal way to measure the Brillouin frequency shifts from Fabry–Perot interferograms. In the above mentioned description we have used the nomenclature as in Williams' small monograph.[19]

Any interferometer has three primary spectral characteristics: (1) resolving power; (2) free spectral range, i.e., interval of frequency over

[20] R. Y. Chiao and B. P. Stoicheff, *J. Opt. Soc. Amer.* **54**, 1286 (1964).

[21] D. H. Rank and J. N. Shearer, *J. Opt. Soc. Amer.* **46**, 463 (1956).

which the nth order will be caused to just overlap the $n + 1$ order; and (3) number of details uniquely observable.

The path difference between successive beams in multiple-beam interference we shall call p, and this quantity $p = 2\mu t/\lambda$. Thus one can show that the free spectral range is given by $\Delta\lambda = \lambda/p$ or $\Delta v = 1/(2\mu t)$. The number of details uniquely observable we shall call q^*. The number of beams producing interference in the Fabry–Perot interferometer is of course infinite. Thus q^* is the number of beams of equal amplitude that would be required to produce the identical interference pattern as observed with the real interferometer. The resolving power $R_p = \lambda/d\lambda = v/dv = pq^*$, where $d\lambda$ and dv are the smallest wavelength and wave number intervals resolvable.

From the theory a very useful quantity appears which Fabry called F, the coefficient of finesse,

$$F = 4R/(1 - R)^2,$$

where R is the reflectivity of the plates. To a good approximation, $q^* = \pi F^{1/2}/2.42$. It also follows approximately that the half-width intensity of the fringes is $d\lambda = \Delta\lambda/q^*$ or dv/q^*.

Finally, the minima in a Fabry–Perot interference pattern are not completely zero intensity. The contrast of the fringes K is given by

$$K = F + 1 = (1 + R)^2/(1 - R)^2.$$

It is seldom reliable to use strict directly measured values of plate reflectivity to calculate F, particularly if R is large. The plates are rarely

TABLE I. Properties of the Fabry–Perot Interferometer

Reflectivity R	Finesse $F = 4R/(1 - R)^2$	Number of effective beams q^*	Contrast $K = (1+R)^2/(1-R)^2$
0.70	31	7.2	32
0.75	48	9.0	49
0.80	80	11.6	81
0.85	151	15.9	152
0.90	360	24.6	361
0.925	658	33.3	659
0.95	1520	50.6	1520

perfect enough to yield the theoretical finesse. A much more reliable effective value of R can be obtained from measurements of K. Of course, the difference of the finesse obtained by the two methods mentioned above is an indication of the lack of perfection of the plates or adjustment of the interferometer. In Table I we have recorded some of the properties of the Fabry–Perot interferometer in tabular form.

Fig. 2. Components of a Fabry–Perot interferometer ready to be assembled.

3.2.3.3.1. CONSTRUCTION OF FABRY–PEROT ETALONS.   When using the Fabry–Perot etalon for work on the Brillouin effect it is often necessary to have a wide variety of etalon spacers. In Fig. 2 are shown the various components of a Fabry–Perot etalon of the type we have used with great satisfaction. The plates are 10 cm in diameter. The Invar plugs in the spacers are tapered conically at each end with the apex of the cone cut off at about half the diameter of the cylindrical plug. The plugs are made a good fit into holes in the spacer rings, which are constructed of brass. In the case of the smaller spacers the plugs are soldered to the brass ring. For larger spacers the plugs are held in place by a single set screw placed at the midpoint of the spacer ring. In Fig. 3 we have shown an etalon barrel and a pair of plates separated by spacers as they would be when inserted into the etalon barrel. For the most satisfactory results the spacer plugs and the etalon barrel should be constructed of Invar.

When quartz plates are used with such an etalon it remains in adjustment for long periods of time and the change of t, the change of length

FIG. 3. Components of a Fabry–Perot interferometer assembled ready for insertion in the barrel.

of the interferometer spacer with temperature, is that occasioned by the small temperature coefficient of Invar. If a brass barrel is used, the temperature change of t is much larger. It must be remembered that in adjusting the etalon for parallelism of the plates, the spacer plugs are unavoidably placed under compression. The Invar plugs will act like springs, and since brass has a much larger coefficient of thermal expansion than Invar, this change of compression with temperature will yield a net coefficient of thermal expansion much larger than that of Invar. It can be seen from Fig. 3 that with well-constructed spacers a number of them can be stacked on top of each other to produce larger values of t.

3.2.3.3.2. CONSTRUCTION OF ETALON SPACERS. The mechanical workshop can easily construct spacer rings with Invar plugs as described above. The machinist can make the plugs the same length to a tolerance of approximately 0.025 mm. The ends of the plugs (mounted in the spacer ring) should then be lapped into a plane so that all points on the ends lie in this plane. This can be accomplished by lapping on a flat cast iron lap with fine emery. During the lapping measurements are made with a good micrometer on all three plugs. The three plugs can easily be brought to identical length to 0.001 or 0.002 mm. We have made use of a method first described by A. H. Pfund more than 60 years ago to finish the correction of the spacer plugs to the necessary tolerance. The plug ends are polished by rubbing on a soft pine board with rouge as the polishing compound.

After polishing the plug ends, the spacer can be assembled into the etalon barrel with a pair of interferometer plates. It is then easy to see which plug is the longer if one attempts to adjust the etalon plates for parallelism. The long plug is then polished further and this procedure repeated until all plugs show identical lengths. If one has a Fizeau "flat tester" (parallel light) it is very easy to test the (parallel) correction of a spacer. One simply places the etalon spacer on the master plane and places another optical plane on top of the spacer. If the plugs are not the same length, one sees a number of straight line fringes crossing the aperture. Again slight pressure applied directly over the plugs allows one to determine the longer plug. Thus we proceed as formerly, but the testing procedure is much simpler and less time consuming than if one has to assemble the etalon and try to put it into parallel adjustment. One, of course, continues the testing and correcting procedure until a single fringe covers the aperture. It is left to the imagination of the reader, but it is easy to make interferometer spacers of identical t using the Fizeau test method.

3.2.3.4. Intensity of Brillouin Components.

When photoelectric methods are used to detect and the Brillouin spectra are recorded on chart recorders, intensity measurements can be made both easily and reliably. Since the photoelectric response can be easily made strictly linear, it is only necessary to obtain the integrals of the lines on the chart recorder in order to obtain relative intensity measurements. A good example of the necessity to take account of the temperature effect on sound velocity and change of intensity with temperature due to change in viscosity is demonstrated in a paper by Rank et al.[22] In glycerine, for example, the sound velocity nearly doubles in changing the temperature from 115 to $-21°C$. The Landau–Placzek[14] ratio $I_c/2I_B$ approaches $\gamma - 1 \approx 0.23$ at $115°C$ but at $-21°C$ this same ratio has a value of 4.9, showing a tremendous attenuation of the hypersonic (Debye) waves as the result of an increase of viscosity of 10^4.

Because of the large variation of sound velocity in different liquids and the large temperature coefficients of velocity, it is necessary to take

---

[22] D. H. Rank, E. M. Kiess, and U. Fink, *J. Opt. Soc. Amer.* **56,** 163 (1966).

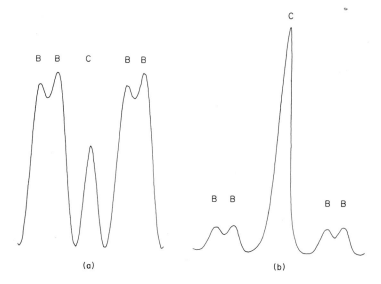

FIG. 4. Brillouin spectra of glycerin at (a) 114°C and (b) −21°C showing the large difference in relative intensities of the central and Brillouin lines. The etalon spacer was 10.534 mm and 8.15 mm, respectively. [Taken from the work of D. H. Rank, E. M. Kiess, and U. Fink, *J. Opt. Soc. Amer.* **56**, 163 (1966).]

certain precautions in using Fabry–Perot etalons. A rather dramatic example of the temperature effect is shown in Fig. 2 of Rank *et al.*[22] Reproduced here as Fig. 4, the two Brillouin components at (a) 114°C are shown nearly superimposed with an etalon for which $t = 10.534$ mm. At (b) −21°C the Brillouin components are shown nearly super-imposed making use of an 8.15-mm etalon spacer. Since it was only desired to obtain the intensity of the central unmodified component and the sum of the intensities of the two Brillouin components, these interference patterns adequately perform their function. If one wished to obtain the sound velocity, it would be necessary to use smaller inter-ferometer spacers for the experiments at the two temperatures. The near superposition of the fringes causes a "pulling or shrinkage effect,"[23] which would seriously affect the accuracy of fringe shifts directly obtained from measurement of peak positions of the interference fringes. There are cases where it is not possible to choose an interferometer spacer so that the shrinkage effect is totally eliminated. In these cases one must go through the painful process of deconvoluting the composite interference

[23] M. S. Sodha, *Indian J. Phys.* **29**, 461 (1955).

pattern. If one wishes to make either good velocity of sound measurements or good intensity measurements on a number of different liquids, it is necessary to have a wide variety of etalon spacers available.

3.2.3.5. Polarization of Brillouin Spectra.

Gross[24] who discovered the Brillouin effect experimentally, noted in his second communication on the subject that both the modified and unmodified Brillouin components were apparently completely polarized. Bhagavantam[6] (p. 84), has assembled a large amount of data on the depolarization of light scattered through 90° by a large number of liquids. The depolarization measurements tabulated are the integrated depolarization values obtained from the total unresolved spectrum (Brillouin modified and unmodified, Rayleigh wing scattering, and the Raman effect). The task of obtaining reliable measurements of the state of polarization of the unmodified and modified Brillouin components was a formidable task before the invention of laser light sources. Even today it is not easy to be certain that Rayleigh wing scattering (sometimes called *rotational wings which are largely depolarized*) has been completely eliminated in making depolarization measurements on the Brillouin components.

Since the invention of the laser, a number of investigators have studied the polarization of Brillouin scattering. Among these are Leite *et al.*[25] and Cummins and Gammon.[26] Rank *et al.*[27] have investigated the polarization of 90° scattering attempting to separate the various types of scattering. The method of measurement of the depolarizations of the Brillouin components used in this paper is essentially the method of polarized incident light described by Douglas and Rank.[28] A schematic diagram of the apparatus used by the above-mentioned authors[27] is shown in Fig. 5. The liquid whose scattering is being investigated is contained in the optical cross C. Parallel light from the laser L is concentrated somewhat by means of a lens which brings the light to a focus in the center of the optical cross C. The light passing through the cross is returned over its original path by means of a spherical mirror indicated in Fig. 5. Optical cross and spherical mirror are enclosed in a blackened box both to avoid parasitic light as much as possible and to provide some temperature

[24] E. Gross, *Nature (London)* **126**, 603 (1930).

[25] R. C. C. Leite, R. S. Moore, and S. P. S. Porto, *J. Chem. Phys.* **40**, 3741 (1965).

[26] H. Z. Cummins and R. W. Gammon, *Appl. Phys. Lett.* **6**, 171 (1965).

[27] D. H. Rank, Amos Hollinger, and D. P. Eastman, *J. Opt. Soc. Amer.* **56**, 1057 (1966).

[28] A. E. Douglas and D. H. Rank, *J. Opt. Soc. Amer.* **38**, 281 (1948).

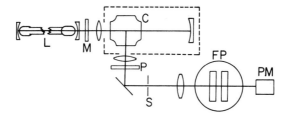

FIG. 5. Experimental arrangement to determine the depolarization ratios for scattering at 90° (see text for details). [Taken from the work of D. H. Rank, A. Hollinger, and D. P. Eastman, *J. Opt. Soc. Amer.* **56**, 1057 (1966).]

stabilization. M is a mica half wave plate which allows the plane of polarization of the incident light to be rotated through 90°.

The above-mentioned method of depolarization measurement is free of convergence error to the third order (which in the case of laser illumination as described above is trivial). The measured depolarization ratio is, however, ϱ_v (the depolarization ratio for vertically polarized incident light). ϱ_v is related to the more commonly used ϱ_u (the depolarization ratio for unpolarized light) by the geometrically derivable Krishnan equation

$$\varrho_u = 2\varrho_v/(1 + \varrho_v).$$

Light that is scattered through 90° is collected by a second lens and passes through a piece of high-quality Polaroid sheet P used as an analyzer and always set to pass only the vertically polarized component, regardless of the polarization of the incident light. The purpose of the diaphragm S is to limit the angular subtense of the beam that falls on the Fabry–Perot etalon FP to a small fraction of the central interference order. The light is detected by the photomultiplier PM and the electrical output recorded on a chart recorder. The Fabry–Perot etalon and its housing are indicated schematically in the figure. Provision is made to linearly vary the air pressure between the etalon plates thus giving a scan of the Fabry–Perot pattern which is linear in wave numbers.

Typical results are portrayed in Fig. 6. The spacer t of the Fabry–Perot etalon was chosen so that the red- and violet-shifted Brillouin components from the nth and $(n + 1)$th orders are superimposed. This superposition has a twofold purpose in that first it allows the depolarization ratio to be more accurately determined than if the measurements were made on single modified components. Secondly, since one can use larger etalon spacers when superposition is used, the resolving power available is considerably increased thus reducing the effects occurring

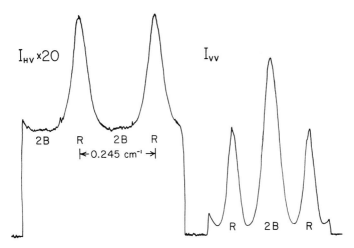

FIG. 6. A continuous interferometer trace of the Rayleigh scattering at 90° in ethylene glycol showing the vertically polarized light I_{VV} and the horizontally polarized light I_{HV}. The gain for I_{HV} has been increased by a factor of 20. The interferometer spacer has been chosen to overlap the Brillouin components which are labeled 2B. [Taken from the work of D. H. Rank, A. Hollinger, and D. P. Eastman, *S. Opt. Soc. Amer.* **56**, 1057 (1966).]

from overlapping of the two types of Brillouin frequencies (modified and unmodified).

It is believed that depolarization of 0.2% of ϱ_v can be detected in the experiments quoted above. For ethylene glycol within the limits of detection, ϱ_u for the modified Brillouin components could not be observed to differ from zero. However, the unmodified Rayleigh component yielded $\varrho_u = 11.2\%$, a value certainly significantly different from zero. The depolarization of the background ϱ_u was found to be 19.8%. The integrated ϱ_u for the unresolved Rayleigh scattering plus the background was 14%.

Nitrobenzene furnishes a very interesting case. At room temperature, 25°C, ϱ_u(Rayleigh) was found to be 100%. ϱ_u(Brillouin modified) = 0.0%, ϱ_u background 75.8%, and integrated unresolved ϱ_u was found to be 75.6%. This integrated ϱ_u is to be contrasted to the directly measured unresolved ϱ_u given by Bhagavantam,[6] page 84, of 68.2%. Measurements were made of ϱ_u(Rayleigh) at 100°C. At this temperature it was found that the unmodified Rayleigh scattering was completely polarized, $\varrho_u = 0.0\%$. This is to be contrasted to complete depolarization observed at 25°C. Aniline behaves in a fashion somewhat similar to the behavior of nitrobenzene but not as dramatically. At 25°C, ϱ_u(Rayleigh) = 15.0%

and at 136°C, $\varrho_u = 8\%$. The reason for the above mentioned behavior of the depolarization of the Rayleigh components of the two above-mentioned liquids is not understood.

3.2.4. Measurements of Sound Speeds and Relaxation Frequency by Means of Light Scattering

3.2.4.1. Liquids. A large amount of work on the measurement of sound speeds in liquids has been accomplished in both the ultrasonic and hypersonic region. Work in the hypersonic region has been accomplished making use of the spontaneous and stimulated Brillouin effect. In this section we will treat only work on the spontaneous Brillouin effect. The stimulated effect will be discussed in a later section. In a paper entitled, "Temperature coefficient of hypersonic sound and relaxation parameters for some liquids," by Eastman et al.[29] more than 45 references pertinent to this subject are listed. One of the more important references for the ultrasonic work on liquids is the book by Herzfeld and Litovitz.[30] Also extensive treatment of this type of work can be found in books by Nozdrev[31] and by Krasil'nikov.[32]

In the paper by Eastman et al.[29] the temperature coefficient of the hypersonic velocity was measured for 10 liquids chosen to be representative of various types of behavior with respect to relaxation frequency, viscosity, etc. It is not possible to determine the velocity of sound as a function of temperature accurately at constant frequency using the Brillouin effect unless a judicious choice of scattering angle is made. If the velocity of sound changes (with temperature), the Bragg coherence condition will demand that the sonic frequency being observed in the measurement will also have changed. In much of the work referred to above hypersonic velocity measurements were made at scattering angles of 25, 90, and 155°. The use of this range of scattering angles produced a spread of hypersonic frequency of 4.5-fold for any given liquid at a given temperature.

[29] D. P. Eastman, A. Hollinger, J. Kenemuth, and D. H. Rank, *J. Chem. Phys.* **50**, 1567 (1969).

[30] K. F. Herzfeld and T. A. Litovitz, "Absorption and Dispersion of Ultrasonic Waves." Academic Press, New York, 1959.

[31] V. F. Nozdrev, "The Use of Ultrasonic Waves in Molecular Physics." Pergamon, Oxford, 1965.

[32] V. A. Krasil'nikov, "Sound and Ultrasound Waves." Israel Program for Scientific Translation, Jerusalem, 1963.

3.2.4.1.1. EXPERIMENTAL METHOD. The arrangement of the experimental apparatus is shown in Fig. 7. The liquid rendered optically empty is contained in the spherical bulb S. For liquids that exhibited small depolarization it was possible to considerably attenuate the Rayleigh and Brillouin components of the light scattered at 25° when it was desired to observe 155° scattering. The plane of polarization of the incident light was rotated into the scattering plane with the $\lambda/2$ plate, and the plane of polarization of the beam returning into the scattering cell from M was rotated perpendicular to the scattering plane by means of the $\lambda/4$ plate. The analyzer A was set to pass the scattered light that is polarized perpendicular to the scattering plane.

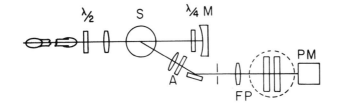

FIG. 7. Schematic diagram of the apparatus to observe Brillouin scattering at 25 and 155°. The liquid is distilled into the spherical container S. For measurements at 155° scattering the analyzer A, $\lambda/2$ and $\lambda/4$ plates are inserted to reduce the intensity of light scattered through 25°. [Taken from the work of D. P. Eastman, A. Hollinger, J. Kenemuth, and D. H. Rank, *J. Chem. Phys.* **50**, 1567 (1969).]

Alignment for 25° scattering was accomplished by adjusting the optical collection system by using a 45° prism set for minimum deviation to initially accurately set up the approximately 25° angle direction. The actual value of the scattering angle was determined from a precise measurement of the angle of minimum deviation of the prism. After the optical system has been adjusted the scattering cell S was set in the position occupied by the prism used for angle adjustment. Scattering at 155° was achieved with the same optical alignment by placing the spherical mirror M in the beam transmitted by the liquid.

In order to measure hypersonic velocities accurately by means of the Brillouin effect extreme care must be exercised in the use of the Fabry–Perot etalon. The "shrinkage effect," first treated by Sodha[23] must not be ignored and corrections for this effect must be made if no alternative solution is possible. If the Brillouin or Rayleigh lines have appreciable breadth, the shrinkage effect can produce grievous errors of measurement. The ideal situation, if Rayleigh and Brillouin components have

TABLE II. The Limiting Speeds, Relaxation Frequency, and Total Dispersion at 20°C for Seven Unassociated Liquids[a]

Liquid	Limiting speed v_0 (m/sec)	Ref.	Relaxation frequency v_R (GHz)	Ref.	Limiting speed v_∞ (m/sec)	Ref.	Total dispersion $[(v_\infty - v_0)/v_0{}^\circ] \times 100$	Ref.
Dibromomethane	963.2	b	0.433	c	996.9	a	3.5	a
							3.7	c
Dichloromethane	1092	c,d,e	0.166	c	1130	a	3.5	a
							3.3	c
Chloroform	1005	f	2.8	a	1087	a	8.2	a
			0.95–1.7	g			9.1	g
							12	t
Carbon tetrachloride	940	b,f,e,g,h,i	2.5	a	1069	a	13.7	a
			2.4	n			10	g
			2.1	j			20	t
			1.8	g				
			1.1	o				
Nitrobenzene	1480	f,j,k	9.1	a	1722	a	23	a
			18.4	n				
Benzene	1324	l	3.0	a	1550	a	17	a
			0.55 and ~10	p			17	t
			1.36	n				
			2.9	q				
			0.064–0.53	g				
			0.69	r				
Acetic acid	1155	m	4.1	a	1206	a	4.4	a
			4.1	g				
			0.56	s				

[a] Taken from the work of D. P. Eastman, A. Hollinger, J. Kenemuth, and D. H. Rank, *J. Chem. Phys.* **50**, 1567 (1969).

[b] R. Mountain, *Rev. Mod. Phys.* **38**, 205 (1966).

[c] J. L. Hunter and H. D. Dardy, *J. Chem. Phys.* **42**, 2961 (1965); **44**, 3637 (1966).

[d] H. W. Leidecker and J. T. La Macchia, *J. Acoust. Soc. Amer.* **43**, 143 (1968).

[e] N. A. Clark, C. E. Moeller, J. A. Bucaro, and E. F. Carome, *J. Chem. Phys.* **44**, 2528 (1966).

[f] K. F. Herzfeld and T. A. Litovitz, "Absorption and Dispersion of Ultrasonic Waves." Academic Press, New York, 1959.

[g] D. I. Mash, V. F. Starunov, E. V. Tiganov, and I. L. Fabelinskii, *Zh. Eksp. Teor. Fiz.* **49**, 1946 (1966); *Sov. Phys. JETP* **22**, 1205 (1966).

[h] D. H. Rank, E. M. Kiess, U. Fink, and T. A. Wiggins, *J. Opt. Soc. Amer.* **55**, 925 (1965).

[i] C. L. O'Connor and J. P. Schlupf, *J. Acoust. Soc. Amer.* **40**, 663 (1966).

[j] I. L. Fabelinskii, *Sov. Phys. Usp.* **8**, 637 (1966).

[k] G. W. Marks, *J. Acoust. Soc. Amer.* **27**, 680 (1955).

[l] "International Critical Tables." McGraw-Hill, New York, 1926.

[m] J. Lamb and J. N. Pinkerton, *Proc. Roy. Soc. (London)* **A199**, 144 (1949).

[n] R. Y. Chiao and P. A. Fleury, *Proc. Quantum Electron. Conf. San Juan, Puerto Rico, 1965* (P. L. Kelley, B. Lax, and P. E. Taunewald, eds.), pp. 241–252. McGraw-Hill, New York, 1965.

[o] R. A. Rasmussen, *J. Chem. Phys.* **48**, 3364 (1968).

[p] J. L. Hunter, E. F. Carome, H. D. Dardy, and J. A. Bucaro, *J. Acoust. Soc. Amer.* **40**, 313 (1966).

[q] J. R. Pellam and J. K. Galt, Jr., *J. Chem. Phys.* **14**, 608 (1946).

[r] S. E. Hakim and W. J. Comley, *Nature (London)* **208**, 1082 (1965).

[s] H. J. McSkimin, *J. Acoust. Soc. Amer.* **37**, 325 (1965).

[t] Assuming all vibrational modes relaxed.

equal intensities, is to choose an etalon spacer $\Delta v_s/Z = 0.33$, where Δv_s is the sound wave frequency in wave number. $Z = 1/2t$, where Z is the free spectral range of the etalon and t is the separation of the etalon plates in centimeters. Because of the shrinkage effect mentioned above, it is necessary to have a considerable number of different etalon spacers if accurate work is to be accomplished.

3.2.4.1.2. RESULTS. We have summarized some of the results obtained by Eastman *et al.*[29] and compared them to those from other sources[33-49] in Table II. The parameters that characterize the relaxation in chloroform and carbon tetrachloride are given here. Hunter and Dardy[34] have suggested that in these liquids the relaxation can be described as a parallel relaxation of all except the lowest vibrational modes at a single frequency. The magnitude of the observed dispersion in dichloromethane is very close to the calculated dispersion. The dispersion calculation of Litovitz and Davis[50] on the basis of all vibrational modes relaxing is 6% higher than the present observed dispersion.

The hypersonic absorption coefficient α is obtainable directly from the broadening of the Brillouin components. $\alpha = \delta v \pi / v$, where δv is the

[33] R. Mountain, *Rev. Mod. Phys.* **38**, 205 (1966).

[34] J. L. Hunter and H. D. Dardy, *J. Chem. Phys.* **42**, 2961 (1965); **44**, 3637 (1966).

[35] H. W. Leidecker and J. T. La Macchia, *J. Acoust. Soc. Amer.* **43**, 143 (1968).

[36] N. A. Clark, C. E. Moeller, J. A. Bucaro, and E. F. Carome, *J. Chem. Phys.* **44**, 2528 (1966).

[37] D. I. Mash, V. S. Starunov, E. V. Tiganov, and I. L. Fabelinskii, *Zh. Eksp. Teor. Fiz.* **49**, 1946 (1966); *Sov. Phys. JETP* **22**, 1205 (1966).

[38] D. H. Rank, E. M. Kiess, U. Fink, and T. A. Wiggins, *J. Opt. Soc. Amer.* **55**, 925 (1965).

[39] C. L. O'Connor and J. P. Schlupf, *J. Acoust. Soc. Amer.* **40**, 663 (1966).

[40] R. Y. Chiao and P. A. Fleury, *Proc. Quantum Electron. Conf. San Juan, Puerto Rico*, 1965 (P. L. Kelley, B. Lax, and P. E. Taunewald, eds.), pp. 241–252. McGraw-Hill, New York, 1965.

[41] I. L. Fabelinskii, *Sov. Phys.–Usp.* **8**, 637 (1966).

[42] R. A. Rasmussen, *J. Chem. Phys.* **48**, 3364 (1968).

[43] G. W. Marks, *J. Acoust. Soc. Amer.* **27**, 680 (1955).

[44] "International Critical Tables." McGraw-Hill, New York, 1926.

[45] J. L. Hunter, E. F. Carome, H. D. Dardy, and J. A. Bucaro, *J. Acoust. Soc. Amer.* **40**, 313 (1966).

[46] J. R. Pellam and J. K. Galt, Jr., *J. Chem. Phys.* **14**, 608 (1946).

[47] S. E. A. Hakim and W. J. Comley, *Nature (London)* **208**, 1082 (1965).

[48] J. Lamb and J. M. Pinkerton, *Proc. Roy. Soc. (London)* **A199**, 144 (1949).

[49] H. J. McSkimin, *J. Acoust. Soc. Amer.* **37**, 325 (1965).

[50] T. A. Litovitz and C. M. Davis, Jr., *J. Chem. Phys.* **44**, 840 (1966).

broadening and v is the hypersonic velocity. We have reproduced the results obtained by Eastman et al.[29] for glycerine in Table III to illustrate the smallness of the broadening and the wide range of interferometer spacers that had to be used to make these determinations. It is pertinent to note here that the hypersound velocity increases dramatically as the temperature of liquids (both associated and viscous and mobile liquids) is decreased. Water is, however, an exception to this rule since the sound velocity increases with temperature in the 0–60°C range.

TABLE III. The Amplitude Absorption Coefficient in Glycerol as a Function of Temperature[a]

Temperature T (°C)	Spacer t (mm)	δv (cm^{-1})	v (GHz)	α (10^4 cm^{-1})
−11.8	5.956	0.003	10.48	0.09 ± 0.07
26.0	5.956	0.034	8.96	1.19 ± 0.11
42.8	6.802	0.029	7.79	1.13 ± 0.12
81.6	8.158	0.044	5.95	2.24 ± 0.18
99.8	8.158	0.033	5.69	1.79 ± 0.15
101.4	8.158	0.033	5.67	1.76 ± 0.09
134.2	9.495	0.016	5.31	0.93 ± 0.13

[a] The amplitude absorption coefficient α is calculated from the broadening δv and the speed. [Taken from the work of D. P. Eastman, A. Hollinger, J. Kenemuth, and D. H. Rank, J. Chem. Phys. **50**, 1567 (1969).]

3.2.4.1.3. MOUNTAIN PEAK. Mountain[51] has shown theoretically that a fourth component should appear in the fine structure of Rayleigh scattering. This component should be unshifted in wavelength (same wavelength as the unmodified central component) and very broad so that it underlays the complete Brillouin pattern. This component should occur when one makes the observations in close proximity to the relaxation frequency. It is very difficult to make an experimental determination to show the existence of the "Mountain peak." One must make scattering

[51] R. D. Mountain, J. Res. Nat. Bur. Std. **70A**, 207 (1966); Rev. Mod. Phys. **38**, 205 (1966).

measurements in which no parasitic light can be tolerated and a complicated deconvolution of the trace of the interference pattern must be consumated. Stoicheff and co-workers[52] have performed these difficult experiments and shown conclusively that the "Mountain peak" is present in the spectrum of CCl_4 and glycerine. Montrose et al.[53] have shown theoretically that in addition to the "Mountain peak," the modified Brillouin components are considerably broadened.

3.2.4.2. Spontaneous Brillouin Scattering in Gases.

Until the advent of the laser, it was considered impossible to detect Brillouin scattering from gases due to the small sound speeds, small density, and the difficulties in obtaining clean samples. However, these problems have now been solved and it is feasible to use this technique for the investigation of acoustical damping and acoustic speeds at frequencies in the gigahertz range in a number of gases.

The first observations[54] were made by Eastman et al. in CO_2 and N_2. Although the gases were not very clean, the work demonstrated the possibility of this type of measurements. Subsequent work has investigated the density and frequency dependence in the speed of sound and some information has been obtained on attenuation.

Experimental techniques used in these measurements are similar to those needed for Brillouin scattering in liquids. However, due to the smaller sound speed in gases compared to liquids and solids, it is necessary to use incident light of small spectral width, constant wavelength, and high stability. Various methods have been used to provide these characteristics. They include Michelson interferometers,[55] nonlinear absorption,[56] in-cavity mode selectors,[57] feedback systems,[58] and the confocal cavity.[59]

[52] W. S. Gornall, G. I. A. Stegeman, B. P. Stoicheff, R. H. Stolen, and V. Voltera, *Phys. Rev. Lett.* **17**, 297 (1966); H. F. P. Knaap, W. S. Gornall, and B. P. Stoicheff, *Phys. Rev.* **166**, 139 (1968).

[53] C. J. Montrose, V. A. Solovyev, and T. A. Litovitz, *J. Acoust. Soc. Amer.* **43**, 117 (1968).

[54] D. P. Eastman, T. A. Wiggins, and D. H. Rank, *Appl. Opt.* **5**, 879 (1966).

[55] P. W. Smith, *IEEE J. Quant. Electron.* **QE 2**, 666 (1966).

[56] P. H. Lee, P. B. Schoefer, and W. B. Barker, *Appl. Phys. Lett.* **13**, 373 (1968); V. N. Lisitsyn and V. P. Chebotaev, *Sov. Phys. JETP* **27**, 227 (1968); G. H. Hanes and C. E. Dahlstrom, *Appl. Phys. Lett.* **14**, 362 (1969).

[57] M. Hercher, *Appl. Opt.* **8**, 1103 (1969).

[58] A. D. White, *Rev. Sci. Instrum.* **38**, 1079 (1967).

[59] D. C. Sinclair, *Appl. Phys. Lett.* **13**, 98 (1968).

Cleanliness and purity of the gas are also important. Since the Brillouin scattering is weak due to the small index of refraction, any dirt in the gas will greatly enhance the Rayleigh line. This can make the tail of the Rayleigh line overlap the Brillouin shifted components and increases the difficulty of making accurate measurements. A means of suppressing the Rayleigh line is discussed by Langley and Ford.[60] In addition, the intensities needed to measure specific heat ratios are meaningless if there is stray light or scattering due to dirt which contributes to only the unshifted component. Successful cleaning has been accomplished by cryogenic techniques and filters.[61]

The purity of the gas is more important than would be expected from the simple weighting of speeds in accord with their concentrations. This is because trace impurities have been shown to markedly change the relaxation frequencies and hence the sound speeds.[30]

The shift of the Brillouin components and the breadth of these components are functions of the scattering angle as well as the density and temperature. This has been discussed by Greytak and Benedek.[62] In addition, corrections to experimental measurements due to a finite linewidth has been treated by Lallemand.[63] Measurements by May et al.[64] and Eastman et al.[65] give additional results on methane and other gases.

Although the critical region is not considered to be part of this topic, interesting work on xenon, sulfur hexafluoride, and carbon dioxide have been reported.[60,66]

In summary, it may be stated that current measurements yield important results concerning the speed of hypersound, at least for the simple gases. Additional work remains for the more complex molecules, particularly with regard to an understanding of the attenuation and the nature of the dispersion.

[60] K. H. Langley and N. C. Ford, Jr., *J. Opt. Soc. Amer.* **59**, 281 (1969).

[61] Millipore filters, Millipore Corp., Bedford, Massachusetts 01730.

[62] T. J. Greytak and G. B. Benedek, *Phys. Rev. Lett.* **17**, 179 (1966).

[63] P. Lallemand, *C. R. Acad. Sci. Paris* **270B**, 389 (1970).

[64] E. G. Rawson, E. H. Hara, A. D. May, and H. L. Welsh, *J. Opt. Soc. Amer.* **56**, 1403 (1966).

[65] M. J. Cardamone, T. T. Saito, D. P. Eastman, and D. H. Rank, *J. Opt. Soc. Amer.* **60**, 1264 (1970).

[66] R. W. Gammon, H. L. Swinney, and H. Z. Cummins, *Phys. Rev. Lett.* **19**, 1467 (1967); R. Mohr, K. H. Langley, and N. C. Ford, Jr., Paper 4J5, ASA meeting (November 1969); D. L. Henry, H. Z. Cummins, and H. L. Swinney, *Bull. Amer. Phys. Soc.* **14**, 73 (1969).

3.2.4.3. Glasses and Plastics Krishnan Effect. Krishnan[67] showed that in the case of critical opalescence when depolarization measurements are made using horizontally polarized light, ϱ_h is less than unity indicating the depolarization is not complete as is the normal situation. Krishnan was also able to show that this phenomenon occurred in light scattering from optical glass and ascribed both the behavior of critical opalescence and the case of optical glass to molecular clustering. Mueller[68] has given a thorough discussion of the "Krishnan effect" (the term commonly used to denote the above mentioned anomalous behavior) and has shown that the reciprocity theorem of Lord Rayleigh is always valid. Krishnan[69] has given the relationship connecting the depolarization factors

$$\varrho_u = (1 + 1/\varrho_h)/(1 + 1/\varrho_v),$$

where ϱ_u, ϱ_v, and ϱ_h refer to depolarization factors measured using unpolarized, vertically polarized, horizontally polarized incident light, respectively. The Krishnan effect in optical glass was also observed by Rank and Douglas[70] using photoelectric measuring methods instead of the visual observation methods used by Krishnan. Somewhat later Yoder[71] and Rank and Yoder[72] made very careful experiments on optical glasses using photoelectric measuring methods. These later experiments achieved excellent agreement between the measured values of the depolarization factors and those predicted from the reciprocity relationship.

Mueller[68] shows that the Krishnan effect in glasses arises from density fluctuations that are frozen in the cooling glass, resulting in internal compression and shear strains. These internal strains do not appear between crossed polarizers because of their random nature, but they cause differences in the dielectric susceptibility fluctuations depending on whether the fluctuations are due to either transverse or longitudinal Debye waves.

Romberger et al.[73] have investigated light scattering from plastics and find even smaller values of ϱ_h than is commonly found in optical glasses. These authors conclude that anomalously high Landau–Placzek ratios obtained are not due to a decrease in modified Brillouin intensity but

[67] R. S. Krishnan, *Proc. Ind. Acad. Sci.* **3**, 211 (1936).
[68] Hans Mueller, *Proc. Roy. Soc.* **166A**, 425 (1938).
[69] R. S. Krishnan, *Proc. Ind. Acad. Sci.* **6**, 21 (1938).
[70] D. H. Rank and A. E. Douglas, *J. Opt. Soc. Amer.* **38**, 966 (1948).
[71] Paul R. Yoder, Jr., MS Thesis, Pennsylvania State Univ. (1950).
[72] D. H. Rank and Paul R. Yoder, Jr., *Mater. Res. Bull.* **5**, 335 (1970).
[73] A. B. Romberger, D. P. Eastman, and J. L. Hunt, *J. Chem. Phys.* **51**, 3723 (1969).

arise from an enhancement of the unmodified Rayleigh component. This enhanced Rayleigh component is consistent with a model of a plastic that contains a random distribution of strains frozen in during the polymerization process and only partially relieved by any annealing process. The observation of the Krishnan effect confirms the above interpretation.

3.3. Spontaneous Raman Scattering

Very extensive work on light scattering was being actively pursued by C. V. Raman and others during the decade 1920–1930. Raman did a great deal of this work using sunlight as a light source. Upon using monochromatic light he observed that $90°$ scattering showed new wavelengths, not present in the light source, to appear in the spectrum of light scattered from liquids. Raman[74] and Raman and Krishnan[75] showed that the frequency displacement of the modified wavelengths was independent of the wavelength of the incident light being scattered. This, of course, was contrary to the case of fluorescence and was recognized by Raman as the discovery of a new phenomenon which now bears his name. Almost simultaneously and independently, Landsberg and Mandelstam[76] in Russia observed the above-mentioned phenomenon in light scattered from crystal quartz.

3.3.1. Historical Introduction and General Resume of Theory

Smekal[77] published a paper on the quantum theory of dispersion in 1923 in which he predicted that the modified wavelengths later observed by Raman would appear in the spectrum of the light scattered from transparent media. It is now well known that the 1925 dispersion theory of Kramers and Heisenberg[78] contains a complete theory of the Raman effect. Placzek,[79] in a monumental piece of work, has given a detailed treatment of the theory of Rayleigh scattering and the Raman effect.

Before development of the complete theory of the Raman effect it

[74] C. V. Raman, *Indian J. Phys.* **2**, 387 (1928).

[75] C. V. Raman and K. S. Krishnan, *Nature (London)* **121**, 501 (1928).

[76] G. Landsberg and L. Mandelstam, *Naturwissenchaften* **16**, 57, 772 (1928).

[77] A. Smekal, *Naturwissenchaften* **11**, 873 (1923).

[78] H. A. Kramers and W. Heisenberg, *Z. Phys.* **31**, 681 (1925).

[79] G. Placzek, "Handbuch der Radiologie" (E. Marx, ed.), Vol. 6. Akad. Verlagsges, 1934.

was thought that the so-called *Raman lines* or *frequency shifts* were identical to the frequency of the infrared absorption bands. It is now well known that, in general, the infrared and Raman frequencies are not identical. The infrared absorption bands appear when the molecule undergoes a change in dipole moment while executing the vibration in question. The Raman frequencies appear when there is a change in polarizability involved in executing the molecular vibration. In general, highly symmetric vibrations make their appearance in the Raman spectrum even for homonuclear molecules which have no infrared dipole absorption spectrum.

The Raman effect has been of tremendous importance in the unraveling of the dynamics of polyatomic molecules. Infrared and Raman spectrum are complementary and both are needed to obtain a complete picture of the behavior of molecular vibrations. General appearance of a Raman spectrum is somewhat more simple than in infrared absorption spectrum, particularly in the case of liquids. Raman lines are really bands but appear to be relatively sharp since the intensity in the widely spaced branches is usually negligible. Because of the appearance of the symmetric vibrations where, in general, only Q branches have appreciable intensity, these Raman "lines" are sharp ~ 1 cm^{-1}. Thus under even moderate dispersion it is possible to make some decisions as to whether a "line" arises from a symmetric or unsymmetrical vibration merely on the basis of line width (1 cm^{-1} versus 5–10 cm^{-1}).

Raman spectra obtained making use of unpolarized exciting light and $90°$ scattering show alteration of polarization depending upon whether the vibration giving rise to the "line" belongs to a symmetric or unsymmetrical class. If the vibration is of a symmetric class, the depolarization factor ϱ_u will vary between 0 and 0.866. If the vibration belongs to an unsymmetrical class, $\varrho_u = 0.866$. Thus by measuring the depolarization ratio one can decide to which class the vibrations giving rise to the Raman line belong.

Further simplification of Raman spectra compared to infrared absorption spectra arises due to the fact that overtones and combination bands are very common in infrared spectra and are usually weak and not observed in the Raman spectrum. A very large amount of work has been done on the vibrational Raman spectrum of polyatomic molecules. Herzberg[80] in his book "Infrared and Raman Spectra" has given a rather

---

[80] G. Herzberg, "Infrared and Raman Spectra." Van Nostrand–Reinhold, Princeton, New Jersey, 1951.

complete treatment of the details of the theory of these molecular vibration spectra. In addition, he has summarized in detail the literature of the subject up to the time of publication of his book.

3.3.2. Experimental Techniques

Raman focused the light of the sun or a mercury arc in a spherical bulb containing the liquid and observed the trace of scattered light at 90° to the direction of propagation. The spectra of the scattered light were weak necessitating long exposures. Wood[81] modified these methods which he has used very successfully in studies of fluorescence spectra, thus greatly increasing the ease with which Raman spectra could be obtained. Wood also devised various combinations of filters which he used to effectively isolate the various strong lines of the mercury arc spectrum. Rank *et al.*[82] have described a modification of Wood's method that is quite suitable for use in making depolarization measurements.

3.3.2.1. Sources. Mercury arcs suffer from a number of shortcomings when used as a source for the excitation of Raman spectra. Intense high-pressure arcs produce relatively broad lines and a continuum that is not negligible. It must be remembered that the strongest vibrational Raman lines are only 1/100 or less as intense as the unavoidable Rayleigh scattering. A great advance in light sources was achieved by H. L. Welsh and his students by the development of the so-called "Toronto" arc. These lamps have internally water-cooled electrodes that maintain a low vapor pressure of mercury, thus reducing the line breadth to essentially that of the central components of the principal mercury lines. (In an arc of even moderate vapor pressure the broadening fills at least the whole region of the hyperfine pattern.) In addition, the low vapor pressure arc has a continuum that is at least an order of magnitude weaker than the conventional high-pressure arc. These low-pressure arcs can be made in various forms and operated at high currents with attendant increase of flux without producing deleterious effects on the spectrum.

Since the invention of the laser, completely new types of light sources are available. The helium–neon laser is to all intents a monochromatic source of parallel light. The line width of a high-power multimode laser is considerably smaller than the widths of the lines produced by the best mercury arcs. Such lasers can easily have a power output of

[81] R. W. Wood, "Physical Optics," p. 447. MacMillian, New York, 1934.
[82] D. H. Rank, R. J. Pfister, and H. H. Grimm, *J. Opt. Soc. Amer.* **33**, 31 (1934).

200 to 300 mW. If the scattering material is placed in the resonant cavity, power levels as high as 10 W have been obtained. Many of the remarks made concerning the helium–neon laser can also be applied to the argon ion laser. In this source, since several wavelengths lase simultaneously, one must first separate these by means of filters or other devices. It is easily possible to obtain considerably higher power with the argon ion laser than with the He–Ne laser. Since this argon laser has its lasing wavelengths in the blue–green region of the spectrum it is much more efficient as a source for light scattering than the He–Ne laser because of the operation of the inverse fourth power of the wavelength law.

3.3.2.2. Scattering Cells. The types of cells used to contain the liquid or gas under investigation by means of light scattering are quite different when different types of light sources are used (arcs or lasers). When one is using illumination by mercury arcs, the type of cell first described by Wood,[81] suitably modified, is indicated. The object is to flood the available scattering volume with the maximum amount of flux, collect the 90° scattered light from this volume, and use it efficiently to illuminate the slit of the spectrograph. It is easy to see that with a given amount of scattered flux produced, efficient illumination of the spectrograph will entail different methods of light collection depending upon whether one uses a photographic spectrograph or a photoelectric spectrograph. In the first case image size is of very small consequence, since it is only flux per unit area on the photographic plate (intensity) that counts. In the case of the photoelectric spectrograph one requires the arrival of the maximum number of photons at the detector, and intensity *per se* is of little or no consequence.

When we deal with lasers as light sources for scattering experiments we have totally different conditions to consider. The laser source produces a parallel beam of light in contradistinction to the divergent light produced by arc sources. With parallel light an "optical cross" type of scattering cell is ideal in many respects. These can be used in such a way that multiple reflections are obtained, thus increasing greatly the amount of light scattered. If one provides such an optical cross cell with Brewster angle windows particularly when scattering from gases is desired, such a cell can be placed in the resonant cavity resulting in a very large increase in scattered light over external usage.

It should be pointed out that the amount of light scattered by a liquid is of the order of 10^{-5} of the incident light. This is the molecular or Rayleigh scattering. The Raman or modified scattering is about 10^{-2}

as intense as the Rayleigh scattering. The weakness of the scattered light puts a stringent requirement on the amount of dust tolerable in the liquid or gas and the necessity of almost complete elimination of stray (parasitic) light from reflections, etc. In the case of the study of Rayleigh scattering the condition for the removal of parasitic light is more severe than in study of the Raman effect. Dust can be removed from liquids effectively by filtration through Millipore filters. If this process is followed by repeated vacuum distillation without ebullition, pouring the liquid back into the vessel after each condensation in the scattering tube, an optically empty liquid can be readily obtained.

3.3.2.3. Spectrographs.

At the time of the discovery of the Raman effect, few if any commercial instruments available were suitable for the observation of such weak spectra. Even with liquids, the ability to resolve 2–3 cm^{-1} is desirable. In order to have reasonable exposure times with the type of illumination described by Wood[81] instruments in the range $f = 2.0$ to $f = 10$ with a prism as the dispersing element were indicated. Typical exposure times would be 0.5 hr at $f = 2.0$ to 15 hr at $f = 10$. It should be noted that the f numbers referred to are the f numbers of the camera lens. (The f number of the collimator only determines the size of the spectral images and not their intensity.) Grating instruments at this time yielded spectra that were too feeble largely because of the low brightness of the spectra produced by the then available speculum diffraction gratings.

Rank et al.[83] were able to observe Raman spectra of liquids making use of photoelectric detection. The spectrograph was a Czerny–Turner plane grating spectrograph of 1-m focal length. The grating was 20 cm in diameter ruled on an aluminized pyrex blank. Later, Rank and Wiegand[84] constructed an improved spectrograph using a 4.5-m radius of curvature 20-cm diameter concave grating in a Wadsworth mounting. This instrument also made use of photoelectric detection and was able to resolve Raman lines at least as close as 2 cm^{-1} apart.

The chlorine isotope effect in CCl_4 and $CHCl_3$ totally symmetric vibration was resolved by Rank and Van Horn[85] using grating spectrographs. One spectrograph was a 3.6-m focus 20-cm plane grating in a Czerny–Turner mounting. The second spectrograph was a 6.4-m radius 15-cm concave grating in a Paschen mounting. In the case of both

[83] D. H. Rank, R. J. Pfister, and P. D. Coleman, *J. Opt. Soc. Amer.* **32**, 390 (1942).
[84] D. H. Rank and R. V. Wiegand, *J. Opt. Soc. Amer.* **36**, 325 (1946).
[85] D. H. Rank and J. A. Van Horn, *J. Opt. Soc. Amer.* **36**, 454 (1946).

instruments, a cylindrical lens was used in front of the plate to increase the intensity of the spectrum lines without altering the dispersion of the spectrograph.

After the development of the "Toronto" type mercury arc and construction of light furnaces at Toronto containing batteries of these powerful lamps, H. L. Welsh and his students were able to obtain Raman spectra under high resolution using both powerful prism and grating spectrographs. Stoicheff at the National Research Council in Ottawa has obtained beautiful pure rotation spectra of gases at a few atmospheres pressure using a 6.4-m concave grating spectrograph. An example of the precision obtained by Stoicheff[86] can be found in his paper on the molecular constants of hydrogen. Since the advent of powerful lasers it is now possible to obtain high-resolution Raman spectra with considerable ease.

At the present time there are many types of spectrographs suitable for the observation of Raman spectra commercially available. Both photographic and photoelectric detection instruments are now manufactured both in the U.S. and abroad. One particularly desirable type of photoelectric instrument makes use of a grating double monochromator which eliminates much of the parasitic light ordinarily present in the scattered light.

3.3.2.4. Depolarization Factors. It was pointed out earlier in this article that if one examines the depolarization of the scattered light, it is possible to obtain information concerning the nature of the molecular vibrations responsible for the Raman line. If the Raman spectrum is excited by unpolarized parallel light and observed by means of $90°$ scattering, the following results are obtained for the depolarization factors. Lines of the symmetric classes will have a depolarization factor ϱ_u that varies from 0 to 0.866. Vibrations that have spherical symmetry will show a $\varrho_u = 0$. Vibrations that belong to the unsymmetrical classes will give rise to Raman lines where $\varrho_u = 0.866$. If one deals with polarized exciting light, $90°$ scattering yields $\varrho_v = 0.75$ for the lines arising from unsymmetrical vibrations, and ϱ_v varies from 0 to 0.75 for lines arising from symmetrical vibrations.

It is well known that there are many difficulties involved in the accurate measurement of depolarization factors, particularly when photographic photometry is used to make the necessary intensity measurements.

[86] B. P. Stoicheff, *Can. J. Phys.* **35**, 730 (1957).

Spectrographs themselves are polarizing devices and they must be calibrated for their transmission of the two kinds of plane polarized light. Care must also be taken to avoid introducing circular or elliptical polarization into the measurements by improper use of totally reflecting prisms or other such devices in the optical system. Care must be taken to avoid introduction of "convergence error" occasioned by the nonparallelity of the incident exciting light.

The better methods of measuring depolarization factors make use of polarized incident light. The method preferred by the author is that described by Douglas and Rank.[28] This particular arrangement is free of "convergence error" to the third-order approximation. The intensity measurements are made using the same kind of polarized light thus eliminating the preferential transmission of the spectrograph for the two kinds of plane polarized light. In the case of this above mentioned method, ϱ_v is the quantity which is measured, and ϱ_v is related to ϱ_u by a purely geometrical relationship $\varrho_u = 2\varrho_v/(1 + \varrho_v)$. Polarized incident light methods are particularly applicable to laser-excited Raman spectra since the laser is almost always completely polarized and the light is parallel thus avoiding "convergence error" automatically.

It is possible to measure depolarization factors with relatively high precision (to a fraction of 1%) if photoelectric detection is used. In the case of Raman spectra it is usually only necessary to know if the Raman line is "polarized" (ϱ_u less than 0.866) or depolarized ($\varrho_u = 0.866$). If there are known to be depolarized lines in the spectrum, the decision of polarized versus depolarized can be made by inspection. It is only necessary to make two photographs of the spectrum using vertical and horizontal polarized exciting light, respectively. The known depolarized lines are brought to the same density on the two photographs. Thus any lines that are weaker on the one photograph than on the other belong to a polarized class.

3.3.3. Gases, Pure Rotation and Vibrational Raman Effect

The Raman effect was first observed in a gas, HCl, by Wood.[87] Wood observed both the pure rotation Raman spectrum of HCl and a single line (vibrational Raman effect) that showed a frequency shift equal to the frequency of the "missing" line (forbidden Q branch) of the fundamental infrared absorption band. Wood's experiments were performed

[87] R. W. Wood, *Nature (London)* **123**, 166 (1992).

with the HCl gas at atmospheric pressure, and the excitation was by the $\lambda 4358$ line of mercury. In the issue of *Nature* the week following Wood's paper, a paper appeared by Rasetti[88] in which he reported observing the Raman effect in gaseous CO and CO_2. Somewhat later, Rasetti[89] reported on the pure rotational Raman spectrum of O_2 and N_2. These molecules were interesting since, having no permanent dipole moment, they have no infrared absorption spectrum and yet exhibit beautiful Raman spectra. Rasetti's work on O_2 and N_2 was performed with the gases at 10 atm pressure, and he made use of the $\lambda 2536$ line of mercury as the exciting radiation.

The structure of the Raman spectrum for the diatomic molecule has been deduced from the quantum-mechanical theory of dispersion by Rasetti.[90] Hill and Kemble[91] have worked out the theory more completely to include a qualitative treatment of intensities and polarization.

It has been pointed out previously that in the hands of Stoicheff[86] and his collaborators, Raman spectra of gases have been obtained by the use of the techniques developed at the University of Toronto that compare favorably in precision with other spectroscopic methods of obtaining molecular constants of nonpolar molecules.

Recently making use of a helium–neon laser, Renschler *et al.*[92] have obtained a very beautiful Raman spectrum of gaseous oxygen at atmosphere pressure. The cell containing the oxygen was placed in the laser cavity. The spectrograph was a 2.5-m Czerny–Turner doubly passed monochromator equipped with a 20-cm echelle grating. Detection was by means of a photoelectric detector. In Fig. 8 we have reproduced the central part of the pure rotation Raman spectrum of oxygen. Satellites appear very prominently on each side of the central Rayleigh line and also on the first Stokes and anti-Stokes Raman lines. This fine structure arises from the multiplicity of a rotational levels of the $^3\Sigma_g^-$ ground state of oxygen. It appears that with the powerful lasers now available, phenomena formerly unobservable can now be studied in the Raman spectra of gases.

[88] F. Rasetti, *Nature (London)* **123**, 205 (1929).

[89] F. Rasetti, *Phys. Rev.* **34**, 367 (1929).

[90] F. Rasetti, *Proc. Nat. Acad. Sci.* **15**, 234, 515 (1929); *Nature (London)* **123**, 757 (1929).

[91] E. L. Hill and E. C. Kemble, *Proc. Nat. Acad. Sci.* **15**, 387 (1929).

[92] D. L. Renschler, J. L. Hunt, T. K. McCubbin, Jr., and S. R. Polo, *J. Mol. Spectrosc.* **31**, 173 (1969).

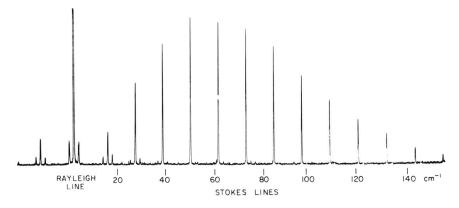

FIG. 8. The rotational Raman spectrum of O_2. [Taken from the work of D. L. Renschler, J. L. Hunt, T. K. McCubbin, Jr., and S. R. Polo, *J. Mol. Spectrosc.* **31**, 173 (1969).]

3.4. Stimulated Scattering

3.4.1. Introduction

Stimulated scattering began shortly after the invention of the giant pulse laser by Maiman.[93] This remarkable device can produce hitherto undreamed of light intensities in well-collimated beams of quite small spectral width. This intensity can be used to investigate nonlinear processes which had previously been unexplored.

The first stimulated process to be investigated (other than that which itself produces the giant pulse) was that of optical harmonic generation.[94] Since this is primarily applied to crystalline materials it will not be discussed here. (Third harmonic generation in gases is discussed by Ward and New.[95]) The second process was that of stimulated Raman scattering, which will be discussed in Section 3.6. The third process, scattering due to longitudinal sound waves, was first observed in liquids by Brewer and Rieckhoff[96] and by Garmire and Townes[97] and is discussed

[93] T. H. Maiman, *Nature (London)* **187**, 493 (1960); *J. Opt. Soc. Amer.* **50**, 1134 (1960); *Phys. Rev.* **123**, 1145 (1961).

[94] P. A. Franken, A. E. Hill, C. W. Peters, and G. Weinrich, *Phys. Rev. Lett.* **1**, 118 (1961).

[95] J. F. Ward and G. H. C. New, *Phys. Rev.* **185**, 57 (1969).

[96] R. G. Brewer and K. E. Rieckhoff, *Phys. Rev. Lett.* **13**, 334 (1964).

[97] E. Garmire and C. H. Townes, *Appl. Phys. Lett.* **5**, 84 (1964).

in the next section. Additional stimulated processes have been observed and are treated in the final section 3.7.

The nature of the stimulated process is such that it can be characterized by a threshold or gain. By a *threshold* is meant the minimum power per unit area required so that a small spontaneous signal I_s in some medium grows in intensity to an arbitrarily set level I that can be easily detected. The gain g can also be defined in terms of these symbols by $I/I_s = e^{glP}$, where l is the length of path over which the light has a power density P. In general, the gain will depend upon molecular parameters of the medium and possibly some laser parameters such as its wavelength and spectral width.

The existence of such a threshold is sufficient evidence that a process is nonlinear or stimulated. However, there may be other indications that are a direct result of the exponential gain. These are: the time duration or the spectral width can be less than the corresponding values for the incident light, and the scattered light will be coherent and collimated.

For this process to be interesting to molecular physicists, the difference in frequency between the incident light and the stimulated light will correspond to parameters of the medium. These parameters range from large vibrational and electronic Raman shifts through rotational Raman shifts, acoustical frequencies, and molecular reorientation times down to very small shifts due to Doppler shifts due to diffusion. In all cases it is necessary that there be some change in the index of refraction of the medium and that this change be dependent upon the electric field. At this date all types of scattering that have been investigated by spontaneous scattering have also been investigated by stimulated scattering. A possible exception is that of Rayleigh scattering for which the stimulated analog may have a very small frequency shift.

3.4.2. Experimental Techniques

3.4.2.1. Giant Pulse Ruby Systems. The characteristics of a laser source that are important in stimulated scattering experiments are: the energy in a single pulse, the duration and intensity of the pulse as a function of time, the wavelength, the spectral width, and the divergence. These characteristics are determined in large part by the arrangement and quality of the optical components. Almost all of the work on stimulated scattering has been done with the ruby laser. For this reason, and because this is the area of competence of the authors, the following discussion will be limited to ruby systems.

FIG. 9. Twyman–Green interferograms of an excellent 1×10-cm ruby rod. One figure shows the result with the interferometer adjusted to show several fringes across the rod aperture, the other for the zero fringe case.

A ruby system consists of four essential parts. The ruby element itself is a chromium-doped sapphire single crystal with about 0.05% Cr^{3+}. It is cut from a boule grown by the Czochralski process with the C axis usually at $60°$ to the axis of the rod. Its outer surface may be rough ground to cause better diffusion of the pump light. The ends are flat and parallel to each other. They may be at the Brewster angle or at $90°$ to the axis in which case they usually are antireflection coated. The quality of the crystal is of great importance in providing a sharp line of low divergence.[98] Twyman–Green interferograms of the rod are useful in predicting their quality. Such an interferogram of an excellent 1×10 cm ruby produced by Linde is shown in Fig. 9.

The second element is the flashlamp or lamps. They are arranged about the rod and fired from a charged capacitor bank through a proper network by a triggering device. A reflector around the lamp is designed to concentrate the light into the rod.[99]

The third element is a pair of mirrors which form the lasing cavity. One has a reflectivity as close to 100% as possible and must be designed

[98] C. M. Stickley, *Appl. Opt.* **2**, 855 (1963); **3**, 967 (1964).

[99] H. Welling and C. J. Bickhart, *J. Opt. Soc. Amer.* **56**, 611 (1966); C. H. Church and I. Liberman, *Appl. Opt.* **6**, 1966 (1967).

to withstand a very high power density. Various types are used. Total internal reflecting prisms can be used if their apex angle is very close to 90° and the front face is parallel to the intersection of the 90° faces. Dielectric coated flat or long radius curved mirrors can also be used. A newer development is the resonant reflector which promises to be quite useful.[100] The other end of the cavity is formed by a mirror of reflectivity between 50 and 65%. This can be a dielectric-coated surface but usually is a solid Fabry–Perot etalon or a two or three plate resonant reflector. The latter is used when it is important to select a single axial mode. One or both of the cavity-forming mirrors can be restricted by means of an aperture to reduce the transverse mode structure thus obtaining a spectral width of a few millikaysers.[101]

The fourth and final element is some sort of Q-switch used to prevent the lase from occurring before the flashlamp has pumped the ruby to a highly excited state. Common forms are saturable absorbers, Pockel or Kerr cells, or a rotating mirror. The one found to give the smallest spectral width is the saturable absorber. For large power requirements and precise timing of a lase, the Pockel cell is usually used.

3.4.2.2. Diagnostic Techniques.

Three characteristics of the light emitted by a giant pulse system should be measured. These are spectral width, power, and divergence. The first can be determined by reflecting a small fraction of the beam into a flat or spherical interferometer and photographing the resulting ring system. Since most laser beams used have small divergence, it is necessary to spread the beam to illuminate a sufficient number of rings of the interferometer to make quantitative measurements. This can be accomplished by a negative lens or by ground or etched glass. The first can be used if a transverse section of the beam is smooth and uniform with regard to intensity. If not, it is convenient to use a random scatterer.[102] Plates of 94 to 96% reflectivity and spacers of up to 150 mm are necessary to assess the transverse mode structure. The use of spherical interferometers for this purpose has been discussed by Bradley and Mitchell.[103] Longitudinal mode structure can be analyzed by lower-resolution interferometers or detected by beating in oscilloscope

[100] J. K. Watts, *Appl. Opt.* **7**, 1621 (1968); T. A. Wiggins, C. E. Procik, and J. Pliva, *ibid.* **10**, 304 (1971).

[101] D. J. Bradley and E. R. Peressini, *Appl. Phys. Lett.* **3**, 203 (1963).

[102] T. A. Wiggins, T. T. Saito, L. M. Peterson, and D. H. Rank, *Appl. Opt.* **9**, 2177 (1970).

[103] D. J. Bradley and C. J. Mitchell, *Phil. Trans.* **263A**, 209 (1968).

traces to be discussed below. The reason that a knowledge of the spectral width is important in much of the scattering work to be described is that for several processes it is necessary to have the spectral width small simply to be able to resolve the expected spectral shifts. Further, the gain of some stimulated processes are directly influenced by the spectral width of the source.

The second characteristic is that of the power and the time dependence of power in the laser pulse. This is usually determined by making two observations; one of the total energy in the pulse by means of a thermocouple which detects the temperature change of an object of known heat capacity, the other of the intensity as a function of time. The latter measurement is made with fast rise-time photo tubes whose output is observed by photographing the trace on a fast sweep-rate oscilloscope. Response times on the order of 0.3 nsec are adequate except for mode-locked systems. Smoothness in the power as a function of time can indicate that only one longitudinal mode is present although many modes can give the appearance of a single mode. The output is perfectly symmetric for truly single mode systems.[104] The average power can be calculated from the total energy and the time between the half-intensity points on the oscilloscopes record.

Divergence measurements are important in assessing the transverse mode structure as well as in being able to compute the size of the focal spot if the laser output is to be focused to increase the power density. A method for measuring the divergence has been given by Winer.[105] It is interesting to note that, while the divergence can be determined directly by the ratio of the size of the beam at the focal point to the focal length of the focusing lens, the presence of a waist in the laser cavity may allow a spot to be formed that is smaller than at the focus.[106] If the laser pulses are very reproducible, the profile can be mapped by moving a pinhole and photodiode across the beam.[104]

These four essential parts, when properly used, can provide a laser source for investigating the various stimulated processes which will now be described. It might be useful to indicate typical characteristics for such a laser system. Using a 1×10-cm ruby rod at full aperture power in the range of 25 to 50 MW can be expected in a single longitudinal

[104] A. J. Alcock, C. DeMichelis, and M. C. Richardson, *J. Quant. Electron.* **QE 6**, 622 (1970).

[105] I. M. Winer, *Appl. Opt.* **9**, 1437 (1966).

[106] H. Kogelnik, *Bell System Tech. J.* **44**, 455 (1965).

mode with a spectral width of 0.03 cm^{-1} and a divergence of 4 to 8 mrad, the half-power angle. With an aperture of 1.5 mm in the cavity, the power will be reduced by a factor of 10 but so will the spectral width, the divergence decreasing to 0.5 to 1 mrad.

3.5. Stimulated Brillouin Scattering

3.5.1. Introduction

As noted above, stimulated Brillouin scattering (SBS) in liquids was first observed by Garmire and Townes[97] and independently by Brewer.[96] They identified the intense back-scattered radiation as being due to scattering from longitudinal sound waves similar to the observations in crystals.[107] SBS in gases was first observed by Rank et al.[108] and by Hagenlocker and Rado.[109]

These sound waves are identical to those investigated in the spontaneous scattering experiments discussed in Section 3.2.2. However, in the stimulated case, the scattered light is not observed directly, but rather what is seen is the result of the stimulated process in which large amounts of power are transferred from the incident beam to the frequency-shifted beam by the process of electro-striction. Unlike the spontaneous case, only the red or Stokes-shifted component of the usual Brillouin doublet is usually observed. This is because the anti-Stokes component depends on the presence of weak thermal noise while the Stokes component has continuously produced phonons as a source.

In the following paragraphs certain experimental techniques, some of which are peculiar to SBS, are described. This is followed by comments on the significant results that have been obtained.

3.5.2. Experimental Techniques

A typical experimental arrangement used to detect SBS is shown in Fig. 10, which is taken from the work of Wiggins et al.[110] The laser with its diagnostic attachments sends a nearly collimated beam onto a lens

[107] R. Y. Chiao, C. H. Townes, and B. P. Stoicheff, Phys. Rev. Lett. 12, 592 (1964).

[108] D. H. Rank, T. A. Wiggins, R. V. Wick, D. P. Eastman, and A. H. Guenther, J. Opt. Soc. Amer. 56, 174 (1966).

[109] E. E. Hagenlocker and W. G. Rado, Appl. Phys. Lett. 1, 236 (1965).

[110] T. A. Wiggins, R. V. Wick, D. H. Rank, and A. H. Guenther, Appl. Opt. 4, 1203 (1965).

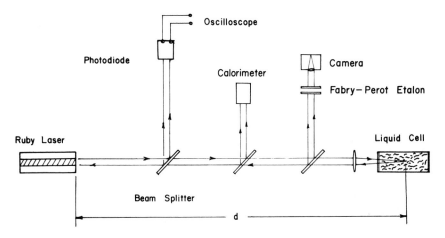

FIG. 10. Experimental arrangement to observe reamplified pulses and to detect the multiple Brillouin scattering. [Taken from the work of T. A. Wiggins, R. V. Wick, D. H. Rank, and A. H. Guenther, *Appl. Opt.* **4**, 1203 (1965).]

that focuses the light into a solid or a sample cell. A beam splitter, flat quartz or glass or a pellicle,[111] sends a portion of the beam into a Fabry–Perot etalon and camera for observing the spectral output of the laser. This beam splitter can also be rotated 90° for observation of the output of the sample cell. This arrangement is the same as used in the first observations.[97]

One of the unique features of stimulated scattering is the collimation of the scattered light. Because only the Stokes component has gain and because the increase in intensity depends upon the interaction length, the intensity increase is at exactly 180° to the incident beam. Thus the back-scattered light retraces its optical path and the beam has a convergence equal to the divergence of the laser.[112,113]

This recollimation provides some advantages as well as some disadvantages in experimental work. The advantages are that the scattering angle has no uncertainty, the same lenses can be used, and there are no aperturing effects. The disadvantage is that the return beam can reenter the laser cavity and may trigger a new output pulse. This can not only cause damage to the components, since it can be very intense even before amplification by the ruby, but can also give false readings of the amount

---

[111] National Photocolor Corp., South Norwalk, Connecticut 06854.

[112] T. A. Wiggins, R. V. Wick, and D. H. Rank, *Appl. Opt.* **5**, 1069 (1966).

[113] R. G. Brewer, *Appl. Phys. Lett.* **5**, 127 (1964); *Phys. Rev.* **140**, A800 (1965).

of energy recorded and confusion in the interferometry. The problem here occurs in identifying which of the ring systems in an interferogram is that due to the laser and which to the scattered light.

The timing of these multiple pulses was investigated by Wiggins *et al.*[110] who showed that there were successive pulses whose time separation depended upon the distance *d* shown in Fig. 10. Further, they showed that the intensity of the amplified back-scattered light could be greater than that of the original pulse. This time behavior is shown in Fig. 11, which is taken from Wiggins *et al.*[110] Thus the most intense ring system in an interferogram is not necessarily due to the laser pulse as was assumed by some authors. Guiliano[114] has done time-resolved spectroscopy to support these findings.

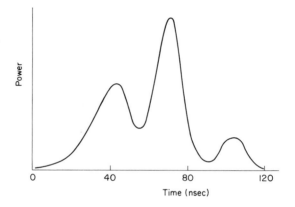

FIG. 11. Output power as a function of time showing the original laser pulse and two reamplified pulses that were back-scattered from a CS_2 cell placed 4 m from the laser. [Taken from the work of T. A. Wiggins, R. V. Wick, D. H. Rank, and A. H. Guenther, *Appl. Opt.* **4**, 1203 (1965).]

Because it is necessary for the ruby to amplify the back-scattered light to obtain intensities comparable to the original laser, there could be some shift between the SBS wavelength and the reemitted wavelength. This is due to the wavelength-dependence of gain in the cavity.[110] The effect may be small but has been observed in one case.[115] A number of schemes have been proposed to eliminate this problem. They include simply increasing the path length so that there is no time overlap of the incident and reamplified pulses from the laser, placing a polarizer and

[114] C. R. Guiliano, *Appl. Phys. Lett.* **7**, 279 (1965).
[115] R. G. Brewer, *Appl. Phys. Lett.* **9**, 51 (1966).

$\lambda/4$ plate in the beam to reject the back-scattered component, and operation so near threshold that the laser can not reamplify the back-scattered light.

The problem of this multiple scattering has been present since the first observations.[96,97] The cause has been found and the remedy is available. However, the situation is not completely clear due to the complication of optical mixing to be discussed in the final section. This has been discussed by Goldblatt and Hercher.[116]

Stimulated scattering has been observed at other than 180° by supplying a cavity[117] or an interaction path[118] at some other angle. This has application for a direct gain measurement, for observing different hypersound frequencies, and for investigating the anti-Stokes as well as the Stokes component.

Problems encountered in SBS are the breakdown of the medium under investigation and self-focusing. In the first case the power density is sufficiently high to produce an ionization of the medium. This was first investigated in gases by Meyerand and Haught[119] and extensively by Alcock et al.[120] The sound wave associated with SBS has been shown to be a source of damage in solids. Studies of crystals[121] and glasses[122,123] indicate that the presence of intense waves have a direct correlation with the damage produced. In plastics this has not been found to be the case.[124]

Self-focusing has been shown to occur in many liquids[125] and in gases.[104] If the nonlinear terms in the index of refraction are large enough, the index in the region of greatest field can be higher than where the field is weak so that the light is refracted into the beam and the medium acts

[116] N. Goldblatt and M. Hercher, *Phys. Rev. Lett.* **20**, 310 (1968).

[117] D. A. Jennings and H. Takuma, *Appl. Phys. Lett.* **5**, 239 (1964).

[118] J. L. Emmett and A. L. Schalow, *Phys. Rev.* **170**, 358 (1968).

[119] R. G. Meyerand, Jr., and A. F. Haught, *Phys. Rev. Lett.* **11**, 401 (1963); **13**, 7 (1964).

[120] A. J. Alcock, C. DeMichelis, and M. C. Richardson, *Appl. Phys. Lett.* **15**, 72 (1969) and references given therein.

[121] D. A. Kramer and R. E. Honig, *Appl. Phys. Lett.* **13**, 115 (1968) and reference given therein.

[122] D. Mash, V. Morozov, V. Starunov, E. Tiganov, and I. Fabelinskii, *JETP Lett.* **2**, 157 (1965).

[123] C. R. Guiliano, *Appl. Phys. Lett.* **5**, 137 (1964).

[124] N. F. Pilipetskii, Y. P. Raizer, and V. A. Upadyshev, *Sov. Phys. JETP* **27**, 568 (1968).

[125] R. Y. Chiao, E. Garmire, and C. H. Townes, *Phys. Rev. Lett.* **13**, 479 (1964).

as its own lens. Obviously, if this happens, the calculated power density and gain have little meaning and thresholds are misleading. Depolarization of the back-scattered light is often a clue that those effects are occurring.

SBS can act as a Q-switch if a liquid cell is added to the ruby cavity.[126] The power builds up in the usual way in the regular cavity until the threshold for SBS is reached. The cell now acts as a good reflector and sends Brillouin-shifted light through the ruby where it is amplified, the process then repeating itself. The pulses produced are short in duration. Multiple pulses or a single pulse can be produced. This process places a limit on the length of a saturable dye cell since the liquid solvent for the dye may act in this manner. The wave number change due to this in-cavity scattering has been shown to be less than if the same liquid is used outside the cavity.[127]

The reason for this difference is not clear. It may be due to mode pulling noted above or to a power dependence on the magnitude of the Brillouin shift. Some evidence for this is seen in gases.[128] It has been treated by Wang.[129] Another explanation may lie in the attenuation present. Solovyev et al.[130] have developed corrections to the measured speed that are dependent upon the acoustic attenuation and show that the resonant frequency can be decreased, particularly if the hypersonic frequency lies near a relaxation frequency of the medium.

Some problems in measurements of stimulated effects are caused by the temporal and spectral narrowing noted in Section 3.4.1. The latter can produce random errors and large scatter, as demonstrated in liquids by Goldblatt and Litovitz.[131] The problem has been shown to be equally important in gases.[132] Near threshold only the most intense temporal and spectral part of the incident beam can produce a stimulated process. Thus the measured shift from the center of the incident pulse, whose structure is not resolved in an interferogram, to the center of the stimu-

[126] A. J. Alcock and C. DeMichelis, *Appl. Phys. Lett.* **11**, 42, 185 (1967).

[127] R. V. Wick and A. H. Guenther, *Appl. Opt.* **7**, 73 (1968).

[128] D. I. Mash, V. V. Morozov, V. S. Starunov, and I. L. Fabelinskii, *Sov. Phys. JETP* **28**, 1085 (1969).

[129] C. S. Wang, *Phys. Rev. Lett.* **24**, 1394 (1970).

[130] V. A. Solovyev, N. R. Goldblatt, and C. J. Montrose, *J. Acoust. Soc. Amer.* **43**, 1170 (1968).

[131] N. R. Goldblatt and T. A. Litovitz, *J. Acoust. Soc. Amer.* **41**, 1301 (1967).

[132] T. T. Saito, L. M. Peterson, D. H. Rank, and T. A. Wiggins, *J. Opt. Soc. Amer.* **60**, 749 (1970).

lated pulse may be the wrong quantity to measure unless the pulse is very uniform. This effect could also cause constant shifts since the wavelength of a ruby laser system has been shown to be a function of time.[133–135] Pohl and Kaiser state there is a constant difference between the wave numbers of the incident and scattered light.[135] Even so, in unresolved interferograms, shifts could be present.

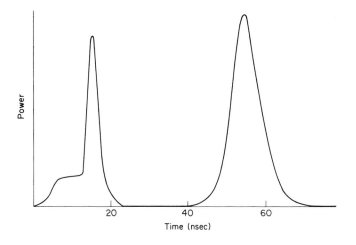

FIG. 12. Time display of power received by a photodiode. The first small peak is due to reflections from the lens and cell windows. The sharp superimposed peak is due to the Brillouin scattered power. The larger peak is used to monitor the laser output power. [Taken from the work of T. A. Wiggins, R. V. Wick, and D. H. Rank, *Appl. Opt.* **5**, 1069 (1966).]

Examples of temporal and spectral narrowing are shown in Figs. 12 and 13 taken from Wiggins *et al.*[112] In Fig. 12, the first large pulse is due to SBS and the second is due to the incident pulse delayed by an optical delay line. The narrowing is obvious; it is most pronounced near threshold. Figure 13 shows an interferogram of SBS in two gases. This positive copy of the original negative shows the inner (dark) SBS rings to be sharper than the rings due to the incident light.

One way to accentuate the production of SBS is to use a temporally long laser pulse, where "long" means the time duration is greater than

[133] D. J. Bradley, G. Magyar, and M. G. Richardson, *Nature (London)* **212**, 63 (1966); V. V. Korobkin, A. M. Leontovich, M. N. Popova, and M. Y. Shchelev, *JETP Lett.* **3**, 194 (1966).

[134] D. Pohl, M. Maier, and W. Kaiser, *Phys. Rev. Lett.* **20**, 366 (1968).

[135] D. Pohl and W. Kaiser, *Phys. Rev.* **B1**, 31 (1970).

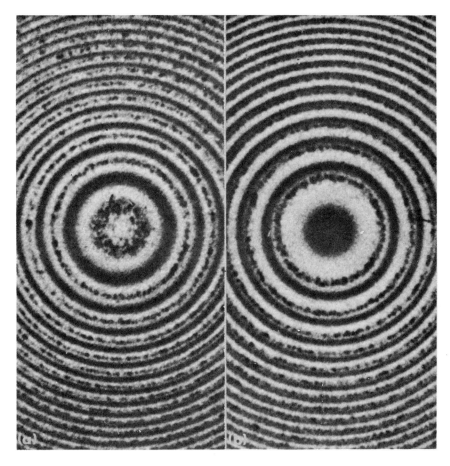

FIG. 13. Interferograms of light scattered from (a) methane at 17 atm and (b) nitrogen at 48 atm illustrating the spectral sharpness of the Brillouin scattered light.

the acoustical relaxation time. This has been treated by Hagenlocker and co-workers for gases[136] and exploited by von der Linde et al. to study Raman scattering using very short pulses to eliminate Brillouin scattering.[137]

An interesting experimental technique for observing the anti-Stokes shifted Brillouin component has been described by Dumartin et al.[138]

[136] E. E. Hagenlocker, R. W. Minck, and W. G. Rado, Phys. Rev. 154, 226 (1967).

[137] D. von der Linde, M. Maier, and W. Kaiser, Phys. Rev. 178, 11 (1969).

[138] S. Dumartin, B. Oksengorn, and B. Vodar, C. R. Acad. Sci. Paris 262B, 1680 (1966); J. Chem. Phys. 64, 235 (1967).

They term this the *inverse Brillouin effect*, in analogy with the inverse Raman process discussed in the next section. The source for this observation is the blue-shifted continuum produced by a laser pulse focused into compressed argon. This back-scattered light traverses a cell containing acetone. The nonlinear interaction in the cell transfers energy from the continuum to the laser beam at a wave number separation that corresponds to the Brillouin shift of acetone. The result is an absorption at the anti-Stokes Brillouin-shifted position.

An experimental technique that is useful in SBS as well as for other nonlinear effects is the direct measurement of gain. This type of experiment makes a direct comparison between the intensity of an oscillator signal and the intensity of this signal after it has traversed a path of length l where there is a high power density P. The difference between the frequency of the high-power light and the frequency of the oscillator is usually adjustable so that the gain can be measured as a function of this frequency difference. This difference can be adjusted by several methods. One is to utilize the frequency sweep of the laser itself and reflect the incident light back through the amplifier cell after traversing a suitable delaying distance. Another is to use the SBS signal from various mixtures of liquids or gases selected to give Brillouin shifts larger or smaller than the frequency shift at which the gain is expected to be maximum. Temperature changes in the oscillator can also be used or crystals can be rotated to give different momentum matching conditions and hence different frequency displacements.

This type of measurement has been used extensively by Lallemand and Bloembergen,[139] Avizonis *et al.*[140] and Grun *et al.*[141] for gain and determination of line widths in stimulated Raman scattering. Work by Kaiser *et al.*[135,137,142,143] Hagenlocker *et al.*,[136] Bret and Denariez,[144] Brewer,[113] and Walder and Tang[145] has yielded an understanding of the

[139] See N. Bloembergen, G. Bret, P. Lallemand, A. Pine, and P. Simova, *IEEE J. Quant. Electron.* **QE 3**, 197 (1967) for an extensive review.

[140] P. V. Avizonis, K. C. Jungling, A. H. Guenther, R. M. Heimlich, and A. J. Glass, *J. Appl. Phys.* **39**, 1752 (1968).

[141] J. B. Grun, A. K. McQuillan, and B. P. Stoicheff, *Phys. Rev.* **180**, 61 (1969).

[142] M. Maier, W. Rother, and W. Kaiser, *Phys. Lett.* **23**, 83 (1966); *Appl. Phys. Lett.* **10**, 80 (1967); M. Maier, *Phys. Rev.* **166**, 113 (1968).

[143] A. Laubereau, W. Englisch, and W. Kaiser, *IEEE J. Quant. Electron.* **QE 5**, 410 (1969); D. Pohl, M. Maier, and W. Kaiser, *Phys. Rev. Lett.* **20**, 366 (1968).

[144] G. Bret and M. Denariez, *C. R. Acad. Sci. Paris* **264B**, 1815 (1967).

[145] J. Walder and C. L. Tang, *Phys. Rev.* **155**, 318 (1967).

Brillouin gain mechanism. Rayleigh wing scattering has been studied by this method by Deraniez and Bret[146] and integrated gains in stimulated thermal Rayleigh scattering were measured by Gires.[147] Stimulated thermal Brillouin scattering was first observed by measurements of gain by Pohl and co-workers.[148]

An experimental technique which can be used to observe weak stimulated scattering processes is to allow the 180° scattered light to reenter the laser and be amplified. As noted above, this is done at the risk of mode-pulling or damage to the components. However, Fabelinskii and his co-workers have used this means of observing the scattering from H_2 at low pressure.[128,149] Arefev and Morozov have studied concentration scattering as well by this technique.[150]

3.5.3. Results

The results of investigations involving stimulated Brillouin scattering can be grouped into two general areas. One of these areas is that of a physical understanding of the stimulated process itself. The gain or efficiency with which power is transferred from the incident beam to the Stokes-shifted beam has been studied by both theorists and experimentalists to the point where there is now reasonable agreement. Kroll[151] has presented a transient theory which accounts for some of the experimental results. However, many of the experiments have actually been performed in a way such that the process is actually proceeding in a steady-state manner. Measurements and experimental results by Maier et al.,[142] Bret and Denariez,[144] Brewer,[113] Walder and Tang,[145] Kaiser et al.,[137,143] and Hagenlocker et al.[136] have contributed to the understanding of the mechanisms of the process. The physical parameters of electrostrictive coupling and phonon lifetime have been deduced from these measurements.

Experimental measurements of gain have been made difficult by saturation, wherein the signals become too large for linear amplification to occur, by self-focusing, by competition with other stimulated pro-

---

[146] M. Denariez and G. Bret, *Phys. Rev.* **171**, 160 (1968).

[147] F. Gires, *C. R. Acad. Sci. Paris* **266B**, 596 (1968).

[148] D. Pohl, I. Reinhold, and W. Kaiser, *Phys. Rev. Lett.* **20**, 1141 (1968).

[149] I. L. Fabelinskii, D. I. Mash, V. V. Morozov, and V. S. Starunov, *Phys. Lett.* **27A**, 253 (1968).

[150] I. M. Arefev and V. V. Morozov, *JETP Lett.* **9**, 269 (1969).

[151] N. M. Kroll, *J. Appl. Phys.* **36**, 34 (1965).

cesses, by breakdown in the medium, and by spatial, wavelength, and temporal inhomogeneities in the light beams. All these things have required separate investigations and so have added to the understanding of the interaction of light with matter.

An example of the results obtained from measurement of gain is given in Table IV taken from the work of Pohl and Kaiser.[135] It gives the experimentally determined gain and phonon lifetimes. The gain is compared to calculated results.

TABLE IV. Experimentally Determined Phonon Life Times, Brillouin Line widths, and Steady-State Gain Factors Calculated from τ_{exp} for Several Liquids[a]

Substance	Phonon life time τ_{exp} (nsec)	Brillouin linewidth $\Gamma_B/2\pi$ (MHz)	Gain factors	
			$g(\infty)_{calc}$ (cm/MW)	g_{exp} (cm/MW)
CS_2 (pure)	2.2	75	140×10^{-3}	130×10^{-3}
CS_2 + 2.5% CCl_4	1.9	85	125	120
Acetone	0.9	180	22	20
n-Hexane	0.72	220	23	26
Toluene	0.33	480	15	13
CCl_4	0.25	640	7	6

[a] Taken from the work of D. Pohl and W. Kaiser, *Phys. Rev.* **B1**, 31 (1970).

The second area is that of determination of sound speeds at giga-herzian frequencies. As pointed out by Garmire and Townes,[97] high-frequency acoustic measurements are difficult to make and some materials do not lend themselves to spontaneous scattering. Hence, measurements by a stimulated process can help determine the dispersion of speed characteristics of many substances. The effect of attenuation due to high-frequency absorption can also be explored in connection with corrections to sound speeds.

The precision that is obtained is not as great as in ultrasonic measurements. Some of the best experiments in liquids show a precision of ± 2 m/sec.[131] In work on water and CS_2, complete agreement to this precision is found when results are compared to ultrasonic values. In gases, the most recent results[132] have a precision of ± 3 m/sec for CO_2.

TABLE V. Hypersound Speeds Determined by Various Observers from Stimulated Scattering from Several Liquids

Material	Speed (m/sec)	Temperature (°C)	Reference
CS_2	1250	21	a
	1242	22	b
	1232	20	c
	1223		d
	1252	20	e
H_2O	1486	21	f
	1471	22	b
	1480	20	g
	1485	21	h
	1464	21	i
CCl_4	1015	20	e
	1020	21	j
	1007	22	b
C_2H_5OH	1170	20	g
CH_3OH	1130	21	h
	1100	22	b
Acetic Acid	1035	21	h
	1105	20	c
$SiCl_4$	780	21	h
Chloroform	1050	21	h
C_6H_6	1500	21	h
	1434	20	c
t-Butyl alcohol	1164	21	i
Nitrobenzene	1546	20	c
Salol	740	20	c
Acetone	1174	22	b

TABLE V (continued)

Material	Speed (m/sec)	Temperature (°C)	Reference
Aniline	1699	22	[b]
Fused quartz	5804	20	[c]
Crown glass	5906	20	[c]
1,2-Propylene glycol	1885	19	[k]
Triacetin	1570	20	[k]

[a] N. R. Goldblatt and T. A. Litovitz, J. Acoust. Soc. Amer. **41**, 1301 (1967).

[b] E. Garmire and C. H. Townes, Appl. Phys. Lett. **5**, 84 (1964).

[c] D. Mash, V. Morozov, V. Starunov, E. Tiganov, and I. Fabelinskii, JETP Lett. **2**, 157 (1965).

[d] D. A. Jennings and H. Takuma, Appl. Phys. Lett. **5**, 239 (1964).

[e] F. Barocchi, M. Mancini, and R. Vallauri, Il Nuovo Cimento **40B**, 233 (1967); F. Barocchi and R. Vallauri, J. Chem. Phys. **51**, 10 (1969).

[f] N. R. Goldblatt and T. A. Litovitz, J. Acoust. Soc. Amer. **41**, 1301 (1967). (Also measured at different temperatures.)

[g] K. Negishi, M. Yamazaki, and Y. Torikai, Jap. J. Appl. Phys. **6**, 1016 (1967).

[h] J. L. Emmett and A. L. Schalow, Phys. Rev. **170**, 358 (1968). (Also data at 58°C.)

[i] I. M. Arefev, V. S. Starunov, and I. L. Fabelinskii, JETP Lett. **6**, 163 (1967).

[j] N. Goldblatt and M. Hercher, Phys. Rev. Lett. **20**, 310 (1968).

[k] I. S. Zlatin, S. V. Krivokhizha and I. L. Fabelinskii, Sov. Phys. JETP **29**, 638 (1969). (Temp. dependence.)

However, the complexities of impurities, relaxation, and pressure dependence make comparison with ultrasonic results more difficult.

Table V gives some results of measurement of sound speeds[152-155] in liquids and noncrystalline solids that are results of ruby laser SBS measurements mainly for 180° scattering. In addition a number of measured results are noted in the section on optical mixing (3.7.1).

[152] F. Barocchi, M. Mancini, and R. Vallauri, Il Nuovo Cimento **40B**, 233 (1967); F. Barocchi and R. Vallauri, J. Chem. Phys. **51**, 10 (1969).

[153] K. Negishi, M. Yamazaki, and Y. Torikai, Jap. J. Appl. Phys. **6**, 1016 (1967).

[154] I. M. Arefev, V. S. Starunov, and I. L. Fabelinskii, JETP Lett. **6**, 163 (1967).

[155] I. S. Zlatin, S. V. Krivokhizha, and I. L. Fabelinskii, Sov. Phys. JETP **29**, 638 (1969).

TABLE VI. Hypersound Speed in Compressed Gases Determined from Stimulated
180° Scattering of Ruby Laser Light

Gas	Temperature (°C)	Pressure range (Atm)	Speed (m/sec)	at	Pressure (Atm)	Reference
N_2	27	70–190	386		125	a
	27	100–1000	356		125	b
	—	250–500	435		250	c
	—	—			125	d
CH_4	27	25–170	448		125	a
SF_6	27	15–25	104		20	a
CO_2	40	40–145	443		100	a
O_2	—	—	330		150	d
H_2	—	—	1130		95	d
Ar	—	300–500	395		300	c

[a] T. T. Saito, L. M. Peterson, D. H. Rank, and T. A. Wiggins, *J. Opt. Soc. Amer.* **60**, 749 (1970).

[b] E. E. Hagenlocker and W. G. Rado, *Appl. Phys. Lett.* **1**, 236 (1965). See W. M. Madigosky, A. Monkewicz, and T. A. Litowitz, *J. Acoust. Soc. Amer.* **41**, 1308 (1965).

[c] S. Dumartin, B. Oksengorn, and B. Vodar, *C. R. Acad. Sci. Paris* **262B**, 1680 (1966); *J. Chem. Phys.* **64**, 235 (1967).

[d] D. I. Mash, V. V. Morozov, V. S. Starunov, and I. L. Fabelinskii, *JETP Lett.* **2**, 349 (1965). (Power dependence.)

Table VI summarizes the results of measurement of sound speed[156,157] in gases by the SBS method. All of the results quoted are for 180° scattered ruby laser light.

3.6. Stimulated Raman Scattering

3.6.1. Introduction

The stimulated Raman process was first observed by accident. Woodbury and Ng, using a nitrobenzene-filled Kerr cell in the cavity of a ruby laser system, observed discrepancies in the power output and

[156] W. M. Madigosky, A. Monkewicz, and T. A. Litovitz, *J. Acoust. Soc. Amer.* **41**, 1308 (1967).

[157] D. I. Mash, V. V. Morozov, V. S. Starunov, and I. L. Fabelinskii, *JETP Lett.* **2**, 349 (1965).

attributed it to infrared radiation not detected by some of their instruments.[158] After a spectral analysis of the output, they concluded that the 1344-cm^{-1} shift which they observed was the same as the spectral shift of the strongest spontaneous Raman scattering in nitrobenzene. Not only was the Stokes shift seen but, in addition, radiation corresponding to twice and thrice this amount of shift were observed. Measurements of wavelength and intensity indicated that the shifts were true harmonics of the fundamental shifts rather than the anharmonic vibrational shifts seen in spontaneous Raman scattering.

This type of scattering was also observed with the scattering cell outside the cavity. The process was shown to have a threshold and all the other characteristics of a stimulated phenomenon.

The theory of this scattering has been developed beginning primarily with the work of Hellwarth.[159] Comparison of the predicted gain with experimental measurements disagreed by orders of magnitude in many cases. This led to the investigation of some new phenomena. Multimode effects in the laser beam have been discussed by Bloembergen and co-workers,[160] and self-trapping described by Chiao et al.[161] A tutorial review of the theory has been written by Bloembergen.[162]

A number of interesting experimental observations have been explained by current theory. They include: only the strongest spontaneous Raman shift is observed in the stimulated spectrum unless saturation occurs; the anti-Stokes lines are much more intense than would be expected and are emitted in the forward direction in a cone;[163] the gain or threshold depends upon line width. Thus they are temperature- and polarization-dependent[164] in addition to being a function of scattering angle.

3.6.2. Experimental Techniques

Some of the necessary techniques have been discussed in Sections 3.4 and 3.5. The giant pulse laser and its diagnostics, as well as the arrange-

[158] E. J. Woodbury and W. K. Ng, *Proc. IRE* **50**, 2367 (1962).

[159] R. W. Hellwarth, *Appl. Opt.* **2**, 847 (1963).

[160] N. Bloembergen and Y. R. Shen, *Phys. Rev. Lett.* **13**, 720 (1964); P. Lallemand and N. Bloembergen, *Phys. Rev. Lett.* **15**, 1010 (1965).

[161] R. Y. Chiao, E. Garmire, and C. H. Townes, *Phys. Rev. Lett.* **13**, 479 (1964).

[162] N. Bloembergen, *Amer. J. Phys.* **35**, 989 (1967).

[163] R. Y. Chiao and B. P. Stoicheff, *Phys. Rev. Lett.* **12**, 290 (1964); R. W. Minck, R. W. Terhune, and C. C. Wang, *Proc. IEEE* **54**, 1357 (1966).

[164] N. Bloembergen, P. Simova, and G. Bret, *Phys. Rev. Lett.* **17**, 1239 (1966); R. W. Minck, E. E. Hagenlocker, and W. G. Rado, *Phys. Rev. Lett.* **17**, 229 (1966).

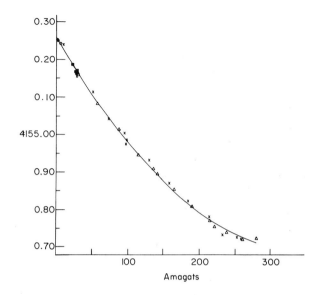

FIG. 14. Frequency shift of the $Q(1)$ line of the 1–0 line of H_2 as a function of pressure. Amagat units of density express the ratio of the density of a gas to the density of that gas at standard conditions. Key: $\bigcirc$, quadrupole; $\blacksquare$, field-induced dipole; $\triangle$, Stokes; $\times$, anti-Stokes. [Taken from the work of J. V. Foltz, D. H. Rank, and T. A. Wiggins, *J. Mol. Spectrosc.* **21**, 203 (1966).]

ment to measure gain, are also important in stimulated Raman scattering (SRS).

One major difference encountered is that diffraction grating spectrographs rather than interferometers are the prime measuring device for observing the large spectral shifts. An example of the accuracy and precision that can be obtained is shown in Fig. 14 taken from the work of Foltz *et al.*[165] The laser output was observed in fifth order. This choice allows the Stokes line as well as the anti-Stokes lines to appear on the same photographic plate even though the dispersion is large (3 cm^{-1}/cm), the first Stokes in fourth order, the first anti-Stokes in sixth order, etc. Precise spectral shifts have also been measured by interferometry.[166]

A different type of experiment has been used by Stoicheff[167] and co-

[165] J. V. Foltz, D. H. Rank, and T. A. Wiggins, *J. Mol. Spectrosc.* **21**, 203 (1966).

[166] P. Lallemand, P. Simova, and G. Bret, *Phys. Rev. Lett.* **17**, 1239 (1966); *J. Mol. Spectrosc.* **26**, 262 (1968).

[167] W. J. Jones and B. P. Stoicheff, *Phys. Rev. Lett.* **13**, 657 (1964); R. A. McLaren and B. P. Stoicheff, *Appl. Phys. Lett.* **16**, 140 (1970).

workers, which they term *inverse stimulated Raman scattering*. The advantage of this type of observation is that all the Raman active modes can be observed, and at the same time, rather than seeing only the one due to the strongest totally symmetric vibrational mode. The experiment consists of causing an intense monochromatic beam and an intense continuum to traverse a sample cell. The continuum was the broad anti-Stokes radiation or fluorescence radiation excited by second harmonic generation. The two beams can be mixed in a capillary containing the sample. The resulting spectrum shows absorption peaks at the wavenumber shifts corresponding to the Raman transitions.

3.6.3. Results

As in the case of SBS, one of the important results of measurements in SRS is the gain of the stimulated process. Most of the measurements have been limited to gases and simple liquids when it was realized that the self-focusing found in many liquids was affecting the results.

It appears that there is now satisfactory agreement between theory and experiment. The results in gaseous H_2 [139,140,166] and in liquid N_2 and O_2 [141] show the importance of line widths in the calculation of gain. Transient gain using picosecond pulses have also been used by Mack *et al.*[168] to study Raman spectra of gases in which the no steady-state SRS is possible. The short pulse also suppresses other stimulated effects such as SBS.

In H_2 with steady-state conditions the $Q(1)$ line of the 1–0 vibrational band is excited. That this is true is evident from the exact agreement found in its spectral shift when compared with known values.[165] Further, it is obvious from photographs of the spectrum that only a single transition is stimulated. This is due to the very nature of the stimulated process. The spontaneous signal is largest for this transition and hence the gain is greatest. It grows at a rate greater than for any other transition and depletion of the laser input prevents the other transitions from becoming intense.

This phenomenon was also shown to exist in the SRS spectrum of $SnCl_4$. In this case a single isotopic species, $SnCl_3^{35}Cl^{37}$, was found to have gain.[169]

---

[168] M. E. Mack, R. L. Corman, J. Reintjes, and N. Bloembergen, *Appl. Phys. Lett.* **16**, 209 (1970).

[169] D. H. Rank, R. V. Wick, and T. A. Wiggins, *Appl. Opt.* **5**, 13 (1966).

A number of interesting uses have been found for SRS. One is in the production of new lasing wavelengths, usually the Stokes-shifted line of a gas or liquid. These new wavelengths have use in that they have a close resonance with some molecular transition or at a wavelength where some medium of interest, for example the atmosphere, is transparent. Dye lasers, not included in this discussion, are also useful in this regard.

An obvious use of SRS would be in measurement of spectral shifts due to Raman transitions of various compounds. It is doubtful, however, as to how much real value these measurements would have. To date, only in H_2 has information[165,166] of interest to molecular spectroscopists been deduced. Even here the results were more in the nature of confirmation rather than discovery.

3.7. Other Stimulated Effects

3.7.1. Optical Mixing

Previous sections have discussed stimulated Brillouin scattering and the stimulated Raman effect. Both of these phenomena arise only because of the nonlinear behavior of optical materials when illuminated by means of very intense radiation. Stimulated Brillouin spectra are observed by means of 180° scattering, which oftentimes produces multiple lasing resulting in the appearance of a number of Brillouin components shifted by multiples of the original stimulated Brillouin shift. To the best of our knowledge, no anti-Stokes stimulated Brillouin components have been observed with certainty.

Wick et al.[170] have illuminated liquids and solids with a single-moded ruby laser beam of high power (70 MW, beam divergence 8 mrad, linewidth 0.03 cm^{-1}). A lens of 15-cm focal length was used to focus the light in the material. The light passing through the material was focused on the slit of a high-resolution grating spectrograph (resolving power 400,000, dispersion 3 cm^{-1}/cm on the photographic plate). The experiment was first performed with CS_2, which gave a truly remarkable result. Instead of observing a single sharp line in the spectrum, a large number of lines were observed equally spaced and shifted in wavelength both to the red and the violet of the laser wavelength. The incident light was polarized and the light traversing the CS_2 was found to be depolarized. Thus this perfectly transparent liquid did not transmit any of

---

[170] R. V. Wick, D. H. Rank, and T. A. Wiggins, *Phys. Rev. Lett.* **17**, 466 (1966).

the incident laser light without modification in the above-mentioned experiment.

Measurements of the spacing of the components in the pattern observed above showed that the spacing was that expected for Brillouin shifts which were back-scattered (180° scattering). The velocity of hypersound calculated on the aforementioned assumptions agreed with the spontaneous Brillouin effect measurements of Chiao and Fleury.[40] The phenomenon observed in these experiments is quite different from the results obtained with 180° back-scattering in the stimulated Brillouin effect. In the stimulated Brillouin effect, multiple lasing provides the source of the various orders of Brillouin shifting. In the case of CS_2, where 40 Stokes and 20 anti-Stokes components have been observed (the total pattern has a width of 12 cm^{-1}), the total spread of the optically pumped ruby excited state is not nearly broad enough to account for the phenomenon on the basis of simple relasing. Apparently, the laser pulse excites the Brillouin effect and the created red-shifted frequency mixes with the laser frequency in the nonlinear medium, thus creating new frequencies which, in turn, are capable of further mixing, thus producing the proliferation of new frequencies. At the same meeting that one of the authors reported on the phenomenon of optical mixing by means of the stimulated Brillouin effect, B. P. Stoicheff showed optical mixing in CS_2 making use of a laser which operated in two modes separated by 0.8 cm^{-1}.

Wiggins et al.[171] have done extensive work on the phenomenon reported above. In this work 23 liquids and two types of optical glass were investigated. In all cases the sound speeds measured in the optical mixing experiments agree with those determined from experiments on spontaneous or stimulated Brillouin scattering. The Brillouin shifts, although observed at 0° scattering, must be attributed to the 180° scattered radiation entering the laser, being amplified, and producing a giant pulse at the shifted frequency. This light, along with the original laser light and multiply shifted components, can be observed in the forward direction.

Figure 15 shows some examples of optical mixing pattern. The (b) and (e) parts of the figure show that mixtures of liquids give rise to the pattern spacings to be expected if a simple additivity theorem would apply to the sound velocity in a mixture of liquids. It should be remarked that some liquids produce a spectrum that shows a continuum up to

[171] T. A. Wiggins, R. V. Wick, N. D. Foltz, C. W. Cho, and D. H. Rank, J. Opt. Soc. Amer. 57, 661 (1967).

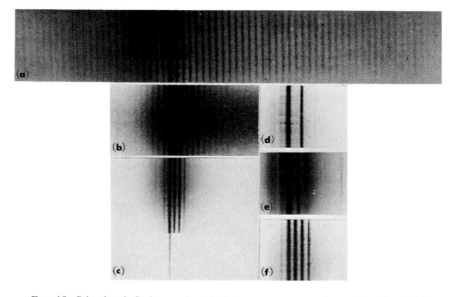

FIG. 15. Stimulated Stokes and anti-Stokes components observed with a high-resolution spectrograph in liquids and mixtures of liquids: (a) CS_2; (b) 50–50 mixture of CS_2 and CCl_4; (c) CCl_4; (d) glycerol; (e) 50–50 mixture of glycerol and water; (f) water. The bottom section of (c) shows the incident laser line. [Taken from the work of T. A. Wiggins, R. V. Wick, N. D. Foltz, C. W. Cho, and D. H. Rank, *J. Opt. Soc. Amer.* **57**, 661 (1967).]

several wave numbers in width, predominantly on the Stokes side. This is presumably the Rayleigh wing observed by Mash *et al.*[172] We shall discuss this phenomenon in detail in the next section of this chapter.

In Table VII we have presented a summary of the work on optical mixing accomplished by Wiggins *et al.*[171] Comparisons to sound speeds from other methods of measurement are included.[173,174] A very wide range of the number of Brillouin components observed in different materials is demonstrated. The range of hypersonic velocity observed in this investigation varies from 1000 to 6000 m/sec depending on the material. The agreement of the sound velocity determined with measurements of other investigators is satisfactory particularly in view of the fact that dispersion of velocity with frequency may be present in some materials.

[172] D. I. Mash, V. V. Morozov, V. S. Starunov, and I. L. Fabelinskii, *JETP Lett.* **2**, 25 (1965).

[173] L. Bergmann, "Der Ultraschall." Hirzel, Zurich, 1949.

[174] G. Benedek and T. Greytak, *Proc. IEEE* **53**, 1623 (1965).

3.7.2. Rayleigh Wing Scattering

Orientational scattering produces a depolarized, frequency-broadened, Rayleigh wing in spontaneous scattering of light by liquids. This phenomenon has been known since the very earliest investigations of the Raman effect, and in the older literature was called *rotational wing scattering*. Broad stimulated Rayleigh wing scattering has been studied by several investigators[172,175] and is attributable to the same cause as in spontaneous scattering.

In 1940 Gross[176,177] observed this broadening of the Rayleigh scattered light from liquids. He attributed this scattering to the finite time required to produce thermal relaxation, the time required being expressed as $\tau = 4\pi\eta a^3/kT$, the Debye relaxation time. Here η is the viscosity of the liquid, a is the effective radius of the liquid molecule, k is the Boltzmann constant, and T is the absolute temperature. Mash *et al.*[172] have observed this broadening in stimulated Rayleigh scattering in certain liquids and have called this the *stimulated Rayleigh wing*. They have indicated that the maximum intensity of this feature should occur at a frequency shift from the laser line of $\Delta\nu$ (in wave numbers), where $2\pi C(\Delta\nu)(\tau) = 1$, where τ is the characteristic reorientation time of the liquid.

In connection with the investigation of optical mixing, Wiggins *et al.*[171] obtained some photographs in which a very sharp line appeared between the unmodified Rayleigh line and the modified Brillouin line in nitrobenzene.

Cho *et al.*[178] have reported on an extensive investigation of this line, which is believed to be the stimulated analog of the Rayleigh wing. Observations were made with a high-resolution grating spectrograph (resolving power 300,000, dispersion 4 cm^{-1}/cm). The light source was a ruby giant pulse laser, single moded with power up to 40 MW and a pulse duration of 10 nsec. The light was focused into the middle of a 10-cm cell of liquid nitrobenzene or *m*-nitrotoluene using a lens of 10-cm focal length. The light transmitted by the cell containing the liquid was focused on the slit of the spectrograph. It was quite clear from the experiments performed that the line observed resulted from a stimu-

[175] N. Bloembergen and P. Lallemand, *Phys. Rev. Lett.* **16**, 81 (1966); P. Lallemand, *Appl. Phys. Lett.* **8**, 276 (1966).

[176] E. Gross, *C. R. Acad. Bulgare* **28**, 786 (1940).

[177] J. Frenkel, "Kinetic Theory of Liquids." Dover, New York, 1955.

[178] C. W. Cho, N. D. Foltz, D. H. Rank, and T. A. Wiggins, *Phys. Rev. Lett.* **18**, 107 (1967).

TABLE VII. Materials in Which Optical Mixing was Investigated[a]

| Material | Number of Brillouin components observed | | $\Delta\nu$ (cm^{-1}) | T (°C) | Present investigation | | | Previous results | | |
	Stokes	Anti-Stokes			ν_B (GHz)	ν_T (m/sec)	ν_{20} (m/sec)	ν_B (GHz)	ν_{20} (m/sec)	Ref.
Carbon disulfide	42	20	0.194	26.0	5.82	1256	1275	4.5	1248	b
								6.4	1255	c
								5.8	1248	d
Carbon tetrachloride	3	1	0.146	26.0	4.38	1049	1069	3.3	1016	b
								4.8	1052	c
								4.2	1014	d
Glycerol	2	—	0.386	26.0	11.6	2734	2851	12.6	2808	b
Water	3	1	0.197	27.0	5.91	1544	1544	6.2	1488	c
								5.7	1471	d
Benzene	2	1	0.211	27.3	6.33	1473	1511	4.9	1481	b
								7.1	1511	c
								6.2	1434	e
Bromobenzene	4	1	0.188	23.5	5.64	1261	1272	~0.005	1170	f
Nitrobenzene	6	2	0.228	26.0	6.84	1541	1563	5.3	1532	b
								7.6	1569	c
								7.0	1546	e
Aniline	10	6	0.259	25.0	7.77	1713	1733	5.9	1682	b
								7.7	1707	d
Toluene	8	4	0.193	27.0	5.79	1355	1389	6.5	1385	c
								6.5	1390	g
Benzaldehyde	7	3	0.224	26.0	6.72	1522	1546	~0.005	1479	f
Cyclohexane	4	1	0.180	25.0	5.40	1321	1341	6.1	1346	c

	n								note
m-Nitrotoluene	5	0.229	26.0	6.87	1553	—	~0.005	1489	f
O-Nitrotoluene	4	0.217	26.0	6.51	1471	—	~0.005	1432	f
p-Xylene	5	0.199	23.0	5.97	1392	1408	~0.005	1330	f
p-Dichlorobenzene	8	0.184	80.0	5.52	1265	—	—	—	
Pyridine	8	0.226	26.0	6.78	1573	1600	~0.005	1445	f
BSC-2 glass	2	0.866	28.0	26.0	5960	—	25.7	5906	e
DF-3	2	0.638	28.0	19.1	4120	—	—	—	
Methanol	4	0.142	23.4	4.26	1111	1124	4.7	1124	c
							4.2	1108	a
Acetic acid	4	0.152	26.0	4.56	1156	1190	3.5	1169	b
							5.0	1191	c
							4.4	1105	e
Methylene iodide	3	0.166	25.0	4.98	1014	1023	~0.005	973	f
Chloroform	3	0.148	23.7	4.44	1070	1083	3.3	1053	b
Acetone	5	0.154	25.0	4.62	1180	1205	3.5	1190	b
							5.1	1198	c
							4.6	1184	d
s-Tetrabromomethane	3	0.173	26.0	5.19	1108	1121	~0.005	1041	f
Octanol	5	0.194	22.5	5.82	1426	1434	4.5	1406	b

[a] The estimated precision of measurement of the frequency shifts is ± 0.002 cm^{-1}. The term v_{20} represents the speed corrected to 20°C. Taken from the work of T. A. Wiggins, R. V. Wick, N. D. Foltz, C. W. Cho, and D. H. Rank, J. Opt. Soc. Amer. 57, 661 (1967).

[b] D. P. Eastman, A. Hollinger, J. Kenemuth, and D. H. Rank, J. Chem. Phys. 50, 1567 (1969).

[c] R. Y. Chiao and P. A. Fleury, Proc. Quantum Electron. Conf. San Juan, Puerto Rico, 1965 (P. L. Kelley, B. Lax, and P. E. Taunewald, eds.), pp. 241–252. McGraw-Hill, New York, 1965.

[d] E. Garmire and C. H. Townes, Appl. Phys. Lett. 5, 84 (1964).

[e] D. Mash, V. Morozov, V. Starunov, E. Tiganov, and I. Fabelinskii, JETP Lett. 2, 157 (1965).

[f] L. Bergmann, "Der Ultraschall." Hirzel, Zurich, 1949.

[g] G. Benedek and T. Greytak, Proc. IEEE 53, 1623 (1965).

TABLE VIII. Observed Frequency Displacement of the Stimulated Rayleigh Line in a Number of Liquids[a]

Liquid	Temperature (°C)	Observed stimulated Rayleigh line			Slope $(\Delta v)/(T/\eta) \times 10^6$ $(\text{cm}^{-1}\,\text{K}^{-1}\,\text{P})$
		Viscosity (cp)	$\Delta v\ (\text{cm}^{-1})$	$\tau(10^{+11}\ \text{sec})$	
Nitrobenzene	20	2.03	0.111	4.78	7.7
o-Nitrophenol	44	2.45	0.078	6.80	6.3
o-Nitroaniline	82	3.07	0.107	4.96	9.1
m-Nitrobenzaldehyde	73	3.30	0.101	5.25	9.3
o-Nitrotoluene	20	2.18	0.133	3.99	8.6
m-Nitrotoluene	20	2.33	0.097	5.47	7.7
p-Nitrotoluene	73	1.03	0.145	3.66	4.3
p-Nitroanisole	74	1.99	0.075	7.07	4.3
m-Nitroacetophenone	82	2.50	0.105	5.05	7.3
1,4-Dimethylnitrobenzene	20	2.54	0.090	5.89	7.9
m-Dinitrobenzene	115	1.78	0.116	4.57	5.2
2,4-Dinitrotoluene	83	3.09	0.098	5.41	8.4
Benzonitrile	20	1.35	0.198	2.68	9.1
Benzoyl chloride	26	1.36	0.124	4.28	5.6
Styrene	26	0.71	0.4	1.0	10.0
Naphthalene	115	0.67	0.5	1.0	8.0
1-Bromonaphthalene	26	4.20	0.076	6.98	10.7
α-Choronaphthalene	26	3.18	0.100	5.31	10.6
Axoxybenzene	44	4.86	0.036	14.7	4.9
n-Benzylideneaniline	67	2.81	0.065	8.16	5.4

[a] Taken from the work of N. D. Foltz, C. W. Cho, D. H. Rank, and T. A. Wiggins, *Phys. Rev.* **165**, 396 (1968).

lated effect, since the line appeared only for powers in excess of 25 MW and is very sharp indicating gain narrowing.

Measured frequency shifts, relaxation times, and molecular radii derived from the measurements[178] are given in Table VIII for nitrobenzene and *m*-nitrotoluene. Figure 16 is a reproduction of stimulated Rayleigh wing spectra of nitrobenzene at three different temperatures.

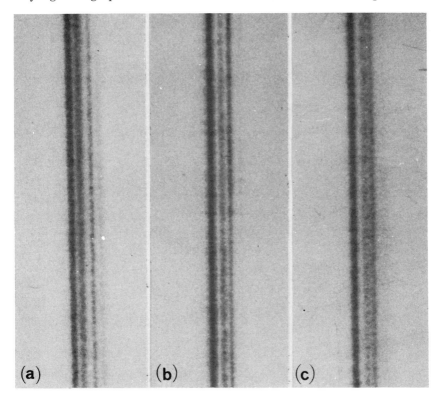

FIG. 16. Stimulated Rayleigh lines in nitrobenzene at three temperatures: (a) 12°C, (b) 16°C, (c) 31°C. The line at the left in each spectrum is the incident laser line, the second is the stimulated Rayleigh line, and the third is the stimulated Brillouin line. [Taken from the work of C. W. Cho, N. D. Foltz, D. H. Rank, and T. A. Wiggins, *Phys. Rev. Lett.* **18**, 107 (1967).]

Figure 17 is a plot of Δv, the frequency shift of the stimulated Rayleigh wing line in nitrobenzene, as a function of T/η. A polarizer and $\lambda/4$ plate were inserted between the laser and the focusing lens to prevent reinitiation of laser action in the ruby due to the reflected Brillouin scattering light.

Sometime after our first highly successful experiments outlined above, we wished to extend our observations on stimulated Rayleigh wing scattering to other types of liquids.[179] In the interval between the work reported above and the new experiments, the ruby formerly used had been replaced by one of much higher quality. We were unable to obtain sharp stimulated Rayleigh wing spectra with the more perfect ruby. It was surmised that the older ruby gave rise to some elliptical polarization

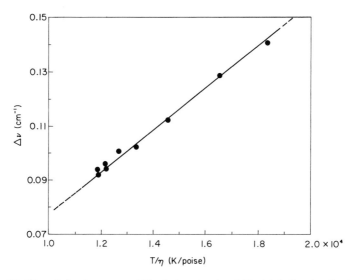

FIG. 17. Plot of the frequency shift of the stimulated Rayleigh line in nitrobenzene as a function of the ratio T/η, where η is the viscosity for the range $T = 285$–304 K.

Thus we arranged our apparatus so that the polarization could be changed continuously from linear to circular by means of a Kerr cell. A diagram of the apparatus used is shown in Fig. 18. The light was again focused into a 10-cm length cell with a lens of 10-cm focal length.

The spectral composition of the back-scattered light was analyzed with a 6.8-mm interferometer. The camera was fitted with analyzing sectors in its focal plane for use with circularly polarized incident light. A low-dispersion spectrograph was used to detect the presence of vibrational Raman spectra in the forward-scattered light. A high-resolution spectrograph (dispersion 4 cm^{-1}/cm and resolving power 300,000) was also used to observe the forward-scattered light. With circularly polarized

[179] N. D. Foltz, C. W. Cho, D. H. Rank, and T. A. Wiggins, *Phys. Rev.* **165**, 396 (1968).

incident light, a $\lambda/4$ plate at the slit and analyzers in the focal plane enabled the state of polarization to be observed.

Very detailed observations were made by Foltz et al.[179] on the thresholds for the appearance of the stimulated Rayleigh line as a function of the polarization of the incident light. The main conclusions arrived at as the result of this work were as follows: the sharpest stimulated Rayleigh line is observed for both forward- and back-scattering when circularly polarized incident light is used; the stimulated Rayleigh line is found to be circularly polarized with a sense opposite to that of the incident light;

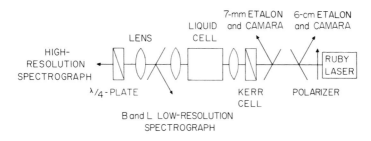

FIG. 18. Block diagram of the optical system used to detect and measure stimulated Rayleigh scattering. [Taken from the work of N. D. Foltz, C. W. Cho, D. H. Rank, and T. A. Wiggins, *Phys. Rev.* **165**, 396 (1968).]

and the stimulated Rayleigh line is not observed with linearly polarized incident light. Only the broad stimulated Rayleigh wing is observed from the forward scattering in liquids when used with linearly polarized incident light. Finally, it was found that the liquids which most readily produce stimulated Rayleigh lines are benzene derivatives particularly those containing NO_2 groups. These molecules possess high anisotropy and polarizability and most of them except styrene and napthalene have a permanent electric dipole moment. Favorite scattering materials like benzene and CS_2 failed to produce stimulated Rayleigh lines in the present investigation. We have tabulated the wide variety of results obtained in this investigation in Table VIII.

Herman[180] has given a theoretical explanation of the effects observed in stimulated Rayleigh wing scattering. Stimulated Rayleigh wing scattering arises by virtue of the ability of intense optical fields to orient liquid molecules having large polarizability anisotropies. Herman argues that for linearly polarized incident light and stimulated radiation fields

[180] R. M. Herman, *Phys. Rev.* **164**, 200 (1967).

in liquids the molecules tend to line up quite rigidly and permanently. Under these circumstances, the liquid fails to exhibit the necessary fluctuations in the index of refraction that are required to provide the gain at the frequency of the maximum intensity ω_R. This so-called *saturation effect* is largely responsible for the nonappearance of the sharp Rayleigh wing scattered line. Herman[180] discusses the case of circularly polarized incident light and shows that the stimulated Rayleigh wave becomes well established before saturation becomes appreciable and thus one obtains the gain-narrowed stimulated Rayleigh line.

3.7.3. Stimulated Thermal Rayleigh Scattering

The origin of spontaneous Rayleigh scattering is well understood. The unmodified Rayleigh line is associated with nonpropagating thermal density fluctuations. Herman and Gray[181] have considered the effect of absorptive heating in causing laser-induced instabilities in fluids. Herman and Gray discuss the combined influence of absorptive heating and electrostriction under the assumption that the absorber electromagnetic radiation is instantly and locally thermalized. The detailed calculations come to the conclusions: (1) a stimulated effect will occur; (2) a critical absorption coefficient for the material α_{cr} must obtain before the stimulated Rayleigh line exhibits a lower threshold than the threshold for the stimulated Brillouin line; and (3) the stimulated thermal Rayleigh line shows an anti-Stokes shift approximately equal to one half of the half intensity width of the laser intensity distribution.

Herman and Gray[181] have given an explicit expression for α_{cr} in their paper in terms of common well-known optical and mechanical parameters

$$\alpha_{cr} = \left[\frac{n^2 - 1}{2}\right]\left[\frac{n^2 + 1}{3}\right]\frac{C_p(\Delta\omega)_B}{\beta n v^2 C} \cdot \frac{\Gamma_L + \Gamma_R}{\Gamma_L + \Gamma_B},$$

where n is the index of refraction, C_p and β are the specific heat at constant pressure and the volume coefficient of expansion of the liquid, respectively, $(\Delta\omega)_B$ is the frequency shift of the Brillouin line, v is the speed of sound, C is the velocity of light in vacuo, and Γ_L, Γ_B, and Γ_R are the full widths at half intensity for the laser, Brillouin, and Rayleigh lines, respectively.

The predictions of the Herman and Gray[181] theory of stimulated

[181] R. M. Herman and M. A. Gray, *Phys. Rev. Lett.* **19**, 824 (1967).

thermal Rayleigh scattering (STRS) have been verified by experiments by Rank *et al.*[182] and by Cho *et al.*[182]

The liquid containing the dye was placed in a 10-cm length cell and a lens of 10-cm focal length was used to concentrate the light in the cell. The light source was a Korad giant pulse ruby laser operated single moded at power up to 50 MW. The pulse duration was 14 nsec and the spectral line width of the laser varied between 0.017 to 0.026 cm^{-1}. The back-scattered light was observed by means of a Fabry–Perot etalon using interferometer spacers up to 101 mm in length. In addition to the back-scattered light the interferometer was also fed with suitably attenuated laser light. The back-scattered light was observed simultaneously with the incident laser light, but the two kinds of light were photographed on separate quadrants of the interferogram by means of a quarter wave plate and analyzers in the plane of the interference fringes. Laser and cell containing the liquid were separated by about 3 m so as to inhibit relasing.

In Fig. 19 we have shown the interference pattern obtained using a solution of I_2 in ethylene glycol (101-mm interferometer spacer). It is easy to see that the (black) rings produced by the back-scattered light (weaker rings) have a larger diameter than the rings produced by the laser itself. This very definitely shows that the STRS line is violet shifted approximately half the laser line width as predicted by the theory of Herman and Gray.[181] Furthermore, it can be noted by referring to Fig. 19 that the STRS line is sharper than the laser line, which is an indication that the effect is a stimulated one and involves the existence of a threshold as demanded by the theory and proved by auxiliary experiments in the present work. A large number of liquid dye combinations were investigated by Cho *et al.*[183] and the results of the experiments in large measure verify fairly quantitatively the theoretical predictions of the Herman and Gray theory of STRS.

Gires and Soep[184] have suggested that STRS can be used as a Q-switch in a manner similar to that for SBS.[126]

A direct measurement of the gain in STRS has been made by Gires.[185] Due to the small anti-Stokes shift, it was not possible to explore the

---

[182] D. H. Rank, C. W. Cho, N. D. Foltz, and T. A. Wiggins, *Phys. Rev. Lett.* **15**, 828 (1967).

[183] C. W. Cho, N. D. Foltz, D. H. Rank, and T. A. Wiggins, *Phys. Rev.* **175**, 271 (1968).

[184] F. Gires and B. Soep, *Proc. IEEE* **56**, 1613 (1968).

[185] F. Gires, *C. R. Acad. Sci. Paris* **266B**, 596 (1968).

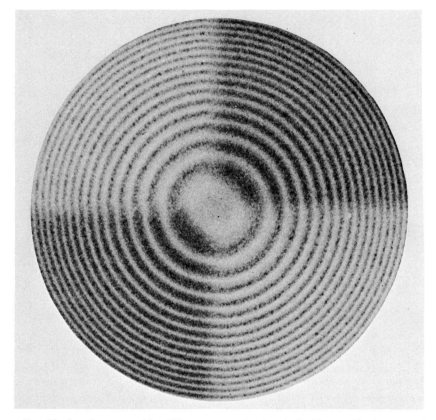

FIG. 19. Interferogram with a 101-mm spacer showing the incident and back-scattered light from ethylene glycol with iodine added to produce an absorption coefficient of 0.045 cm^{-1}. The darker quadrants are due to the incident laser light. The sharp rings in the lighter quadrants are due to the anti-Stokes shifted STRS. [Taken from the work of C. W. Cho, N. D. Foltz, D. H. Rank, and T. A. Wiggins, *Phys. Rev.* **175**, 271 (1968).]

gain as a function of frequency displacement. Rather the gain was integrated over the spectral width of the laser. Results were in fair agreement with the work of Cho *et al.*[183]

Herman and Gray predicted in conjunction with their work[181] on STRS that there should be a change in the frequency at which the SBS gain maximizes if an absorber is present in a liquid. This change in gain has been observed by Pohl *et al.*[186] by measuring the gain in the neighborhood of the Brillouin shifted line as a function of the amount of absorber

[186] D. Pohl, I. Reinhold, and W. Kaiser, *Phys. Rev. Lett.* **20**, 1141 (1968).

present. The Stokes shift of the maximum gain was clearly seen and agreed well with the theoretical prediction.

A distinction has been made by Rother et al.[187] between the type of gain to be expected in transient and steady-state conditions for STRS. Using the frequency drift of the laser during the pulse they were able to explore the gain for small changes in frequency and found that for transient conditions the gain was not zero for zero frequency difference as required by the steady-state theory. This gain at small frequency displacements could explain why the measured frequency shifts in STRS were smaller than predicted.[182,183] In additional work Rother et al.[188] have confirmed the previous findings[187] by exploring further the gain at zero frequency shift.

A thermal scattering has been observed by Mack and attributed to induced temperature waves using picosecond pulses.[189] The theory has also been discussed by Pohl.[190] Two beams crossing at a small angle θ in an absorber can interact to give gain to the weaker beam and produce additional beams showing no frequency change at angles $-\theta$, $\pm 2\theta$, etc. The medium thus acts as a refraction grating. It appears[189] that a true scattering from temperature waves is a very special case.

Bespalov et al.[191] claim that the frequency shift in STRS depends upon the incident power, the length of the laser pulse, and the temperature or relaxation time. Very small anti-Stokes frequency shifts are predicted for their experimental conditions and none were observed.

Anti-Stokes shifts have been observed by Kyzylasov et al.[192] in absorbing liquids. They also report that Stokes shifts are found when no absorber is present.

STRS has been observed in compressed gases as well as in liquids. Using NO_2 as an absorber, the back-scattered light was observed to have similar characteristics to those from liquids.[193] Values of α_{cr} were measured but found to be generally lower than theoretical values. An exception to

[187] W. Rother, D. Pohl, and W. Kaiser, *Phys. Rev. Lett.* **22**, 915 (1969).

[188] W. Rother, H. Meyer, and W. Kaiser, *Phys. Lett.* **31A**, 245 (1970).

[189] M. E. Mack, *Phys. Rev. Lett.* **22**, 13 (1969).

[190] D. Pohl, *Phys. Rev. Lett.* **23**, 711 (1969).

[191] V. I. Bespalov, A. M. Kubarev, and G. A. Pasmanik, *Phys. Rev. Lett.* **24**, 1274 (1970).

[192] V. S. Starunov, *Sov. Phys. JETP* **30**, 553 (1970); Y. P. Kyzylasov, V. S. Starunov, and I. L. Fabelinskii, *JETP Lett.* **11**, 66 (1970).

[193] T. A. Wiggins, C. W. Cho, D. R. Dietz, and N. D. Foltz, *Phys. Rev. Lett.* **20**, 831 (1968).

this was found using I_2 vapor as an absorber. The large value of α_{cr} needed appears to reflect the inefficiency of conversion of energy into thermal motion due to the low dissociation energy of I_2.[194]

In additional work[195] the threshold for STRS in gases was measured as a function of density. The variation in gain is due to the density dependence of the relaxation time τ_1, the time for the absorbed heat to be deposited in the bulk gas at a density of one amagat. This relaxation time can be deduced from

$$\tau_1 \sim \frac{\mu\varrho}{\mu^2 - 1} \frac{v^2\beta}{\omega_B C_p \Gamma_L} \alpha_{cr}.$$

The times determined are 60 nsec for CO_2, 100 nsec for N_2, and 8 nsec for CH_4.

A stimulated back-scattering has been observed from argon[196] under circumstances similar to those in which STRS was observed in other gases. However, the anti-Stokes shifted component was present without the added absorber used with CO_2, N_2, and CH_4. The explanation of this scattering is not known but possibly is associated with the breakdown in these experiments.

One additional kind of stimulated scattering in which small frequency changes are seen is that of concentration scattering.[150,197] The frequency shifts depends upon the diffusion coefficient for the motion of a gas through another gas of different molecular weight. Small Stokes shifts are predicted and observed.

[194] M. A. Gray and R. M. Herman, *Phys. Rev.* **181**, 374 (1969).

[195] D. R. Dietz, C. W. Cho, and T. A. Wiggins, *Phys. Rev.* **182**, 259 (1969).

[196] D. R. Dietz, C. W. Cho, D. H. Rank, and T. A. Wiggins, *Appl. Opt.* **8**, 1248 (1969).

[197] N. Bloembergen, W. H. Lowdermilk, M. Matsuoka, and C. S. Wang, *Quantum Electron. Conf., Kyoto.* Paper 19.6 (September 1970).

AUTHOR INDEX

Numbers in parentheses are footnote numbers. They are inserted to indicate that the reference to an author's work is cited with a footnote number and his name does not appear on that page.

A

Abe, K., 298
Abouaf-Marquin, L., 289
Agar, D. M., 300
Akitt, D. P., 336, 375, 387
Alcock, A. J., 433, 437, 438
Allegretti, J. M., 268
Allen, H. C., 15, 27, 28, 29, 30, 32, 129, 148(5), 261, 274, 275
Allen, J. C., 252
Allison, R., 240
Alpert, B. D., 140
Amat, G., 61, 148, 149, 150, 169
Anacker, F., 222
Anderson, E. D., 252
Andrews, D. H., 196
Angell, C. L., 289
Antić-Jovanović, A. M., 296
Arai, S., 237
Arefev, I. M., 442, 445, 464(150)
Ashley, E. J., 189
Atherton, N. M., 298
Avery, D. G., 172, 188(67), 189(67), 192 (67), 197(67)
Avizonis, P. V., 441, 449(140)

B

Bair, E. J., 300
Baker, A. D., 291
Baker, C., 291
Ballard, S. S., 189, 242
Ballhausen, C. J., 270
Ballik, E. A., 289

Barbrow, L. E., 222
Barger, R. L., 289
Barker, E. S., 240
Barker, W. B., 418
Barnes, R. B., 147
Barnett, T. L., 158, 169(43, 44, 45)
Barocchi, F., 445
Barrell, H., 217
Barrow, R. F., 288
Barrow, T., 299
Basco, N., 297
Bass, A. M., 235
Bassler, J. M., 297
Bates, B., 244
Battaglia, A., 116
Beckel, C. L., 288
Becker, J. A., 195, 196
Bedo, D. E., 231
Bell, E. E., 311
Bell, S., 299
Bellamy, L. J., 147
Bender, C. F., 252, 294
Benedek, G. B., 401, 419, 452, 455
Benedict, W. S., 307, 317(18), 318(18), 319, 331(18), 354, 381, 382, 383
Benjamin, F., 212
Bennett, H. E., 189, 244
Bennett, J. M., 189
Bergmann, L., 452, 455
Berman, P. R., 370
Bernard, H. W., 252
Bernheim, R. A., 252
Berning, P. H., 246
Bernstein, H. J., 247
Berthou, J. M., 294

SUBJECT INDEX

Date Due